OPTICAL STUDIES IN LIQUIDS AND SOLIDS

RAYLEIGH AND RAMAN SCATTERING, METAL OPTICS,
AND LUMINESCENCE OF ALKALI IODIDES

RELEEVSKOE I KOMBINATSIONNOE RASSEYANIE, METALLOOPTIKA,
LYUMINESTSENTSIYA SHCHELOCHNYKH IODIDOV

РЕЛЕЕВСКОЕ И КОМБИНАЦИОННОЕ РАССЕЯНИЕ, МЕТАЛЛООПТИКА,
ЛЮМИНЕСЦЕНЦИЯ ЩЕЛОЧНЫХ ИОДИДОВ

The Lebedev Physics Institute Series

Editor: Academician D. V. Skobel'tsyn

Director, P. N. Lebedev Physics Institute, Academy of Sciences of the USSR

Volume 25   Optical Methods of Investigating Solid Bodies
Volume 26   Cosmic Rays
Volume 27   Research in Molecular Spectroscopy
Volume 28   Radio Telescopes
Volume 29   Quantum Field Theory and Hydrodynamics
Volume 30   Physical Optics
Volume 31   Quantum Electronics in Lasers and Masers
Volume 32   Plasma Physics
Volume 33   Studies of Nuclear Reactions
Volume 34   Photomesonic and Photonuclear Processes
Volume 35   Electronic and Vibrational Spectra of Molecules
Volume 36   Photodisintegration of Nuclei in the Giant Resonance Region
Volume 37   Electrical and Optical Properties of Semiconductors
Volume 39   Optical Studies in Liquids and Solids

In preparation

Volume 38   Wideband Cruciform Radio Telescope Research
Volume 40   Experimental Physics: Methods and Apparatus
Volume 41   The Nucleon Compton Effect at Low and Medium Energies
Volume 42   Electronics in Physics Experiments
Volume 43   Nonlinear Optics
Volume 44   Nuclear Physics and Interaction of Particles with Matter
Volume 45   Programming and Computer Techniques in Physics Experiments
Volume 46   Cosmic Rays and Interaction of High-Energy Particles
Volume 47   Radio Astronomy Instruments and Observations
Volume 48   Surface Properties of Semiconductors and Dynamics of Ionic Crystals
Volume 49   Quantum Electronics and Paramagnetic Resonance

Proceedings (Trudy) of the P. N. Lebedev Physics Institute

Volume 39

OPTICAL STUDIES IN LIQUIDS AND SOLIDS

Edited by
Academician D. V. Skobel'tsyn
Director, P. N. Lebedev Physics Institute
Academy of Sciences of the USSR, Moscow

Translated from Russian

SPRINGER SCIENCE+BUSINESS MEDIA, LLC
1969

ISBN 978-1-4684-8726-8   ISBN 978-1-4684-8724-4 (eBook)
DOI 10.1007/978-1-4684-8724-4

The original Russian text, published by Nauka Press in Moscow in 1967
for the Academy of Sciences of the USSR as Volume 39 of the Proceedings
(Trudy) of the P. N. Lebedev Physics Institute, has been corrected by
the editor for this edition.

Library of Congress Catalog Card Number 69-12523

CONTENTS

STUDY OF SECOND-ORDER LINES IN THE VIBRATIONAL
SPECTRA OF MOLECULES

Z. MULDAKHMETOV

INTRODUCTION 3

CHAPTER I. Review of Literature and Presentation of the Problem 5

CHAPTER II. Experimental Method 8
    1. Study of Second-Order Lines in the Raman Spectra of Liquids 8
    2. Study of Infrared Absorption Spectra 10
    3. Calculation of the Anharmonicity Coefficients of the Molecular Vibrations. 12

CHAPTER III. Experimental Results...................................... 14
    1. Second-Order Lines in the Raman Spectrum of Deuterocyclohexane...... 14
    2. Study of Overtones and Composite Frequencies in the Vibrational Spectra
       of Benzene and Hexadeuterobenzene 17
    3. Second-Order Vibrational Spectra of Chloroform and Deuterochloroform.. 26
    4. Anharmonicity of Some Characteristic Molecular Vibrations 32

CHAPTER IV. Discussion of Results and General Conclusions................ 35
    1. General Conclusions Regarding the Anharmonicity Coefficients 35
    2. Electro-Optical Parameters of Second-Order Lines................... 38
    3. Use of Anharmonicity Coefficients for Finding the Zero Vibration
       Frequencies 41

CONCLUSION 43

LITERATURE CITED 45

STUDY OF THE ROTATIONAL OSCILLATION OF MOLECULES
IN LIQUIDS BY THE RAMAN METHOD

U. A. ZIRNIT

INTRODUCTION... 51

CHAPTER I. Review of Literature 52
    1. Review of Experimental Data Relating to the Low-Frequency Vibrational
       Spectra 52
    2. Methods of Interpreting Low-Frequency Spectra 54
    3. Presentation of the Problem 58

CHAPTER II. Methods of Obtaining and Analyzing Low-Frequency
Raman Spectra.   59

CHAPTER III. Appearance of the Rotational Oscillations of Methyl Groups
in Low-Frequency Raman Spectra.   62
    1. Experimental Results on the Low-Frequency Raman Spectra of Methyl-
       Substituted Cyclohexanes.   62
    2. Referring the Lines in the Raman Spectra of Methyl-Substituted
       Cyclohexanes to Rotational Oscillations   65

CHAPTER IV. Appearance of the Rotational Oscillations of Ethyl and Heavier
Groups in the Low-Frequency Raman Spectra   71
    1. Experimental Results on the Low-Frequency Raman Spectra of Methyl-
       Substituted Butanes   71
    2. Experimental Results on the Low-Frequency Raman Spectra of Methyl-
       Substituted Pentanes.   75
    3. Experimental Results on the Low-Frequency Raman Spectra of Ethyl-
       Substituted Naphthenes   80

CONCLUSION   83

LITERATURE CITED   85

STUDY OF THE OPTICAL AND ELECTRICAL PROPERTIES
OF CERTAIN FOURTH-GROUP METALS

A. I. GOLOVASHKIN

INTRODUCTION.   89

CHAPTER I. Theoretical Study of the Optical Properties of Metals   91
    1. Classical Theory of the Skin Effect.   91
    2. Anomalous Skin Effect   92
    3. Frequency of Electron Collisions   93
    4. Quantum Kinetic Equation   93

CHAPTER II. Measuring Method   96
    1. Method of Measuring Optical Constants   96
    2. Experimental System for Measuring Optical Constants   101
    3. Measurement of the Real and Imaginary Parts of the Complex
       Refractive Indices n and $\varkappa$.   102
    4. Possible Systematic Errors.   104
    5. Samples for Optical Investigations   106
    6. Measurement of the Thickness, Density, Conductivity, Critical
       Temperature, and Hall Effect.   107

CHAPTER III. Results of the Measurements.   112
    1. Results of Measuring the Optical Constants of Tin.   112
    2. Results of Measuring the Optical Constants of Lead.   113
    3. Effect of Oxidation and Annealing on Optical Constants   116
    4. Results of Measuring the Density, Conductivity, Critical Temperature,
       and Hall Effect of Tin.   117
    5. Results of Measuring the Density, Conductivity, Critical Temperature,
       and Hall Effect of Lead.   119

CHAPTER IV. Analysis of Experimental Results................................... 121
    1. Obtaining Microcharacteristics.................................... 121
    2. Micro-Characteristics of Tin.................................... 123
    3. Micro-Characteristics of Lead.................................... 125
    4. Character of the Skin Effect for Tin and Lead..................... 126
    5. Concentration of Conduction Electrons............................ 129
    6. Experimental Proof of the Gurzhi-Holstein Theory 132
    7. Velocity of Electrons on the Fermi Surface and Other Micro-
       characteristics 135
    8. Effect of Interelectron Collisions on the Optical Constants of Tin and Lead 137
    9. Separation of the Effects of Interband Transitions 138

CONCLUSION 141

LITERATURE CITED 143

STUDY OF THE SPECTRUM OF THE THERMAL AND STIMULATED MOLECULAR SCATTERING OF LIGHT IN LIQUIDS

V. S. STARUNOV

INTRODUCTION.. 149

CHAPTER I. State of Theoretical and Experimental Knowledge
Regarding the Spectral Composition of the Molecular Scattering of Light 152
    1. Scattering of Light at Fluctuations of Anisotropy in Liquids (Thermal
       Scattering) 152
    2. Fine Structure of the Molecular-Scattering Lines of Liquids 156
    3. Mandelshtam—Brillouin Stimulated Scattering...................... 159

CHAPTER II. Theory of the Spectral Composition of the
Depolarized Scattering of Light in Liquids 161
    1. Relation Between the Spectral Composition of the Scattered Light and the
       Rotational Thermal Motion of the Molecules in the Liquid............. 161
    2. Effect of Rotational Diffusion of Molecules on the Spectral Composition
       of the Depolarized Scattering of Light 163
    3. Modulation of Scattered Light by the Rotational Oscillations of the
       Molecules in a Liquid... 166
    4. Stimulated Scattering of Light on the Wing of the Rayleigh Line 168

CHAPTER III. Methods of Studying the Spectral Composition
of Scattered Light... 174
    1. Method of Studying the Spectral Composition of the Wing of the Rayleigh
       Line (Ordinary Thermal Scattering) 174
    2. Experimental Apparatus for Studying the Fine-Structure Components
       of the Rayleigh Line (Thermal Scattering) 178
    3. Method of Studying the Stimulated Scattering of Light................ 181

CHAPTER IV. Results of an Experimental Study of the Spectrum
of Depolarized Scattered Light in Low-Viscosity Liquids................... 184
    1. General Characteristics of the Spectrum of the Depolarized Scattering
       of Light. Diffusion Wing of the Rayleigh Line...................... 184
    2. Results of a Study of the High-Frequency Part of the Rayleigh-Line Wing
       in Low-Viscosity Liquids at Room Temperature 186
    3. Study of the Remote Region of the Rayleigh-Line Wing in Liquids
       at Various Temperatures 189

CHAPTER V. Determination of the Velocity and Absorption of
Hypersound in Liquids from the Width of and Spacing Between
the Mandelshtam—Brillouin Components. 193
  1. Measured Values of the Absorption and Velocity of Hypersound 193
  2. Analysis of the Results from the Point of View of the Phenomenological
     Relaxation Theory 195
  3. Analysis of the Results from the Point of View of the Molecular Theory
     of Relaxation. 198
  4. Intensity Ratio of the Fine-Structure Components 202

CHAPTER VI. Stimulated Molecular Scattering of Light. 206
  1. Stimulated Mandelshtam—Brillouin Scattering in Liquids and Glasses 206
  2. Results of an Investigation into the Stimulated Scattering of Light
     on the Wing of the Rayleigh Line. 210

LITERATURE CITED 214

GAMMA- AND PHOTOLUMINESCENCE OF ALKALI IODIDES

N. N. VASIL'EVA

CHAPTER I. Presentation of the Problem 219

CHAPTER II. Study of γ-Luminescence in Alkali Iodides 222
  1. Review of Published Data Relating to the Spectral Characteristics
     of Alkali Iodides 222
  2. Choice of Subjects for Study. 224
  3. Method of Studying the γ-Luminescence Spectra of Alkali Iodides 224
  4. Effect of Foreign Impurities on the Unactivated Fluorescence of
     Alkali Iodides 225
  5. Effect of Structural Defects in the Crystals on the Unactivated
     Luminescence. 230
  6. Temperature Behavior of the Activator Emission Bands of Phosphors. . . . 232
  7. Thermoluminescence of Alkali Iodides. 233

CHAPTER III. Study of the Spectral Characteristics of the Photoexcitation
of Phosphors Based on Alkali Iodides. 235
  1. Review of Published Data Relating to the Absorption of Alkali Halides. . . . 235
  2. Method of Studying Excitation Spectra 237
  3. Experimental Results. Photoexcitation Spectra of the Ultraviolet
     Fluorescence of Alkali Iodides 239
  4. Photoexcitation Spectra of the Intermediate Emission Band of Alkali Iodides 241
  5. Photoexcitation Spectra of Alkali Iodides Subjected to Heat Treatment. . . . 242
  6. Photoexcitation of the Activator Fluorescence of Phosphors Based on
     Alkali Iodides 244

CHAPTER IV. Concentration Dependence of the Time Characteristics
of the Phosphors CsI—Tl and KI—Tl 246
  1. Attenuation Time of CsI—Tl and KI—Tl Phosphors as a Function of
     Activator Concentration 246
  2. Concentration Dependence of the Growth Time of Scintillations in a
     Number of CsI—Tl and KI—Tl Phosphors 250

CHAPTER V. Discussion of Results   254
    1. Discussion of the Nature of the Fluorescence in the Ultraviolet Emission
       Band of Pure Alkali Iodides   254
    2. Discussion of the Nature of the Fluorescence in the Intermediate Emission
       Band of Alkali Iodides   256
    3. Discussion of the Mechanism of Energy Transfer from the Main
       Substance to the Activator   257

CONCLUSION   262

LITERATURE CITED   263

STUDY OF SECOND-ORDER LINES IN
THE VIBRATIONAL SPECTRA OF MOLECULES*

Z. MULDAKHMETOV

*Dissertation in pursuit of the degree of Candidate of Physicomathematical Sciences. Defended January 8, 1964. Scientific director: Professor M. M. Sushchinskii.

INTRODUCTION

The study of the vibrations of polyatomic molecules has recently turned into one of the most widespread and powerful methods of studying molecular structure. These vibrations appear directly in the infrared absorption spectra and Raman spectra of gases, liquids, and solids. A measurement of the number of bands in addition to their positions (frequencies or wavelengths) offers the possibility of obtaining a great deal of important information regarding the geometric and mechanical properties of the molecules, the types of chemical bonds, and so forth. It is now quite difficult to list the vast number of specific problems solved by measuring vibrational frequencies.

As a result of the successful development of research methods and the widespread application of vibrational spectra in analyzing the structures of molecules and the constitution of materials, it now becomes necessary to develop the theory of molecular vibrations further.

Existing theory, of course, is based on the assumption of the harmonicity of molecular vibrations, which, strictly speaking, is not justified experimentally. The anharmonicity of the molecular vibrations has therefore to be taken into account by introducing appropriate approximations. Thus, in carrying out calculations on the vibrations of polyatomic molecules, one uses the force constants calculated from the observed frequency values. However, as a result of the anharmonicity of the vibrations, the values of the observed frequencies differ from the harmonic values, and the force constants used therefore differ from the true ones, i.e., the constants which would be obtained on the basis of harmonic vibrations. In a number of investigations anharmonicity has been taken into account by the introduction of "spectroscopic masses" for hydrogen [1]. This method assumes that the vibrations involving hydrogen atoms give the greatest contribution to the total anharmonicity. However, this assumption indisputably constitutes an approximation, since experience shows that the contribution of other atoms to the total anharmonicity of the vibrations is also substantial [2-4].

The most reliable method of allowing for anharmonicity would be to correct the observed first-order line-frequency values by means of anharmonicity coefficients obtained experimentally by reference to second-order line data. For this purpose we need extensive experimental material on overtones and composite frequencies in the vibrational spectra of molecules.

A study of second-order lines in vibrational spectra offers the possibility of obtaining more complete information regarding the structure of the molecule.

The resultant quantitative data (frequencies, intensities, degrees of depolarization, and anharmonicity coefficients) may be used for calculating quantities characterizing the molecular vibrations in various ways; such quantities include the second derivative of the polarizability and the dipole moment of the molecule with respect to the normal coordinates, and also the "zero" frequencies.

Vol'kenshtein expressed an interesting idea regarding the dependence of the intensity of the overtones in the Raman spectra on the polarity of the bond vibrating at the frequency in question. As the polarity of the bond increases, the ratio of the intensity of the overtone to the intensity of the fundamental becomes greater [5].

The second-order lines in the Raman spectrum of liquids have not been studied very much, no doubt because of the experimental difficulties associated with the low intensity of these lines. Published experimental data relating to second-order lines refer principally to infrared spectra, for which the overtone range is more accessible to study.

However, on passing to comparatively complex molecules, the second-order spectra become so complicated that their interpretation on the basis of line frequencies and intensities only becomes too indefinite. From this point of view, the deciphering of the infrared spectrum is more awkward than that of the Raman spectrum. Additional data relating to the degree of polarization in the Raman spectra greatly aids interpretation, although even in this case it is not always absolutely specific.

The complexity of analyzing second-order spectra is due to the fact that it is almost always possible to set up several combinations of fundamental frequencies compatible with the selection rules and fairly close to the frequency of a given line. In all such cases it is usual to take the relation which gives the lowest anharmonicity coefficient [6, 7]. This principle cannot of course really be justified, but if we neglect it there are no guiding lines at all for interpretation of the data.

Naturally, in order to secure a well-founded interpretation of the observed second-order lines and also to verify the principle of lowest anharmonicity coefficients, we require the widest possible experimental material, such as may be obtained by measuring other parameters of the Raman lines as well as their frequencies, by incorporating infrared data, and in some cases by studying the deuteroderivatives.

No systematic work of this kind has yet been done.

In this investigation we shall study the second-order lines in the vibrational spectra of polyatomic molecules.

CHAPTER I

REVIEW OF LITERATURE
AND PRESENTATION OF THE PROBLEM

Data relating to the frequencies of second-order lines in infrared spectra are quite plentiful; however, they have been obtained in connection with widely differing problems and are often of a random nature. Data relating to the second-order infrared spectral lines of molecules which we have studied will be included in the appropriate sections of this treatment.

Recently considerable interest has been aroused in the second-order lines of the Raman spectra of liquids. Long exposures are usually required for observing overtones and composite frequencies in such spectra. However, individual overtones in the Raman spectra of certain compounds are comparable in intensity, for various reasons, with the fundamental frequencies, and may be easily observed when studying the main vibrations. Thus, the strong overtone $\nu = 1550$ cm^{-1} of the degenerate vibration in CCl_4 has been observed by many authors. The intensity and degree of polarization of this vibration were measured in [8, 9].

The first investigations into second-order lines in Raman spectra were carried out by Landsberg and Malyshev [10], and also by Ananthakrishnan [11]. It was found that the intensity of the overtones and composite frequencies was $\sim 3 \cdot 10^{-3}$ of that of the fundamental line.

The second-order lines in the Raman spectra of CCl_4, $SnBr_4$, and $SnCl_4$ were studied quite fully in [4]; the frequencies and intensities of the observed lines were determined, and in some cases the degrees of depolarization were measured. The experimental data relating to the degree of depolarization agreed with the theory of polarizability for molecules with tetrahedral symmetry. The first- and second-order Raman spectra of the tetrahedral molecules CCl_4, $SiCl_4$, $GeCl_4$, $SnCl_4$, and $TiCl_4$ were studied in [12]; quantitative data were obtained for the intensities of the observed lines. The intensity of the overtones increased much more rapidly in the series of compounds cited than that of the fundamental vibrations. This appeared particularly in the completely symmetrical vibrations ν_1 and $2\nu_1$. A comparison of the results for $GeCl_4$ and $TiCl_4$, for example, showed that the fundamental vibration ν_1 increased by about 3 times, while the overtone $2\nu_1$ increased by almost 18 times. Bobovich associated this behavior of the vibrations with the properties of the second derivative of the polarizability with respect to the normal coordinate $\partial^2 \alpha / \partial g^2$. By comparison with the first derivative $\partial \alpha / \partial g$ this depended more sharply on the various parameters of the excited states. By analyzing the results, certain conclusions were drawn regarding the magnitude and sign of the ratio of the first and second derivatives of the polarizability with respect to the vibration coordinate. The question as to the effect of the chemical bond on the intensity of the spectrum in various molecules was also discussed.

The second-order Raman spectra of comparatively complex molecules were studied in [2, 6, 13, 14]. The first of these [2] was devoted to a study of the anharmonicity of the vibrations of the CH groups in normal paraffins. The most interesting result was the fact that the anharmonicity coefficients of the deformation (strain) and valence vibrations differed very little from each other.

Some very sparse data relating to overtones in $CHCl_3$, $C_2H_2Cl_2$, $C_2H_4Cl_2$, CH_3OH, C_6H_5Cl, CS_2, and $TiCl_4$ molecules were presented in [13, 14]. Two cases of Fermi resonance were noted. The second-order Raman spectra of cyclohexane and tetramethylethylene were studied in more detail in [6]. Not only were the intensities of the observed second-order lines measured, but the degrees of depolarization were determined as well; this greatly facilitated interpretation of the lines.

Certain questions relating to the theory of second-order line intensities in vibrational spectra were considered in [15-17]. A general formula was derived in [15] for the intensities of the overtones and composite frequencies in the infrared spectra of polyatomic molecules. The general formula given for the second derivatives of the dipole moment of the molecule with respect to the normal coordinates explicitly contained all the molecular parameters on which the values of these derivatives depended. It should be emphasized that the formula was derived on the assumption that there was no mechanical anharmonicity.

The theory of the intensities of overtones in Raman spectra is considered in [16]. Attention is concentrated on the frequency dependence of this intensity. The authors conclude that the frequency dependence of such lines is more significant than for the fundamental vibrations. On the other hand, it is shown in [17] that the frequency dependence of the fundamental vibrations should be the same as for the overtones.

We see from this brief review of papers devoted to the second-order lines in the Raman spectra of liquids that work has chiefly been carried out on very simple molecules, and that the actual number of investigations is small. We note that in these papers it is chiefly the frequencies of the second-order lines which have been measured. Yet the intensity and degree of depolarization of the second-order lines in the Raman spectra, as well as the intensity of the infrared spectra, are of considerable independent interest. Data relating to the intensities and degrees of depolarization also reduce the difficulties associated with interpreting second-order spectra.

It is in fact almost always possible to set up several combinations of fundamental frequencies satisfying the selection rules and giving results fairly close to the frequency of a given line. Hence the interpretation of second-order lines often remains uncertain. This uncertainty is greatly alleviated on using information relating to the intensities and degrees of depolarization (although even so some lines allow ambiguous interpretations).

As mentioned earlier, a study of second-order lines in the vibrational spectra of molecules is of great importance from the point of view of studying the anharmonicity of molecular vibrations. The anharmonicity of the vibrations may be taken into account by correcting the observed first-order line frequencies by means of an anharmonicity coefficient obtained experimentally from second-order line data.

It is an important point that the "zero" frequencies, which in principle characterize the vibrations of the molecules more accurately, may differ considerably from the observed frequencies in polyatomic molecules. This is because each observed frequency differs from the "zero" one by a sum containing a large number of anharmonicity coefficients. If a large number of the anharmonicity coefficients have the same sign, then even if each is by itself quite small, the sum will be appreciable and the "zero" frequencies will deviate severely from the experimental values. We note that the question of errors in the calculation of the force constants of the molecules resulting from the use of the observed frequencies instead of the "zero" values does not arise in the literature owing to the absence of experimental data.

However, errors of this kind committed by extending the force constants of simple molecules to more complicated systems may lead to uncontrollable errors in the values of the calculated constants. Hence, the building up of experimental material relating to second-order lines in both Raman and infrared spectra is a matter of undoubted interest.

In the present investigation, we made a detailed study of three pairs of molecules: (a) cyclohexane (C_6H_{12}) and deuterocyclohexane (C_6D_{12}); (b) benzene (C_6H_6) and hexadeuterobenzene (C_6D_6); and, (c) chloroform ($CHCl_3$) and deuterochloroform ($CDCl_3$). These molecules were chosen as subjects for study because they had quite a high symmetry, while calculations of the vibrational characteristics had already been carried out for some of them, greatly facilitating the interpretation of the resultant spectra.

The second- and third-order spectra of eleven molecules with selected characteristic bonds were studied in less detail. For the molecules studied we obtained quantitative data relating to the intensity and degree of depolarization as well as the frequencies, and also calculated the anharmonicity coefficients. The quantitative measurements of the parameters of the second-order lines in the Raman spectra, together with a parallel study of the overtones and composite frequencies in the Raman and infrared spectra, yielded a fairly clear and unambiguous interpretation of the second-order spectra of the molecules studied.

CHAPTER II

EXPERIMENTAL METHOD

1. Study of Second-Order Lines in the
Raman Spectra of Liquids

In studying second-order lines in the Raman spectra, the main difficulty is the separation of these from the continuous background. At the present time, when the use of low-pressure tubes has become quite general, this difficulty may be regarded as largely overcome. The appearance of low-pressure tubes does not mean that the high-pressure type cannot be used for observing second-order lines. These may in fact still be used on condition that reliable filtration is provided. In the present investigation, low-pressure tubes were chiefly employed. Hence, the customary method may be used in order to measure isolated lines with an intensity of the order of 0.5 units in the scale of [18] (where the intensity at the maximum of the 802 cm^{-1} line of cyclohexane is taken as 250).* We note that even this intensity is not the limiting one, since, in a number of practical cases the background had no interfering effect on the measurements.

A serious difficulty in our measurements was the separation of neighboring (sometimes overlapping) lines. In particular, certain spectral ranges were almost inaccessible for study owing to the presence of a large number of first-order lines.

For measuring the intensities and degrees of depolarization we used a two-prism Hilger E 612/3 spectrograph with a relative aperture of 1:1.5 and a dispersion of 64 Å/mm in the region of 4368 Å. For measuring the frequencies we used a Huet spectrograph of the B-II type. In taking the photographs we used a two-lamp illuminating system with elliptical surfaces. This system was adjusted in accordance with the results of [19]. The exciting lines were the mercury lines 4358 and 4047 Å, which were separated by glass or liquid filters. The spectra were photographed on Kodak OaG and Raman Orth Plates. Under these conditions the time for measuring the intensities of the second-order lines was 2-3 h without filters and 8-12 h with filters; for measuring the frequencies with filters 5-8 h were required. The frequencies were determined by reference to the iron spectrum. The accuracy of the determination was 1-2 cm^{-1} for the sharp, strong lines, and 2-3 cm^{-1} for weak, wide lines.

The intensities were estimated by the ordinary means of photographic photometry [20]. The fluorescence of quinine sulfate was used for creating the reference marks (the photometric density scale). We made use of the fact that the Schwartzchild constant of our plates equalled unity, so that the marks could be created with a brief period of exposure (15 min).

In order to avoid errors associated with variations in the sensitivity of the plates over the spectrum, the intensities of the second-order lines were measured by reference to neighboring fundamental lines, the intensities of which were known from earlier measurements.

---

*We shall always use the scale of [18] in this investigation.

Measurement of the second-order line intensities was impeded by the fact that standard lines with intensities similar to those of the second-order lines were in general absent. In order to obtain quantitative data regarding the intensity and degree of depolarization of the overtones and composite frequencies, we covered part of the slit with a neutral filter, so that the photograph showed three spectra, two of which were 10.5 times weaker than the third. The intensities of the fundamental lines were measured on a common scale by reference to the weakened spectra, while the intensities of the second-order lines were determined relative to the former by reference to the unattenuated spectrum.

In the photographic recording of the spectra, the degree of depolarization is normally found as the ratio of the intensities at the maxima of the π and σ components of the line in question.

In order to measure the degree of depolarization of the second-order lines, we tried a variety of modifications of the method ordinarily used for measuring this parameter. First of all the measurements were made with plane-parallel polaroids, placed inside the illuminating system (condenser) between the sample and the light source. However, we found that, owing to the long duration of the exposure, the polaroids were damaged and the results of the measurements were substantially distorted. We therefore decided to use polarization prisms, placing these in front of the spectrograph slit.

In view of the fact that the high-transmission spectrograph gave a greatly reduced image when measuring the degree of depolarization, we were unable to obtain good-quality photographs with two polarization prisms placed simultaneously in front of the slit and had to use a simplified measuring method.

We used the depolarization data chiefly in order to relate the observed spectral lines to specific types of molecular vibrations. For this purpose high accuracy in determining the degree of depolarization was not required, it being sufficient to know whether a particular line in the spectrum were polarized or depolarized. Hence, in view of the long exposures (50-100 h) required for the photographs used in measuring the degree of depolarization of the second-order lines, we confined our attention simply to the σ component of the scattered light, assuming that the relative intensity of the π component and the intensities on the ordinary photographs were approximately the same. The coefficient converting the intensities of the lines on the photographs of the σ component to a common scale was determined by referring the intensity of the depolarized first-order lines on these photographs to the intensity at the maximum of the corresponding lines on the ordinary photographs, equating the resultant ratio to $^6/_7$. This method leads to errors in determining the true value of the degree of depolarization of the strongly polarized lines, but this is not important for present purposes.

It is well known that

$$\rho_{tr} = \frac{I_x}{I_z},\tag{1}$$

where I_x and I_z are the intensities at the maxima of the σ and π components of the scattered light, respectively.

In our own method we measured the quantities

$$\rho = k\,\frac{I_x}{I_x + I_z},\tag{2}$$

where k is a coefficient converting the intensities of the lines of the σ-component photographs to a common scale; $I_x + I_z = I_0$ (I_0 is the intensity of the lines on the common scale).

It follows from (1) that $I_X = \rho_{tr} I_z$; then

$$\rho_{obs} = \frac{k\rho_{tr} I_z}{\rho_{tr} I_z + I_z} = k\frac{\rho_{tr}}{1 + \rho_{tr}}. \tag{3}$$

In determining the coefficient k from this formula we expressed the normalizing conditions in the following way. For $\rho_{tr} = {}^6/_7$, $\rho_{obs} = \rho_{tr} = {}^6/_7$. Then, from (3), ${}^6/_7 = k[{}^6/_7/ (1 + {}^6/_7)]$, whence $k = {}^{13}/_7$. Finally, we obtain

$$\rho_{obs} = 1.86 \frac{\rho_{tr}}{1 + \rho_{tr}}. \tag{4}$$

Specifying a number of values of ρ_{tr} for ρ_{obs} in succession, we obtain the following:

ρ_{tr}	0	0.1	0.2	0.3	0.4	0.5	0.6	0.7	0.86
ρ_{obs}	0	0.17	0.31	0.43	0.53	0.62	0.7	0.77	0.86

We see from these results that our values of ρ_{obs} for the strongly polarized lines are too high. The values of ρ_{tr} may be obtained if necessary from formula (4).

In measuring degrees of depolarization, one sometimes employs polaroids similar in parameters to PPU polaroids of type 4, which were developed by the Institute of Crystallography of the Academy of Sciences of the USSR [21].

The second-order lines may only be separated when they lie at a fair distance from first-order lines. Hence the 600-1650 cm^{-1} range is very difficult to study in the spectra of hydrocarbons, as it usually contains a large number of very strong lines. On the other hand, a suitable region for studying second-order lines in the spectra of hydrocarbons is the 1650-2800 cm^{-1} range, which cannot contain any fundamental frequencies; for the same reason, any possible impurities cannot reveal themselves in this region.

Special attention was paid to the separation of weak lines of other origins from the overtones and composite frequencies. By photographing the spectrum of the lamps employed, the weak mercury lines and the lines of other elements accompanying these were eliminated. The Raman lines from other excited lines were eliminated by means of filters. Glass Schott filters were mainly used: GG-3 for separating out the mercury line at 4358 Å and BG-15 for the 4047 Å line. It was found that the total frequencies of the CH valence vibrations and the low frequencies excited by the violet line were superimposed on the overtones excited by the blue mercury line. However, comparison of the photographs obtained with and without the filter showed that the intensity of the interfering lines was low and could be neglected. Hence, in order to economize on time, the polarization photographs were taken (as far as possible) without filters.

The samples of the compounds studied were carefully purified; however, traces of impurities remaining in some of them appeared in the spectrum, and this will be considered in more detail in the corresponding sections.

2. Study of Infrared Absorption Spectra

The infrared absorption spectra were recorded on a two-beam automatic IKS-14 spectrometer in the range 650 to 9000 cm^{-1}. The accurate measurement of frequencies in the second-order spectra is very important for the determination of the anharmonicity coefficients of the molecular vibrations. The values of these anharmonicity coefficients usually lie between 10 and 15 cm^{-1}. Hence, a measuring error of a few cm^{-1} may lead to a considerable distortion of the anharmonicity coefficients.

In the Raman spectrum, by averaging data obtained from several photographs, quite accurate values of the frequencies of the observed lines can be obtained (within 1-2 cm^{-1}). In the infrared spectrum, however, it is practically impossible to obtain this accuracy for the frequencies. This fact is largely associated with the calibration of the apparatus. Great attention thus had to be paid to the calibration of the spectrometer.

In the range lying between 650 and 2000 cm^{-1} an NaCl prism was used, the calibration being carried out by reference to the indene spectrum [22]. In the indene spectrum there are 29 standard points in the range up to 2000 cm^{-1}, and with these a calibration accuracy of 2-4 cm^{-1} can be achieved.

The infrared spectrum between 1800 and 4500 cm^{-1} was covered by means of a LiF prism. Special attention was paid to the exact calibration of this range, since the second-order lines of present interest usually extended up to 4000 cm^{-1}. This range was calibrated by reference to the spectra of ammonia, polystyrene, and 1,2,4-trichlorbenzene [23]. There were about 30 standard frequencies which could be specified by reference to these three substances; however, these lay mainly above 2800 cm^{-1}. Hence, in order to calibrate the range up to 2800 cm^{-1}, we used additional data obtained from Raman spectra. These data included the line frequencies of the chloroform (3018, 2443, 2440 cm^{-1}), deuterochloroform (2253 cm^{-1}), methyldichlorsilane (2221 cm^{-1}), ethyldichlorsilane (2210 cm^{-1}), and acetonitrile (2246 cm^{-1}) molecules.

In the range between 4900 and 9000 cm^{-1}, the spectrometer was calibrated with an F-1 prism by reference to the spectra of benzene, chloroform, and trichlorethylene [24].

In all cases the working parts of the calibration curves constituted straight lines. Within the limits of measuring errors, the end of the calibration curve relating to the NaCl prism overlapped the origin of the graph relating to LiF; the same occurred for the graphs of the LiF and F-1 prisms.

Thus, as a result of the careful calibration of the apparatus, the errors introduced into the frequency values on account of this factor were never greater than 2-4 cm^{-1}. Other factors causing inaccuracies in the frequency measurements (insufficiently accurate reproducibility of the spectrum, indeterminacy of the position of the band maxima) also introduce an additional error. Hence, the over-all error committed in measuring the frequencies used for studying the absorption bands equalled 3-6 cm^{-1}.

In unfavorable cases (poorly resolved lines, presence of a strong perturbing band near a weak one) we took photographs with different cuvette thicknesses (between 0.01 and 10 cm) and also with different recording speeds. In all cases measurements were made for at least three photographs.

By comparing the results with published data, we see that the discrepancies in measuring frequencies up to 4000 cm^{-1} lie within 3-6 cm^{-1}. No systematic errors were observed.

In the range 4000 to 9000 cm^{-1} we obtained the complete spectrum of deuterochloroform. As there were no published data for the vibrational spectra of deuterochloroform we were unable to make any comparisons.

The overall errors committed in measuring the frequencies and anharmonicity coefficients in the infrared spectra were higher than in the Raman spectra, despite all the precautions taken. Hence, although the anharmonicity coefficients were calculated in the infrared spectra, they were only used for qualitative comparisons. However, this does not mean that the data obtained from the infrared spectra cannot be used for studying second-order lines in the vibrational spectra of molecules. On the contrary, a study of the infrared spectrum greatly eases the interpretation of the second-order Raman spectra, which is very important when studying vibrational spectra, particularly for calculating the anharmonicity coefficients of the corresponding vibration. In addition to this, the use of data relating to the infrared spectrum is important for determining the general laws relating to second-order spectra.

3. Calculation of the Anharmonicity Coefficients

of the Molecular Vibrations

For calculating the anharmonicity coefficients we used the following formula for the vibrational energy levels of the molecule [25, 26]

$$G(v_1, v_2, v_3, \ldots) = \sum_i \omega_i \left(v_i + \frac{d_i}{2}\right) + \sum_i \sum_{k > i} x_{ik} \left(v_i + \frac{d_i}{2}\right)\left(v_k + \frac{d_k}{2}\right) + \sum_i \sum_{k > i} g_{ik} l_i l_k + \cdots \quad (5)$$

Here v_i are the vibrational quantum numbers for the normal vibrations, ω_i are the frequencies of the normal vibrations, d_i is the degree of degeneracy, x_{ik} is the anharmonicity coefficient, l_i are whole quantum numbers which take the values $l_i = v_i, v_i - 2, v_i - 4, \ldots, 1,$ or 0, g_{ik} are coefficients characterizing the coupling of the vibrations and rotations.

For nondegenerate vibrations, $l_i = 0$ and $g_{ik} = 0$. The last term, which has to be taken into account for degenerate vibrations, we shall in general neglect, considering it to be very small. This was checked for the case of the $CHCl_3$ and $CDCl_3$ molecules.

Let us write formula (5) in expanded form for a case with three vibrational quantum numbers

$$G(v_1, v_2, v_3) = \omega_1\left(v_1 + \frac{d_1}{2}\right) + \omega_2\left(v_2 + \frac{d_2}{2}\right) + \omega_3\left(v_3 + \frac{d_3}{2}\right) + x_{11}\left(v_1 + \frac{d_1}{2}\right)^2 + x_{22}\left(v_2 + \frac{d_2}{2}\right)^2 + x_{33}\left(v_3 + \frac{d_3}{2}\right)^2$$

$$+ x_{12}\left(v_1 + \frac{d_1}{2}\right)\left(v_2 + \frac{d_2}{2}\right) + x_{13}\left(v_1 + \frac{d_1}{2}\right)\left(v_3 + \frac{d_3}{2}\right) + x_{23}\left(v_2 + \frac{d_2}{2}\right)\left(v_3 + \frac{d_3}{2}\right). \quad (6)$$

The vibrational energy is not usually referred to the minimum of the potential surface, as in (6), but to the lowest vibrational state ($v_1 = 0$, $v_2 = 0$, $v_3 = 0$), the energy of which, referred to the energy minimum, equals

$$G(0, 0, 0) = \frac{d_1}{2}\omega_1 + \frac{d_2}{2}\omega_2 + \frac{d_3}{2}\omega_3 + \frac{d_1^2}{4}x_{11} + \frac{d_2^2}{4}x_{22} + \frac{d_3^2}{4}x_{33} + \frac{d_1 d_2}{4}x_{12} + \frac{d_1 d_3}{4}x_{13} + \frac{d_2 d_3}{4}x_{23}. \quad (7)$$

It follows from this that

$$G(v_1, v_2, v_3) - G(0, 0, 0) = \omega_1 v_1 + \omega_2 v_2 + \omega_3 v_3 + x_{11} v_1^2 + x_{22} v_2^2 + x_{33} v_3^2 + v_1 x_{11} d_1 + v_2 x_{22} d_2 + v_3 x_{33} d_3 +$$

$$+ x_{12} v_1 v_2 + x_{13} v_1 v_3 + x_{23} v_2 v_3 + \frac{1}{2} x_{12}(v_2 d_1 + v_1 d_2) + \frac{1}{2} x_{13}(v_3 d_1 + v_1 d_3) + \frac{1}{2} x_{23}(v_2 d_3 + v_3 d_2). \quad (8)$$

Substituting the values of the vibrational quantum numbers for the transitions corresponding to the fundamental frequencies, overtones, and composite frequencies successively into (8), we obtain

$$G(1, 0, 0) = v = \omega_1 + x_{11} + x_{11} d_1 + \frac{1}{2} x_{12} d_2 + \frac{1}{2} x_{13} d_3, \quad (9)$$

$$G(2, 0, 0) = v_1 = 2\omega_1 + 4x_{11} + 2x_{11} d_1 + x_{12} d_2 + x_{13} d_3, \quad (10)$$

$$G(3, 0, 0) = v_2 = 3\omega_1 + 9x_{11} + 3x_{11} d_1 + \frac{3}{2} x_{12} d_2 + \frac{3}{2} x_{13} d_3, \quad (11)$$

$$G(0, 1, 0) = v' = \omega_2 + x_{22} + x_{22} d_2 + \frac{1}{2} x_{12} d_1 + \frac{1}{2} x_{23} d_3, \quad (12)$$

$$G(0, 0, 1) = v'' = \omega_3 + x_{33} + x_{33} d_3 + \frac{1}{2} x_{13} d_1 + \frac{1}{2} x_{13} d_2, \quad (13)$$

$$G(1, 1, 0) = v_{12} = \omega_1 + \omega_2 + x_{11} + x_{22} + x_{11} d_1 + x_{22} d_2 + x_{12} + \frac{1}{2} x_{12}(d_1 + d_2) + \frac{1}{2} x_{13} d_3 + \frac{1}{2} x_{23} d_3, \quad (14)$$

$$G(1, 1, 1) = \nu_{123} = \omega_1 + \omega_2 + \omega_3 + x_{11} + x_{22} + x_{33} + x_{11}d_1 + x_{22}d_2 + x_{33}d_3 + x_{12} + x_{23} + x_{13} + \frac{1}{2}x_{12}d_1 +$$

$$+ \frac{1}{2}x_{12}d_2 + \frac{1}{2}x_{13}d_1 + \frac{1}{2}x_{13}d_3 + \frac{1}{2}x_{23}d_3 + \frac{1}{2}x_{23}d_2. \tag{15}$$

Carrying out some simple calculations, we then find:

$$G(2, 0, 0) - 2G(1, 0, 0) = \nu_1 - 2\nu = 2x_{11},$$

$$x_{11} = \frac{\nu_1 - 2\nu}{2}; \tag{16}$$

$$G(3, 0, 0) - 3G(1, 0, 0) = \nu_2 - 3\nu = 6x_{11},$$

$$x_{11} = \frac{\nu_2 - 3\nu}{6}; \tag{17}$$

$$G(1, 1, 0) - [G(1, 0, 0) - G(0, 1, 0)] = \nu_{12} - (\nu + \nu') = x_{12},$$

$$x_{12} = \nu_{12} - (\nu + \nu'); \tag{18}$$

$$G(1, 1, 1) - [G(1, 0, 0) + G(0, 1, 0) + G(0, 0, 1)] = \nu_{123} - (\nu + \nu' + \nu'') = x_{12} + x_{13} + x_{23},$$

$$x_{12} + x_{13} + x_{23} = \nu_{123} - (\nu + \nu' + \nu''). \tag{19}$$

In this way we obtained working formulas for determining the anharmonicity coefficients from experimental data relating to lines of the first, second, and third orders. In general form these formulas appear as

$$x_{ii} = \frac{1}{2}(\nu_1 - 2\nu) = \frac{1}{6}(\nu_2 - 3\nu) = \frac{1}{n(n+1)}[\nu_n - (n+1)\nu],$$

$$x_{ik} = \nu_{12} - (\nu + \nu') = \nu_{123} - (\nu + \nu' + \nu'') = \nu_{ik\ldots} - \nu - \nu^i - \nu^k - \ldots$$

Here, $k > i = n = 1, 2, 3, \ldots$, ν is the fundamental frequency, ν_1 is the frequency of the first overtone, ν_2 is the frequency of the second overtone, and ν_{12} is the frequency of the composite tone.

CHAPTER III

EXPERIMENTAL RESULTS

1. Second-Order Lines in the Raman Spectrum of Deuterocyclohexane

It is well known that cyclohexane is the simplest representative of the six-member naphthenes, an important class of hydrocarbons. Hence, a study of the vibrational spectrum of this hydrocarbon is of great significance for understanding the structure of the molecules in other six-membered naphthenes. In addition to this, the cyclohexane molecule has a high symmetry; vibrational calculations have already been carried out for it [27-29], so that it is much easier to interpret the second-order lines in the vibrational spectra of this molecule. The second-order lines in the Raman spectrum of cyclohexane were accordingly studied in [6].

For the deuterocyclohexane molecule also, calculations of the vibrational spectrum have been carried out [30]. We felt it particularly desirable to study the second-order lines in the Raman spectrum of C_6D_{12} in order to compare with the cyclohexane spectrum and establish some general laws.

The first-order Raman spectrum of C_6D_{12} was studied in [31]. Some weak lines were also observed, and these were interpreted as overtones and composite frequencies. Unfortunately, the interpretation of the lines given in [31] was carried out, not in order to understand the second-order spectra of deuterocyclohexane, but in order to verify the D_{6h} model for the cyclohexane molecule. This model was later disproved. For this reason we shall not discuss the interpretation of the spectrum given by Langseth and Bak [31] here.

Our sample of deuterocyclohexane was a colorless liquid with a boiling point of 81°C and a refractive index of 1.4221. The C_6D_{12} content according to specification was 97%, the possible hexadeuterobenzene content being no more than 0.5%.

During the investigation we observed that drops were formed in the working part of the vessel, not mixing with the main mass of the liquid; the continuous background accordingly became much greater, impeding precision measurements of the Raman lines. The sample was therefore subjected to special additional purification by the method of [32]. First the sample was washed several times in a cold mixture of concentrated nitric and sulfuric acids in order to nitrify the benzene possibly present in the specimen. After repeated washing with distilled water, the deuterocyclohexane was subjected to fractional distillation over sodium. After this purification of the deuterocyclohexane the continuous background in the spectrum was eliminated, and it became possible to measure the intensity of the second-order lines quantitatively to 0.5 units.

As already mentioned, our sample of deuterocyclohexane may have contained not more than 0.5% of hexadeuterobenzene. However, in the C_6D_{12} spectrum none of the strongest lines of hexadeuterobenzene known from published data [33] and our own measurements were observed.

Table 1. Interpretation of the Fundamental Frequencies of Cyclohexane
and Deuterocyclohexane

Active frequencies, cm^{-1}						Inactive frequencies, cm^{-1}				
Raman spectrum			Infrared spectrum				C_6H_{12}		C_6D_{12}	
Symmetry	C_6H_{12}	C_6D_{12}	Symmetry	C_6H_{12}	C_6D_{12}	Symmetry	calc. in [30]	refined	calc. in [30]	refined
A_{1g}	384	298	A_{2u}			A_{2g}				
	802	724		523						
	1158	1015		905	783		1117	1150	801	820
	1445	1120		1450	1092					
	2941	2153		2854	2150		1355		1080	
	2924	2194		2927	2250					
E_g	427	373	E_u			A_{1u}	1121		822	
	785	637		861	678					
	1029	796		905	710		1150		865	
	1267	937		1256	992					
	1348	1214		1350	1176		1377		1276	
	1445	1071		1450	1076					
	2852	2082		2927	2190					
	2886	2106		2927	2220					
	2895	2119								
	2924	2170								

The spectrum of our sample showed two weak lines at 2886 and 2924 cm^{-1}, which may be ascribed to the CH valence vibrations of the partly deuterized derivatives of cyclohexane. We note that the strongest lines of such compounds should be distributed between 802 cm^{-1} (cyclohexane) and 724 cm^{-1} (deuterocyclohexane). In this range we observed only one line (746 cm^{-1}) in the spectrum; this line was also noted in [31], where the deuterocyclohexane sample was prepared by a different method. Hence, the 746 cm^{-1} line can hardly belong to an impurity.

Thus we may suppose that our sample of deuterocyclohexane contained slight traces of certain partially deuterized derivatives of cyclohexane. The amount of each of these impurities was too small to show its characteristic lines. The lines in the range 2886 to 2924 cm^{-1}, however, were common for all the impurities in question and thus appeared in the spectrum. From this point of view it is not impossible that our 1440 cm^{-1} line was also due to α vibrations of the CH groups in these compounds.

In interpreting the spectra we used the calculations of [30] relating to the form and frequencies of the vibrations of cyclohexane and deuterocyclohexane. The interpretation of the fundamental frequencies of cyclohexane and deuterocyclohexane is given in Table 1.

In order to interpret the second-order spectral lines of deuterocyclohexane we set up a table of all possible combinations of frequencies of first-order lines allowed by the selection rules. In the range between 250 and 2900 cm^{-1} there were 182 of these. The difference transitions which required that the molecule should have a high vibrational energy before interaction with the photon were not considered. The experimentally observed frequencies of the second-order lines of deuterocyclohexane (25 of them) were compared with the nearest ones appearing in Table 1, allowing for the type of vibrations, the degree of depolarization, and the intensity. It was thus found that 17 of the second-order lines of deuterocyclohexane and cyclohexane could be interpreted unambiguously as the combinations of fundamental vibration frequencies, similar,

Table 2. Interpretation of the Second-Order Lines in the Raman Spectrum of Deuterocyclohexane

C_6D_{12}						C_6H_{12}			
$\Delta\nu$, cm^{-1}	I_0	ρ	Interpretation	Type of vibrations	Δx_{im}, cm^{-1}	$\Delta\nu$, cm^{-1}	Δx_{im}, cm^{-1}	I_0	ρ
598	1.3	—	2·298=596	A_{1g}	1	771	−1	0.1	p
746	9	—	2·373=746	$A_{1g}+E_g$	0	852	−1	0.5	p
1257	0.9	0.8	298+937=1235	E_g	22	1642	−9	0.1	d
1286	1.0	0.66	2·637=1274	$A_{1g}+E_g$	6	1574	2	1.2	0.65
1301	7.0	0.5	373+937=1310	$A_{1g}+A_{2g}+E_g$	−8	1691	3	2.0	0.67
1354	0.8	—	2·678=1356	$A_{1g}+E_g$	−1	1772	−1	1.2	0.47
1440	1.0	0.4	373+1071=1444	$A_{1g}+A_{2g}+E_g$	−4	1770	−2	1.0	0.63
1489	1.0	0.6	390+1092=1482	A_{1g}	—	1978	−1	1.9	0.67
1544	0.9	—	724+801=1525	E_g	—	1952	—	1.5	0.67
1566	0.6	—	2·783=1566	A_{1g}	0	1810	0	0.5	0.42
1591	0.5	—	2·796=1592	$A_{1g}+E_g$	−0.5	2056	−1	1.4	0.54
1612	0.4	—	373+1214=1587	$A_{1g}+A_{2g}+E_g$	25	1777	2	0.4	0.18
1637	0.3	—	678+992=1670	$A_{1g}+A_{2g}+E_g$	−33	2119	−2	1.0	0.53
1701	2.0	0.8	637+1071=1708	$A_{1g}+A_{2g}+E_g$	−3	2227	−3	2.7**	0.78
1748	3.3	0.7	1120+637=1757	E_g	−9	2227	−3	2.7**	0.78
1800	1.0	0.6	710+1076=1786	$A_{1g}+A_{2g}+E_g$	14	2374	16	0.7	0.78
1846	2.2	0.4	724+1120=1844	A_{1g}	2	2254	7	2.7**	0.78
1874	1.3	0.6	2·937=1874	$A_{1g}+E_g$	0	2529	−5	1.3	0.58
1909	1*		1120+796=1916	E_g	−7	2458	−16	3.5	0.82
—	—	—	785+1348=2133	$A_{1g}+A_{2g}+E_g$	—	2162	−29	1.5	0.62
1941	3.1	0.5	2·992=1984	$A_{1g}+E_g$	21	—	—	—	—
2062	10*		992+1076=2068	$A_{1g}+E_g+A_{2g}$	−6	2700	−11	16	0.33
2230	18		2·1120=2240	A_{1g}	−5	—	—	—	—
2313	27	0.1	2·1176=2352	$A_{1g}+E_g$	−20	2667	−18	32	0.17
2380	12.8	0.15	298+2106=2404	A_{1g}	−24	—	—	—	—
2415	11	0.12	2·1214=2428	$A_{1g}+E_g$	−6	2689	−3	2*	—

Note.   p = polarized, d = depolarized.
*Intensity estimated visually.
* *Intensity and polarization given are the total values for the 2227 and 2254 cm^{-1} lines.

as regards symmetry and form, in both compounds.   The resultant data are presented in Table 2, together with the data for cyclohexane [34].

If we follow the principle of "minimum anharmonicity coefficients" strictly (see [6]), we cannot find any analogs in the cyclohexane spectrum for certain of the second-order lines of deuterocyclohexane and vice versa. However, if we deviate slightly from this principle for one of the compounds, the second-order lines of both compounds may be given a better interpretation. We did this for the 1612, 1637, 1748 cm^{-1} lines of deuterocyclohexane and the 2374, 2700, 2667 cm^{-1} lines of cyclohexane, while still preserving the principle of minimum anharmonicity for their respective analogs.

We interpreted the 1489 cm^{-1} line as a composite frequency by using the earlier-calculated [30] frequency of 390 cm^{-1} (A_{2u} type), which is allowed by the selection rules but not found experimentally. Corresponding to the two deuterocyclohexane lines at 1701 and 1748 cm^{-1} there was one cyclohexane line at 2227 cm^{-1}. In the case of the 2230 and 2380 cm^{-1} lines of deuterocyclohexane, the cyclohexane lines corresponding to these fell in the region of 2900 cm^{-1}, which was inaccessible for study owing to the nearness of the strong lines of the CH valence vibrations. For the same reason (overlapping by lines belonging to the CD valence vibrations excited by the violet lines) we were unable to study the low-frequency range of deuterocyclohexane.

In interpreting the spectra we tried as much as possible not to employ combinations involving forbidden frequencies of the A_{1u} and A_{2g} type, calculated in [30]. It was considered risky to rely on frequencies so obtained, although our refusal to use them was not really justified.

We were forced to use the calculated frequency 801 cm^{-1} (A_{2g} type) in order to interpret the line at 1544 cm^{-1} (deuterocyclohexane), and the corresponding combination with the frequency 1117 cm^{-1} in order to explain the 1952 cm^{-1} line of cyclohexane. Agreement between the observed and calculated frequencies improved somewhat on considering that the calculated frequencies were too low and that the actual values of the forbidden frequencies were 820 and 1150 cm^{-1}, respectively. In this way we obtained the following values of the composite frequencies:

$$820 + 724 = 1544 \text{ cm}^{-1} \text{(deuterocyclohexane)},$$
$$1150 + 802 = 1952 \text{ cm}^{-1} \text{(cyclohexane)}.$$

Using these values of forbidden frequencies, we may obtain a more satisfactory interpretation of the line at 1612 cm^{-1} than that given in Table 2, namely:

$$820 + 796 = 1616 \text{ cm}^{-1} \text{(deuterocyclohexane)},$$
$$1150 + 1029 = 2179 \text{ cm}^{-1} \text{(cyclohexane, experimental line 2162 cm}^{-1}\text{)}.$$

The interpretation given in Table 2 (without using forbidden frequencies) leads to considerable anharmonicity coefficients.

For the 1257 cm^{-1} line of deuterocyclohexane and the corresponding 1642 cm^{-1} line of cyclohexane, the anharmonicity coefficients were opposite in sign. This may be avoided if we use a combination involving forbidden frequencies, for example, $390 + 865A_{1u} = 1255E_g$.

Thus, the second-order lines enable us to refine the values of the forbidden fundamental frequencies.

As shown by the data presented, the anharmonicity coefficients of the cyclohexane and deuterocyclohexane lines of analogous origin are the same in sign and similar in magnitude. The exceptions may be explained if we employ forbidden fundamental frequencies. The intensities of the analogous lines in the two spectra are, as a rule, of the same order. The considerable intensity of the 746 cm^{-1} line may be explained by reference to Fermi resonance with the nearby frequency of 724 cm^{-1}. The intense 2313 cm^{-1} line of deuterocyclohexane and the corresponding 2667 cm^{-1} line of cyclohexane should be particularly noted.

We see from the catalog [18] that the spectra of all cyclohexane derivatives show a line with a frequency of the order of 2660 cm^{-1} and a considerable intensity. We may therefore suppose that the strong lines in the cyclohexane and deuterocyclohexane spectra mentioned above are overtones of lines with the characteristic frequencies of the CH_2 (CD_2) groups due to β deformation (strain) vibrations.

2. Study of Overtones and Composite Frequencies in the Vibrational Spectra of Benzene and Hexadeuterobenzene

The detailed study of the vibrational spectra of benzene and its derivatives is of considerable interest, since the problem of the structure of benzene is one of the classical problems of organic chemistry. Despite the large number of papers devoted to benzene [35-44], detailed studies of the benzene structure (including the study of its second-order lines) are few and far between. The main problem treated in these papers was that of establishing the model of the benzene molecule and the corresponding complete system of fundamental frequencies, including inactive frequencies, forbidden by the selection rules. We note that a number of "forbidden" frequencies in fact appear in the spectra (we shall return to this question later).

In view of the great importance and complexity of the problem in question, we shall give some more detailed consideration to certain papers devoted to the study of the spectra arising from benzene and its deuterium derivatives.

Ingold and his colleagues [45, 46] studied the Raman and infrared absorption spectra of benzene and its deuterium derivatives, and also studied the ultraviolet absorption spectra of C_6H_6 and C_6D_6. By analyzing the results, the authors came to the conclusion that the benzene molecule had D_{6h} symmetry, and on this basis they gave the complete system of fundamental frequencies for the two substances.

The values of the active frequencies were obtained directly from observation of the spectra, and the "forbidden" frequencies by using the rule of products, subsequently checking the values obtained by reference to the observed second-order lines.

Despite the fact that the method described for finding the inactive frequencies in in principle quite correct, serious errors have been committed in its practical use. These errors are associated, first, with the indeterminacy in the interpretation of the lines in the vibrational spectra, and, secondly, with the insufficient accuracy of the "rule of products," which should be applied not to the experimental frequencies, but to the "zero" oscillation frequencies. As a result of such errors, certain inactive frequencies obtained by Ingold were not subsequently confirmed. Mair and Hornig [47] gave some data relating to the infrared spectrum of liquid benzene at 28°C and crystalline benzene at −12, −65, and −170°C. The work was carried out over the range 600–4600 cm^{-1}. The frequency data of [47] relating to liquid benzene agreed in general with the results of [46].

In the spectrum of crystalline benzene * appeared such effects as a change in the vibration frequencies and the development of new vibrational frequencies, absent in the liquid state. It was characteristic of the spectra of crystalline benzene that the fundamental lines were very narrow. This factor was allowed for by the authors in interpreting the inactive fundamental frequencies. To these the authors ascribed the frequencies 975, 987, 1010, 3070, 1147, and 1312 cm^{-1}, forbidden in the infrared spectrum in accordance with the D_6 model, but appearing in practice as very narrow lines. The remaining three inactive frequencies were found in [47] from the observed second-order lines.

In this way all the fundamental frequencies of benzene for the D_{6h} model were determined in [47]. These frequencies in general coincided with the values given by Ingold, except for the frequencies of vibrations of the B_{2u} type (in cm^{-1}): Mair, 1310, 1150; Ingold, 1648, 1110. The interpretation of Mair et al. was set in doubt by [49], in which the infrared spectrum of liquid benzene was studied. In this paper the frequencies of the B_{2u} type given by Ingold were confirmed. However, in view of the fact that the investigation was carried out in the range 4000 to 10,000 cm^{-1}, where measuring errors are usually large, the reliability of the results presented in [49] is doubtful.

In a paper by Miller [50], the infrared spectra of liquid and gaseous C_6D_6 were studied by means of a Perkin—Elmer spectrometer in the range 300 to 3700 cm^{-1}. The C_6D_6 spectrum was thus studied in two states, and Miller was able to give a more definite interpretation of the lines observed. It is well known that forbidden lines breaking the selection rules may sometimes appear in the spectrum of a liquid, owing to intermolecular interactions, whereas in a gas the selection rules are obeyed more strictly. In Miller's paper a new interpretation was given for the 1552 cm^{-1} line in the infrared spectrum of liquid C_6D_6. This line was formerly

---

*The infrared spectra of crystalline benzene were studied earlier by Halford and Schaffer [48] but with low-resolution apparatus.

Table 3. Interpretation of Certain Lines in C_6H_6 and C_6D_6 [7]

Interpreta-tion	Type of vibra-tions	C_6H_6			C_6D_6		
		Frequency, cm^{-1}			Frequency, cm^{-1}		
		calc.	exptl.	$-x_{ik}$, cm^{-1}	calc.	exptl.	$-x_{ik}$, cm^{-1}
$\nu_{17} + \nu_{14}$	E_{1u}	2213	2211	2	1681	1680	1
$\nu_{11} + \nu_{13}$	A_{2u}	2329	2326	3	1994	1994	0
$\nu_{17} + \nu_{13}$	E_{1u}	2653	2653	0	2199	2198	1

Table 4. Frequency Differences in the Spectra of the Vapor
and Liquid Forms of Benzene and Benzene Derivatives, cm^{-1}

C_6H_6			$C_6H_3D_3$			C_6D_6		
Vapor	Liquid	Δ	Vapor	Liquid	Δ	Vapor	Liquid	Δ
1037	1035	2	833	833	0	814	812	2
1482	1479	3	1414	1412	2	1333	1330	3
3064	3053	11	3063	3053	10	2288	2276	12

considered as a fundamental line of an E_{2g} vibration, the appearance of which infringed the selection rule. However, in Miller's work the line appeared in the spectrum of the vapor as well as the liquid. Hence the assumption that the selection rule had been broken with respect to the 1552 cm^{-1} line was refuted and the line was interpreted as a composite frequency $\nu_4 + \nu_{12}$.

Giving preference to the values of B_{2u}-type frequencies for C_6H_6 found in [47], Miller [50] determined the corresponding frequencies 825 and 1571 cm^{-1} for C_6H_6, which differed from the values of 838 and 1287 cm^{-1} given by Ingold.

The observed bands 709, 1862, 2156, 2831 cm^{-1} for which no other explanation could be found were interpreted unambiguously as combinations involving the frequency 1287 cm^{-1}. The line 2391 cm^{-1} was explained as a combination involving the frequency 838 cm^{-1}, which constitutes an additional confirmation of the frequency values found.

The infrared spectra of liquid and gaseous C_6D_6, $C_6H_3D_3$, and C_6H_6 were studied in [7]. For the interpretation of the observed bands of these molecules, the values of the fundamental frequencies inactive in the infrared spectrum from [47, 50] were taken as a first approximation. Then all the binary combinations active in the infrared spectrum were calculated and provisionally interpreted by comparison with the observed spectra. The authors observed that for combinations consisting of fundamental lines appearing in the Raman or infrared spectra the discrepancies between the observed and calculated frequencies were small (1-3 cm^{-1}) as may be seen from Table 3. This fact was extended as a general principle to combinations involving fundamental frequencies forbidden by the selection rules in Raman and infrared spectra.

The study of fundamental lines appearing in the spectra of the vapor and liquid forms of the molecules under consideration showed that the differences Δ between the frequencies of one type of vibration in the two states were almost the same for all three molecules (Table 4).

This experimental fact was extended to forbidden frequencies also, as a second general principle. The values of the fundamental frequencies forbidden in the infrared spectrum were determined on the basis of these two principles; in the view of the authors this gave a better

Table 5. Interpretation of the Fundamental Frequencies of Benzene
and Hexadeuterobenzene

Active frequencies, cm^{-1}				Inactive frequencies, cm^{-1}			
Symmetry	Interpret.	C_6H_6	C_6D_6	Symmetry	Interpret.	C_6H_6	C_6D_6
A_{1g}	ν_1	3062	2291	A_{2g}	ν_3	1346	1055
	ν_2	992	943	B_{1u}	ν_5	3060	2290
A_{2u}	ν_4	676	496		ν_6	1010	963
E_{1g}	ν_{11}	849	662	B_{2g}	ν_7	985	827
					ν_8	703	601
E_{1u}	ν_{12}	3054	2280	B_{2u}	ν_9	1309	1287
	ν_{13}	1480	1330		ν_{10}	1150	830
	ν_{14}	1037	811				
E_{2g}	ν_{15}	3047	2265	E_{2u}	ν_{19}	970	793
	ν_{16}	1596	1550		ν_{20}	405	349
	ν_{17}	1177	864				
	ν_{18}	606	576				

reproduction of the observed spectrum. The values of the fundamental frequencies in general coincide with those of [47].

Thus, the results obtained from a number of investigations on the infrared spectra of benzene and its deuterium derivatives confirm the original model of D_{6h} proposed by Ingold. However, the interpretation of the inactive frequencies of certain types of vibrations given by Ingold was not confirmed. This interpretation is even now ambiguous with respect to certain lines.

The study of lines in the second-order spectra may provide extensive material for confirming one interpretation or the other. However, up to the present time the second-order lines in the spectra of benzene and its deuterium derivatives have not been fully studied. The infrared spectra of these molecules have been mainly considered. The second-order lines in the Raman spectra have not been investigated since Ingold's papers. We note that in Ingold's papers the degrees of depolarization were not measured, while the intensities were determined visually. In view of this it is of special interest to make a broadly based study of the second-order lines in the vibrational spectra of benzene and hexadeuterobenzene.

In the present investigation the benzene and hexadeuterobenzene samples were purified carefully by the method proposed in [51]. The samples were washed with several portions of concentrated sulfuric acid, water, a solution of caustic soda, and again water, until a neutral reaction was obtained, after which they were dried with roasted calcium chloride. The compounds thus purified were redistilled over metallic sodium. The Raman and infrared absorption spectra of hexadeuterobenzene showed lines at 3050 and 709 cm^{-1}, evidently due to the presence of C_6HD_5 in our sample. The frequency 709 cm^{-1} in the Raman spectrum of C_6D_6 we interpreted simulataneously as $2\nu_{20}$, assuming the overlapping of an overtone of C_6D_6 and a fundamental frequency of C_6HD_5, since the line 709 cm^{-1} had the form of a wide band. No other impurities were found in the compounds.

The second-order spectra of C_6H_6 and C_6D_6 were interpreted on the basis of the generally accepted model with D_{6h} symmetry. In interpreting our data by reference to overtones and composite frequencies, we tried a number of different variations of the fundamental frequencies. The most convincing results were obtained for the values given in Table 5.

As regards the most discussed frequencies of the A_{2g} and B_{2u} type, our interpretation coincides with that given by Broderson and Langseth [7]. This interpretation is also accepted in the theoretical treatments of Kovner [52] and Bogomolov [53, 54].

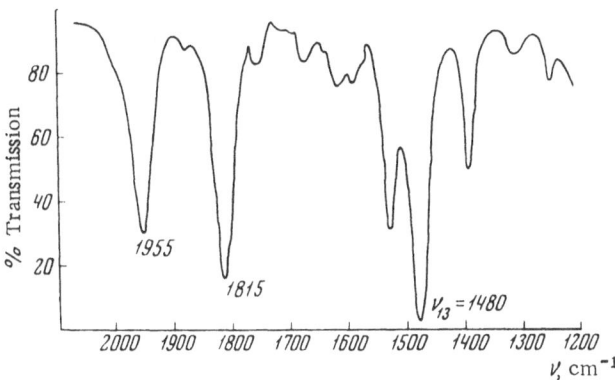

Fig. 1. Infrared absorption spectrum of benzene. Range
1400 to 2000 cm^{-1}. NaCl prism, cell thickness 0.5 mm.

In the Raman spectrum of benzene we observed 32 lines, and in the hexadeuterobenzene spectrum 33 lines of the second and third orders. Quantitative characteristics of the overtones and composite frequencies in the Raman spectra of C_6H_6 and C_6D_6 are given in Table 6. In the benzene spectrum the overtones and composite frequencies are mainly situated in the range 1600-2800 cm^{-1}, convenient for study owing to the absence of fundamental frequencies. In hexadeuterobenzene the second-order spectrum is slightly displaced in the direction of lower frequencies and falls in the range of fundamental frequencies. However, this did not prevent us from making some quantitative measurements, since, as a result of high dispersion of the apparatus our lines were sharply separated from one another. We interpreted the observed lines principally as belonging to the second order. Certain frequencies (1033 and 1092 cm^{-1} in C_6D_6 and 2543 cm^{-1} in C_6H_6) had to be interpreted as third-order lines. We note that as regards intensity these did not differ from in any way from second-order lines. Seventeen second-order lines of C_6D_6 and C_6H_6 are interpreted unambiguously as combinations of fundamental vibration frequencies, identical as regards type of symmetry in both compounds. The intensities of these lines in the spectra of C_6H_6 and C_6D_6 are as a rule quite close. The anharmonicity coefficients of the vibrations of the molecules in question are low. In the Raman spectrum of benzene there are neighboring (in frequency) groups of lines 2923, 2947 and 3164, 3185, 3208 cm^{-1} with anomalously high intensities. The rise in the intensity of these lines is eviddently due to Fermi resonance at the fundamental frequencies 3062 cm^{-1} (A_{1g}) and 3047 cm^{-1} (E_{2g}). In the hexadeuterobenzene spectrum the lines 2123 and 2417 cm^{-1} also have an increased intensity.

Data relating to the infrared spectra of C_6H_6 and C_6D_6 are given in Table 7 and in Figs. 1 and 2. We see that these are in good agreement with the results of [7]. It is interesting that individual composite frequencies in the infrared spectrum, for example 1815 and 1955 cm^{-1} in C_6H_6 and, correspondingly, 1448 and 1619 cm^{-1} in C_6D_6, are comparable in intensity with the fundamental frequencies. The band 4052 cm^{-1} in C_6H_6, which is interpreted as a composite frequency $\nu_6 + \nu_{15}$, is comparable in intensity with the fundamental frequencies of the CH vibrations. The very weak bands 2430 and 2453 cm^{-1} in C_6H_6 are interpreted (according to us) as difference frequencies ($\nu_{12} - \nu_{18}$, $\nu_5 - \nu_{18}$, respectively).

We see from the experimental results obtained and our study of the benzene and hexadeuterobenzene spectra, that there is no clear relationship between the intensity of the observed bands and the type of molecular vibrations. In fact, for example, the above-mentioned intense 1815 cm^{-1} band in C_6H_6 is interpreted as $\nu_{11} + \nu_{19}$, and the weak 1250 cm^{-1} band as $\nu_{11} + \nu_{20}$. Here, ν_{19} and ν_{20} belong to the same type of molecular vibration, E_{2u}, but the intensities of the composite frequencies at 1815 and 1250 cm^{-1}, which are explained as combinations of fundamental frequencies relating to a single type of vibrations, are entirely different.

Table 6. Interpretation of the Overtones and Composite Frequencies in the Raman Spectra of Benzene and Hexadeuterobenzene

Interpretation	Symmetry	C_6H_6 calc. frequency cm⁻¹	$-x_{ik}$ cm⁻¹	obs. frequency cm⁻¹	I_0	ρ	C_6D_6 calc. frequency cm⁻¹	$-x_{ik}$ cm⁻¹	obs. frequency cm⁻¹	I_0	ρ
$\nu_{13} - \nu_{19}$	E_{1g}	510	0	510	0.2	—	—	—	—	—	—
$\nu_{10} - \nu_{20}$	E_{1g}	745	5	740	0.2	—	—	—	—	—	—
$2\nu_{20}$	$A_{1g} + A_{2g} + E_{2g}$	810	4	802	4	p	698	−5	709	16	0.6
$\nu_4 + \nu_{20}$	E_{2g}	—	—	—	—	—	845	4	841	10	d
$2\nu_{18}$	$A_{1g} + A_{2g} + E_{2g}$	1212	−2	1217	0.5	—	1152	−3	1159	3	p
$2\nu_4$	A_{1g}	1352	−6	1365	2	p	992	−3	999	4	p
$\nu_{18} + \nu_{14} - \nu_{20}$	E_{1g}	—	—	—	—	—	1098	5	1033	3	—
$\nu_{14} + \nu_{18} - \nu_3$	$A_{1g} + A_{2g} + E_{2g}$	—	—	—	—	—	1086	−6	1092	2	d
$2\nu_8$	A_{1g}	1406	1	1400	7	p	1202	2	1197	4	p
$\nu_{19} + \nu_{20}$	$A_{1g} + A_{2g} + E_{2g}$	—	—	—	—	—	1142	5	1137	6	0.7
$\nu_{18} + \nu_8$	E_{1g}	—	—	—	—	—	1177	4	1173	3	d
$\nu_{18} + \nu_{11}$	E_{1g}	—	—	—	—	—	1238	6	1232	1	—
$\nu_4 + \nu_{19}$	E_{2g}	—	—	—	—	—	1289	6	1283	2	d
$\nu_2 + \nu_{18}$	E_{1g}	1597	—	1605	c.	d	1519	5	1514	1	—
$2\nu_{11}$	$A_{1g} + E_{2g}$	1698	2	1693	3	0.6	1324	0	1324	8	0.6
$\nu_{11} + \nu_7$	E_{1g}	1834	2	1832	0.1	—	—	—	—	—	—
$\nu_{17} + \nu_8$	E_{1g}	1880	8	1872	0.2	—	1465	6	1459	0.5	—
$\nu_7 + \nu_8$	A_{1g}	—	—	—	—	—	1428	7	1421	1	p
$2\nu_{19}$	$A_{1g} + E_{2g}$	1940	2	1936	2	p	1586	4	1578	3	p
$2\nu_2$	A_{1g}	1984	−1	1986	1	p	1886		1905	Полоса	
$2\nu_6$	A_{1g}	2020	−1	2022	0.9	p	—	—	—	—	—
$\nu_6 + \nu_{14}$	E_{2g}	2047	2	2045	0.5	—	1774	—	1792	Полоса	
$\nu_9 + \nu_{20}$	E_{1g}	—	—	—	—	—	1639	2	1637	3	d
$\nu_{18} + \nu_{20}$	E_{1g}	—	—	—	—	—	1679	1	1678	1	—
$2\nu_7$	A_{1g}	—	—	—	—	—	1654	1	1651	3	p
$\nu_{19} + \nu_{10}$	$A_{1g} + A_{2g} + E_{2g}$	2120	7	2113	0.7	d	—	—	—	—	—
$\nu_{18} + \nu_{16}$	$A_{1g} + A_{2g} + E_{2g}$	2191	6	2185	2	d	—	—	—	—	—
$\nu_{18} + (\nu_{18} + \nu_2)$	E_{1g}	2242	1	2211	1	0.7	—	—	—	—	—
$\nu_9 + \nu_{10}$	E_{1g}	2280	3	2277	0.4	—	2080	+8	2072	0.5	—
$\nu_8 + \nu_{16}$	E_{1g}	2289	−4	2293	3	d	—	—	—	—	—
$\nu_{14} + \nu_9$	E_{2g}	2347	−1	2348	2	—	2098	0	2098	0.4	—
$\nu_{19} + \nu_{13}$	E_{1g}	2450	−3	2453	6	d	—	—	—	—	—
$2\nu_{17}$	$A_{1g} + E_{2g}$	2356	−9	2375	1	p	1788	0	1728	0.4	—
$\nu_{14} + \nu_{13}$	$A_{1g} + A_{2g} + E_{2g}$	2517	9	2508	1	d	—	—	—	—	—
$\nu_{20} + \nu_{19} + \nu_{17}$	$A_{1g} + A_{2g} + E_{2g}$	2552	9	2543	7	p	—	—	—	—	—
$2\nu_9$	A_{1g}	2620	1	2619	10	p	2574	1	2571	8	p
$2\nu_3$	A_{1g}	2692	0	2692	0.7	—	2110	−6	2123	15	0.7
$\nu_2 + \nu_8 + \nu_{18}$	E_{1g}	—	—	—	—	—	2116	3	2108	3	—
$\nu_{16} + \nu_{17}$	$A_{1g} + A_{2g} + E_{2g}$	—	—	—	—	—	2414	−3	2417	6	0.7
$\nu_3 + \nu_{16}$	E_{2g}	2931	8	2923	8 } 0.7			—	—	—	—	—
$2\nu_{13}$	$A_{1g} + E_{2g}$	2952	5	2947	23 }		2660	1	2657	3	—
$\nu_3 + (2+18)$	E_{2g}										
$\nu_{17} + \nu_{15}$	$A_{1g} + A_{2g} + E_{2g}$	—	—	—	—	—	3129	6	3123	2	—
$2\nu_{16}$	$A_{1g} + E_{2g}$	3172	4	3164	7	p	—	—	—	—	—
$\nu_{16} + (\nu_2 + \nu_{18})$	$A_{1g} + E_{2g}$	3192	8	3184	20	p	3100	2	3095	4	—
$2 (\nu_2 + \nu_{18})$	$A_{1g} + E_{2g}$	3212	4	3208	4	p	—	—	—	—	—

Note. p = polarized, d = depolarized.

*Intensity estimated visually.

Table 7. Interpretation of the Overtones and Composite Frequencies of the Infrared Spectrum of Benzene and Hexadeuterobenzene

Interpretation	Symmetry	C_6H_6			C_6D_6		
		Calc. freq., cm⁻¹	$-x_{ik}$, cm⁻¹	Observed frequency, cm⁻¹	Calc. freq., cm⁻¹	$-x_{ik}$, cm⁻¹	Observed frequency, cm⁻¹
$\nu_{17} - \nu_{20}$	A_{2u}	773	4	769 m			
ν_{11}	E_{1g} forb			848 m			656 w
ν_{19}	E_{2u}			970 sh			
ν_2	A_{1g}			992 sh			
$\nu_{18} + \nu_{20}$	$A_{1u} + A_{2u} + E_{2u}$	1011	−6	1017 sh	925	−3	928 s
$\nu_{16} - \nu_{14}$	E_{1u}				739	5	734 w
$\nu_{11} + \nu_{20}$	E_{1u}	1254	4	1250 w	1011	4	1007 w
ν_9	B_{2u}			1309 vw			
$\nu_7 + \nu_{20}$	E_{1u}	1390	−3	1393 m	1176	−9	1185 w
ν_3	A_{2g}						1057 vw
?							1100 vw
$\nu_{11} + \nu_4$	E_{1u}	1525	−7	1532 m	1158	−8	1166 m
$\nu_{17} + \nu_{20}$	A_{2u}	1582	−7	1589 w	1213	−13	1226 m
$\nu_2 + \nu_{18}$	E_{2u}			1610 w			
$\nu_8 + \nu_{19}$	E_{1u}	1673	2	1671 w	1394	1	1393 s
$\nu_8 + \nu_6$	A_{2u}	1713	3	1710 vw	1564	2	1562 w
$\nu_{18} + \nu_{10}$	E_{1u}	1756	3	1753 vw			1502 vw
$\nu_{11} + \nu_{19}$	E_{1u}	1819	4	1815 s	1455	7	1448 s
$\nu_{13} + \nu_{20}$	E_{1u}	1885	8	1879 vw	1679	0	1879 vw
?							1715 vw
$\nu_7 + \nu_{19}$	E_{1u}	1955	0	1955 s	1619	0	1619 s
$\nu_{16} + \nu_{20}$	$A_{1u} + A_{2u} + E_{2u}$	2001	13	1988 w	1899	−7	1906 m
$(\nu_2 + \nu_{18}) + \nu_{20}$	$A_{1u} + A_{2u} + E_{2u}$	2003		2005 w			
$\nu_{18} + \nu_{13}$	E_{1u}	2086	1	2085 vw			
$\nu_{17} + \nu_{14}$	E_{1u}	2204	−2	2203 m			
$\nu_{16} + \nu_4$	E_{2u}	2272	−6	2278 w	2048	−11	2057 vw
$\nu_{13} + \nu_{11}$	A_{2u}	2330	4	2326 m	1992	0	1992 w
$\nu_6 + \nu_7$	A_{2u}				1790	1	1789 w
$\nu_{18} + \nu_9$					1858	2	1856 m
$\nu_{13} + \nu_8$	E_{2u}				1931	5	1926 m
$\nu_{14} + \nu_8$	E_{1u}	2383	2	2384 m			
?				2409 w			
$\nu_{12} - \nu_{18}$	E_{1u}	2430	0	2430 vw			
$\nu_5 - \nu_{18}$	E_{1u}	2456	1	2453 vw			
$\nu_2 + \nu_{13}$	E_{1u}	2472	2	2470 vw			
$\nu_9 + \nu_{17}$	E_{1u}	2487	2	2485 w	2151	0	2151 vw
$\nu_{17} + \nu_3$	E_{2u}	2523	3	2520 vw			
$\nu_{16} + \nu_{19}$	$A_{1u} + A_{2u} + E_{2u}$	2555	1	2554 vw	2343	10	2333 w
$(\nu_2 + \nu_{18}) + \nu_{19}$	$A_{1u} + A_{2u} + E_{2u}$	2576	2	2574			
$\nu_{16} + \nu_6$	B_{1u}	2596	0	2596 m	2513	3	2510 w
$\nu_2 + 2\nu_{19}$	E_{1g}				2547	−7	2554 w
$\nu_{18} + \nu_{17} + \nu_{13}$	E_{1u}				2770	−2	2772 vw
$\nu_1 + \nu_4$	E_{2u}				2787	−5	2792 vw
$(\nu_2 + \nu_{18}) + \nu_6$	E_{1u}	2616	2	2614			
$\nu_{17} + \nu_{18}$	E_{1u}	2657	4	2653 m			
$\nu_{10} + \nu_{16}$	E_{1u}				2380	5	2375 w
$\nu_{18} + \nu_9$	E_{1u}	2826	6	2820 m	2385	−2	2387 m
?				2850 w			
$\nu_{16} + \nu_9$	E_{1u}	2895	10	2885 w	2837	7	2830 vs
$(\nu_2 + \nu_{18}) + \nu_9$	E_{1u}	2916	10	2906 w			
ν_{12}				3030 s			
$\nu_{16} + \nu_{13}$	E_{1u}	3073	0	3073 s			
$(\nu_2 + \nu_{18}) + \nu_{13}$	E_{1u}	3084		3090 s			
?				3407 sh			
$\nu_{15} + \nu_{20}$	E_{1u}	3452	2	3450 w	2614	1	2613 w
?				3488 vw			
$\nu_{13} + \nu_{18}$	E_{1u}	3642	−2	3644 m	2861	4	2857 w
$\nu_{11} + \nu_5$	E_{1u}				2953	0	2953 vw
$(\nu_2 + \nu_{18} + \nu_{19}) + \nu_{18}$	E_{1u}	3697	0	3697 m			
$\nu_8 + \nu_{15}$	E_{1u}	4057	5	4052 s	3228	−6	3234 m
$\nu_1 + \nu_{14}$					3102	−6	3108 w
$\nu_{13} + \nu_3$					3335	7	3328 w
							3366 vw
							3430 vw
							3456 vs
							4024 m

Note. vw, very weak; sh, shoulder; w, weak; m, medium; s, strong; vs, very strong; forb, forbidden by the selection rules.

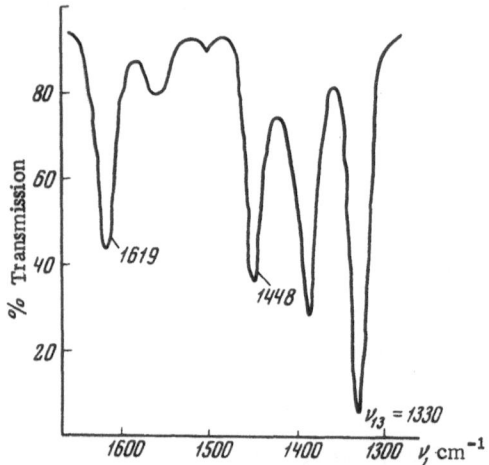

Fig. 2. Infrared spectrum (absorption) of hexadeuterobenzene. Range 1250–1650 cm^{-1}. NaCl prism, cell thickness 0.5 mm.

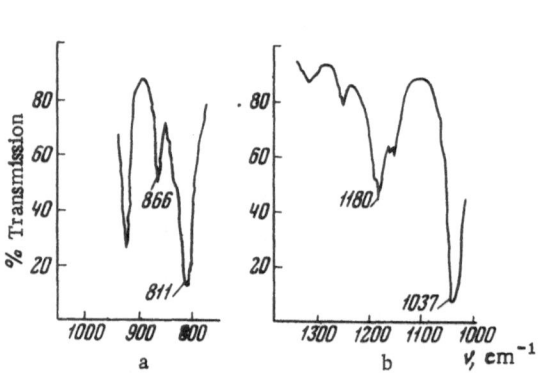

Fig. 3. Part of the infrared spectrum. a) Hexadeuterobenzene; b) benzene.

Our study of second-order lines simultaneously in the Raman and infrared absorption spectra and our derivation of quantitative data relating to the intensity and degree of polarization of the lines in the Raman spectrum for the molecules in question enabled us to give a clearer interpretation of the observed lines than that given by other authors. Thus, the interpretation given in [46] for the 1936, 1988, and 2030 cm^{-1} lines in the Raman spectrum of benzene was indefinite. Knowing the degree of depolarization of these lines, we were able to interpret them unambiguously as frequencies of the A_{1g} type. This in turn made it possible to determine the anharmonicity coefficients for these vibrations.

Let us trace the behavior of the overtones and composite frequencies involving the 1606 and 1586 cm^{-1} lines observed experimentally in the benzene spectrum. The first of these is interpreted as a composite frequency, $\nu_2 + \nu_{18}$, intensified as a result of Fermi resonance with the second fundamental frequency, $\nu_{16} = 1586$ cm^{-1}.

Combinations of these frequencies individually with other fundamental frequencies should respectively form a second-order line (involving $\nu_{16} = 1586$ cm^{-1}) and a third-order line (involving $\nu_2 + \nu_{18} = 1605$ cm^{-1}), while the overtone $\nu_2 + \nu_{18} = 1606$ cm^{-1} should form a fourth-order line. We initially expected that the intensity would fall by factors of 10 and 100 as the order of the lines increased. However, this is not found experimentally. Thus in the Raman spectrum the intensities of the lines of the third-order $\nu_{16} + (\nu_2 + \nu_{18}) = 3184$ cm^{-1} ($I_0 = 20$ units) and fourth-order $2(\nu_2 + \nu_{18}) = 3208$ cm^{-1} ($I_0 = 4$ units) differ only slightly from the intensity of the second-order lines $2\nu_{16} = 3164$ cm^{-1} ($I_0 = 7$ units).

Just as close are the intensities of the second- and third-order lines 2185 and 2211 cm^{-1}.

Combinations involving ν_{16} and $\nu_2 + \nu_{18}$ also explain the neighboring bands in the infrared spectrum:

$$
\begin{aligned}
1988 &= \nu_{20} + \nu_{16} & (A_{1u} + A_{2u} + E_{2u}), \\
2005 &= \nu_{20} + (\nu_2 + \nu_{18}) & (A_{1u} + A_{2u} + E_{2u}), \\
2554 &= \nu_{19} + \nu_{16} & (A_{1u} + A_{2u} + E_{2u}), \\
2574 &= \nu_{19} + (\nu_2 + \nu_{18}) & (A_{1u} + A_{2u} + E_{2u}), \\
2885 &= \nu_9 + \nu_{16} & (E_{1u}), \\
2906 &= \nu_9 + (\nu_2 + \nu_{18}) & (E_{1u}).
\end{aligned}
$$

Table 8. Interpretation of Forbidden Lines in the Vibration Spectra of Benzene and Hexadeuterobenzene

Interpretation	Symmetry	C_6H_6						C_6D_6					
		Calc. freq., cm^{-1}	$-x_{ik}$, cm^{-1}	Raman spectrum			Infrared spectrum	Calc. freq., cm^{-1}	$-x_{ik}$, cm^{-1}	Raman spectrum			Infrared spectrum
				obs. frequency cm^{-1}	I_0	p	obs. frequency cm^{-1}			obs. frequency, cm^{-1}	I_0	p	obs. frequency, cm^{-1}
ν_{20}	B_{2u}	—	—	405	10	d	—	—	—	349	8	d	—
ν_4	A_{2u}	—	—	682	5	d	—	—	—	—	—	—	—
$\nu_{17} - \nu_{20}$	A_{2u}	772	—2	774	0.3	—	—	—	—	—	—	—	656 m
ν_{11}	E_{1g}	—	—	—	—	—	848 m	—	—	—	—	—	—
ν_{10}	B_{2u}	—	—	1150	1	—	1158 w	—	—	—	—	—	—
ν_{17}	E_{2g}	—	—	—	—	—	1180 m	—	—	—	—	—	866 m
$\nu_{13} - \nu_{20}$	E_{1g}	—	—	—	—	—	—	981	—6	—	—	—	987 w
ν_9	B_{2u}	—	—	—	—	—	1309 w	—	—	—	—	—	—
$\nu_{18} + \nu_{14}$	$A_{1u} + A_{2u} + E_{2u}$	—	—	—	—	—	—	1387	3	1384	1	—	—
$\nu_{13} + \nu_3$	E_{2u}	—	—	—	—	—	—	2385	—4	2389	0.3	—	—
$\nu_{17} + \nu_3$	E_{2g}	2523	3	—	—	—	2520 vw	—	—	—	—	—	—
$\nu_4 + \nu_{11} + \nu_{17}$	E_{1u}	2710	—2	2712	—	—	—	—	—	—	—	—	—

Note. d = depolarized, m = medium, w = weak, vw = very weak.

A qualitative estimate shows that the intensities of each pair of lines are also of the same order. Quantitative intensity data given by Langseth [7] in no way contradict this conclusion.

Thus the composite frequency $\nu_2 + \nu_{18} = 1605$ cm^{-1} behaves as a fundamental line in the C_6H_6 spectrum. In the hexadeuterobenzene spectrum, however, the composite frequency $\nu_2 + \nu_{18} = 1514$ cm^{-1} differs in no way from other second-order lines.

Table 8 presents some quantitative data for lines forbidden in the spectra of the molecules under consideration. We note that the forbidden fundamental lines in the low-frequency parts of the Raman spectra of benzene and hexadeuterobenzene, 405 and 349 cm^{-1}, respectively, appear in the form of wide bands, in contrast to the difference frequencies in this region, which appear as weak, narrow lines. The A_{2u} line at 682 cm^{-1}, which is only allowed by the selection rules in the benzene infrared spectrum, also appears in the Raman spectrum with a considerable intensity (5 units).

In addition to this, there are quite a few reasonably intense lines which are forbidden in the spectra by the selection rules for the D_{6h} model. By way of illustration, Fig. 3 shows parts of the infrared spectra of C_6H_6 and C_6D_6 with frequencies of 1180 and 866 cm^{-1}, respectively, active in the Raman spectrum (frequencies 1177 and 849 cm^{-1}) and forbidden in the infrared spectrum. We see from the figure that these forbidden lines appear in the form of fairly strong bands. The appearance of "forbidden lines" in the spectra of liquids is also observed for many other molecules and is usually explained as being a result of intermolecular interactions. However, such lines always have a very low intensity, and their number is extremely limited. Hence, the appearance of a large number of "forbidden" lines with fair intensities in the benzene and hexadeuterobenzene spectra compels us to look more carefully at the original D_{6h} model (at any rate for liquids). We would emphasize that we are not thinking here of returning to a less symmetrical model, for example D_{3d}, since, in general, the spectra are interpreted very well on the basis of the D_{6h} model. It is more likely that there is a slight distortion in the geometrical configuration of the C_6H_6 and C_6D_6 molecules, associated, for example, with slight distortions in the plane of the ring. In view of this it is of great interest to make a detailed study of the vibrational spectra of benzene and hexadeuterobenzene in the solid and liquid phases.

3. Second-Order Vibrational Spectra of
Chloroform and Deuterochloroform

A knowledge of the anharmonicity coefficients offers the possibility of calculating the "zero" frequencies of the vibrations; these are the frequencies which are in principle most justifiable to use in solving the vibrational problem, and the use of these frequencies increases the accuracy of the calculations. However, for this purpose we must obtain the complete system of anharmonicity coefficients, and it is quite difficult to make the required measurements for complex molecules.

In order to obtain the complete system of anharmonicity coefficients, we studied the comparatively simple molecules $CHCl_3$ and $CDCl_3$, the second-order spectra of which each consist of 21 lines. Despite the simple structure, data relating to the vibrational spectra of these molecules are rather sparse.

Some quantitative data for the lines in the Raman spectrum of chloroform are presented in [55]. In addition to the fundamental frequencies, six weak lines were observed: 497, 622, 1027, 1501, 1518, and 1888 cm^{-1}. These lines were interpreted as overtones and composite frequencies.

Earlier results taken from [56-81] were also systematized in [55] with respect to the Raman spectrum of chloroform. Among these the 1442 cm^{-1} line was mentioned without interpretation in [57-61], while Rao mentioned two lines in [62], at 1442 and 1420 cm^{-1}, interpreting the latter as a $\nu_2 + \nu_5$ line of symmetry E. Other second-order lines observed by Sirkar [61], particularly the overtone $2\nu_5$, appearing in the form of a fairly intense, wide band, were not found in [62]. There is little doubt therefore that the 1442 and 1420 cm^{-1} lines were due to acetone and ethyl alcohol impurities in the sample, as indicated in [55].

Dabadghao [63] mentioned a 3072 cm^{-1} line, which was not confirmed in the other papers. This line nevertheless appears to have been a genuine one, as we observed it ourselves.

The infrared spectrum of chloroform was studied in [82] by Lisitsa and Tsyashchenko. Earlier papers on the infrared absorption of liquid $CHCl_3$ were also systematized in [82].

The measurements in [82] were carried out with an IKS-6 spectrometer over the spectral range 470-11,500 cm^{-1}. The bands observed were interpreted and compared with published data. The comparison showed that in the range 500-3000 cm^{-1}, which had only been properly studied earlier in [83], agreement with the measured data was quite satisfactory.

However, in the ternary combination range between 3000 and 11,500 cm^{-1}, which had been studied by a large number of authors [24, 84-87], the results sometimes diverged sharply.

The Raman spectrum of $CDCl_3$ was studied by Cleveland and colleagues in [88] in order to obtain data for calculating the thermodynamic parameters of deuterochloroform. In addition to the fundamental frequencies, three weak lines were observed; these were interpreted as overtones and composite frequencies. Data relating to the infrared spectrum of liquid $CdCl_3$ and its solution in carbon disulfide were also given in [88]. In the liquid a 1095 cm^{-1} line was observed as well as the fundamental frequencies 910 and 2245 cm^{-1}. In the spectrum of the solution the authors observed two weak lines 743 and 1400 cm^{-1} and explained these as $2\nu_3$ and $\nu_2 + \nu_5$, respectively.

We know of no other investigations into the vibrational spectra of deuterochloroform.

In view of the fact that the literature contains no complete and systematized data relating to $CHCl_3$ and $CDCl_3$, we made an attempt at obtaining fuller quantitative data regarding the Raman and infrared absorption spectra of these molecules and at interpreting the resultant bands.

Table 9. Interpretation of the Fundamental Frequencies of Chloroform
and Deuterochloroform

Symmetry	Interpret.	Freq. $\Delta\nu$, cm^{-1}		Symmetry	Interpret.	Freq., $\Delta\nu$, cm^{-1}	
		CHCl$_3$	CDCl$_3$			CHCl$_3$	CDCl$_3$
A_1	ν_1	3018	2254	E	ν_4	1219	908
A_1	ν_2	669	649	E	ν_5	764	736
A_1	ν_3	366	365	E	ν_6	264	260

*According to [25].

Table 10. Interpretation of the Observed Overtones and Composite Frequencies
in the Raman Spectrum of Chloroform

Interpretation	Symmetry	Calc. frequency cm^{-1}	$-x_{ik}$, cm^{-1}	Obs. frequency cm^{-1}	I_0	ρ
$2\nu_6$	$A_1 + E$	528	3	521	0.8	d
$\nu_3 + \nu_6$	E	630	1	629	4	d
$2\nu_6 + \nu_3$	$A_1 + E$	894	16	862	2	d
$\nu_2 + \nu_6$	E	933	3	930	0.3	—
$\nu_2 + \nu_3$	A_1	1035	0	1035	1	—
$\nu_5 + \nu_2 + \nu_3$	E	1067	3	1064	0.3	—
$\nu_3 + \nu_5$	E	1130	13	1117	1	—
$\nu_2 + \nu_3 + \nu_6$	E	1299	—4	1303	6	—
$2\nu_2$	A_1	1336	1	1334	2	—
$2\nu_3 + \nu_5$	$A_1 + E$	1496	1	1495	12	0.7
$2\nu_5$	$A_1 + E$	1528	4	1519		
$\nu_5 + \nu_4$	$A_1 + A_2 + E$	1983	19	1964	1	0.7
$2\nu_4$	$A_1 + E$	2438	19	2400	5	p
$2\nu_5 + \nu_2 + \nu_6$	$A_1 + A_2 + 2E$	2461	18	2443	2	d
$\nu_2 + \nu_3 + \nu_4 + \nu_5$	$A_1 + A_2 + 2E$	3018	—	3047	10	—
$2\nu_4 + \nu_2$	$A_1 + E$	3107	19	3069	5	—

Note. d = depolarized.

The chloroform and deuterochloroform samples were first purified by the method proposed in [89]. The sample was washed for an hour with a solution of caustic soda and twice with distilled water. Then the sample was treated in three portions with concentrated sulfuric acid, twice with distilled water, once with mercury, and finally with distilled water again. After this the sample was dried over calcium chloride and redistilled in a column 1 m long.

The deuterochloroform sample with which we were working contained 98.6% of the deuterized product; this was checked by analysis when obtaining the product. The boiling point of the sample was 61°C. The chloroform apparently contained a slight trace of CCl$_4$, since the Raman spectrum of the chloroform showed a weak polarized line coinciding with the fundamental frequency of carbon tetrachloride, 459 cm^{-1}. The rest of the CCl$_4$ lines remained unobserved, since they coincided with intense chloroform lines or repetitions of these.

In the Raman spectrum of deuterochloroform there was a very weak line at a frequency of 446 cm^{-1}, which was probably due to the presence of phosgene in our specimen. We were unable to detect any other phosgene lines, since these were superimposed on intense lines

Table 11. Interpretation of the Observed Overtones and Composite Frequencies in the Raman Spectrum of Deuterochloroform

Interpretation	Symmetry	Calc. frequency cm^{-1}	$-x_{ik}$, cm^{-1}	Obs. frequency cm^{-1}	I_0	ρ
$2\nu_6$	$A_1 + E$	520	0	520	0,2	—
$\nu_3 + \nu_6$	E	625	6	619	0,4	—
$\nu_5 + \nu_6$	$A_1 + A_2 + E$	996	6	990	0,2	—
$\nu_2 + \nu_3$	A_1	1014	—1	1015	0,5	—
$\nu_3 + \nu_5$	E	1101	3	1098	Band	
$\nu_4 + \nu_6$	$A_1 + A_2 + E$	1168	1	1167	0,7	—
$\nu_3 + \nu_4$	E	1272	—8	1280	0,1	—
$2\nu_2$	A_1	1298	2	1297	2	p
$\nu_2 + \nu_5$	E	1385	5	1380	3	
$2\nu_5$	$A_1 + E$	1472	9	1453	8	
$\nu_5 + \nu_4$	$A_1 + A_2 + E$	1644	16	1628	6	0,70
$2\nu_4$	$A_1 + E$	1816	13	1790	5	
$\nu_6 + \nu_4 + \nu_2$	$A_1 + A_2 + E$	1817	3	1814	3	
$\nu_2 + \nu_4 + \nu_5$	$A_1 + A_2 + E$	2293	—4	2297	0,3	—
$\nu_4 + 2\nu_5$	$A_1 + A_2 + 2E$	2380	20	2339	1	—

Note.   p = polarized.

in the $CDCl_3$ spectrum. In the infrared spectrum (cuvette thickness 0.3-1 mm) the 1217 and 3022 cm^{-1} lines appeared; this indicated a slight $CHCl_3$ impurity.

We might have expected ethyl alcohol to be among the impurities, but we failed to find any lines of this compound in the spectrum. No signs of the decomposition of chloroform and deuterochloroform were observed while the photographs were being taken.

The chloroform and deuterochloroform molecules belong to the C_{3v} type of symmetry. All six fundamental frequencies are active in both the Raman and infrared spectra. These frequencies are given in Table 9.

The observed overtones and composite frequencies were interpreted as combinations of these frequencies. The resultant characteristics of the observed lines in the Raman spectra of chloroform and its deutero-analog are given in Table 10.

Individual overtones appear with quite a high intensity. We must mention in particular the 1519 cm^{-1} line of $CHCl_3$ and the 1453 cm^{-1} line of $CDCl_3$, which are interpreted as overtones of the degenerate vibration $2\nu_5$. However, we observed two real lines, as in [55]. The intensities of these could not be determined quantitatively, although visual estimation indicated that the 1495 line was rather stronger than the 1519.

In the Raman spectrum of $CHCl_3$, we observed sixteen lines (apart from the fundamental ones), nine of which were interpreted as second-order lines, while the other seven were of both the third and fourth orders.

In the Raman spectrum of $CDCl_3$ we found fifteen overtones and composite frequencies (Table 11), twelve of which were explained quite satisfactorily as second-order lines, while the 1814, 2297, and 2339 cm^{-1} lines were explained as composite frequencies of the third order.

It was a characteristic feature that the lines interpreted as third-order composite frequencies were no different as regards intensity from the second-order composite frequencies. Thus the 862 cm^{-1} line with an intensity of 2 units in the Raman spectrum of chloroform is interpreted as a composite third-order composite $2\nu_6 + \nu_3$, while the 1035 cm^{-1} line with an intensity of 1 unit is interpreted as a second-order composite frequency.

The 1495 cm^{-1} line of CHCl$_3$ and the 2297 cm^{-1} line of CDCl$_3$ were interpreted as third-order lines; however, it might also have been considered that they were due to the splitting of the levels of the degenerate vibrations ν_4 and ν_5 as a result of the coupling of the vibrations and rotations.

Table 12. Interpretation of Overtones and Composite Frequencies
in the Infrared Spectrum of Deuterochloroform

Interpretation	Symmetry	Calc. freq., cm^{-1}	$-x_{ik}$, cm^{-1}	Obs. freq., cm^{-1}
$2\nu_3$	A_1	730	4	722
$\nu_2 + \nu_6$	E	909	0	909
$\nu_5 + \nu_6$	$A_1 + A_2 + E$	996	7	989 w
$\nu_2 + \nu_3$	A_1	1014	9	1023 w
$\nu_3 + \nu_5$	E	1101	1	1100 m
$\nu_4 + \nu_6$	$A_1 + A_2 + E$	1168	1	1167 w
$\nu_3 + \nu_4$	E	1272	5	1276 vw
$2\nu_2$	A_1	1298	3	1295 vw
$\nu_2 + \nu_5$	E	1385	5	1380 m
$2\nu_5$	$A_1 + E$	1472	7	1465 m
$\nu_2 + \nu_4$	E	1557	0	1557 vw
$\nu_5 + \nu_4$	$A_1 + A_2 + E$	1644	9	1635 vsw
$2\nu_5 + \nu_3$	$A_1 + E$	1632	-15	1762 m
$2\nu_4$	$A_1 + E$	1816	12	1793 w
$\nu_6 + \nu_4 + \nu_2$	$A_1 + A_2 + E$	1817	—	1817 w
$\nu_1 + \nu_6$	E	2514	5	2509 vsw
$\nu_1 + \nu_3$	A_1	2619	0	2619 m
$\nu_1 + \nu_2 - \nu_6$	E	2643	-11	2654 m
$3\nu_2 + \nu_5$	$A_1 + E$	2683	0	2684 m
$\nu_1 + \nu_6 + \nu_3$	E	2879	6	2873 m
$\nu_1 + \nu_2$	A_1	2903	1	2902 m
$\nu_1 + \nu_5$	E	2990	3	2987 m
$\nu_1 + \nu_4$	E	3162	3	3159 s
?				3481 *
$\nu_1 + \nu_2 + \nu_4$	$A_1 + A_2 + E$	3527		3522 *
$2\nu_2 + \nu_1$	A_1	3552		3555 *
$\nu_1 + \nu_2 + \nu_5$	$A_1 + A_2 + E$	3639		3611 *
$2\nu_1 - \nu_5$	E	3772		3670 *
$2\nu_1 - \nu_3$	A_1	4143		4020 m
$2\nu_1 - \nu_6$	E	4248		4209 m
$2\nu_1$	A_1	4508	34	4439 s
$2\nu_1 + \nu_6$	E	4768		4619 m
$2\nu_1 + \nu_3$	A_1	4873		4795 w
				4918 w
$2\nu_1 + \nu_2$	A_1	5157		5070 vw
$2\nu_1 + \nu_5$	E	5244		5159 w
$2\nu_1 + \nu_4$	E	5416		5340 m
$2\nu_1 + \nu_4 + \nu_6$	$A_1 + A_2 + E$	5676		5486 vww
$3\nu_1 - \nu_4$	E	5854		5616 vww
$3\nu_1 - \nu_2$	A_1	6113		5906 m
$3\nu_1 - \nu_6$	A_1	6502		6228 vww
$3\nu_1$	A_1	6762	32	6569 m
				6950 } wd
				7036 }

*Region of water vapor.

Note. w, weak; vw, very weak; vww, very weak and wide; s, strong; vsw, very strong and wide; m, medium; wd, wide and double.

In the $CHCl_3$ spectrum we observed an intense line at 3047 cm^{-1}, not mentioned earlier in the literature. This was evidently due to the fact that, owing to the low resolution of former equipment, the line merged with the fundamental $\nu = 3018$ cm^{-1}. We interpret it as a fourth-order line. The increased intensity of this line is due to Fermi resonance with the completely symmetrical line at 3018 cm^{-1}.

Since the infrared spectrum of liquid $CHCl_3$ was studied over a fairly wide range in [82], we shall confine attention to the range of the first overtones and composite frequencies.

In view of the absence of data relating to the vibrational spectrum of $CDCl_3$, we studied the infrared spectrum of this molecule over a wide range, 600-9000 cm^{-1} (Table 12).

The number of observed overtones and composite frequencies in the range studied (including the third- and fourth-order lines) was 43. Of all the possible second-order lines (21) in the infrared spectrum, 19 appeared. The overtone $2\nu_6$ and the composite frequency $\nu_3 + \nu_6$ could not be recorded, since they lay in the neighborhood of 550 cm^{-1}, inaccessible for our apparatus. These lines were observed in the deuterochloroform spectrum.

In the region of the first overtones and composite frequencies of the infrared spectrum of deuterochloroform we observed some fairly intense bands and interpreted these as belonging to the third order. For comparison we may indicate the 1762 cm^{-1} ($2\nu_5 + \nu_3$) and 1793 cm^{-1} ($2\nu_4$) bands. The third-order composite frequency 1762 cm^{-1} was stronger than the second-order line 1793 cm^{-1}. These effects, as mentioned earlier, also occurred in the spectra of other molecules. Hence, the widely held view that third-order lines must necessarily have a lower intensity than lines of the second order is only true "on average." The values of the frequencies of the 3481, 3522, 3555, 3611, and 3670 cm^{-1} bands in the infrared spectrum of deuterochloroform should be regarded as approximate, since they fall in a range in which some of the radiation is absorbed by water vapor, and may therefore be displaced or suffer a change of structure.

The infrared spectrum of deuterochloroform was photographed between 4500 and 9000 cm^{-1} with a cuvette 10 cm in thickness. The resultant spectrum was resolved almost completely. In the 7000 cm^{-1} region there was a wide band, probably unresolved by our apparatus, and this remained uninterpreted.

The strongest band in this range corresponds to the second overtone ($2\nu_1$) of the vibrations of CD (analogously to the first overtone in the corresponding range).

In the case of these comparatively simple molecules, chloroform and deuterochloroform, we were able to obtain the complete second-order spectra, calculated as containing 21 lines each. Some data relating to these spectra are presented in Table 13. Although by the selection rules all 21 lines in the second order of chloroform and deuterochloroform should be active both in the Raman and infrared spectra, in practice individual overtones and composite frequencies only appeared in one of the spectra. Thus, in the Raman spectrum of chloroform between 1000 and 3000 cm^{-1} there should be 11 second-order lines, but only six appeared.

For example, we failed to observe the composite frequencies $\nu_2 + \nu_4 = 1888$ cm^{-1}, $\nu_2 + \nu_5 = 1433$ cm^{-1}, $\nu_3 + \nu_4 = 1585$ cm^{-1}, which lay in a range very convenient for study. In the infrared spectrum, however, these appeared quite clearly. Thus, full information regarding the second-order spectra may only be obtained by studying the Raman and infrared spectra simultaneously.

We see from the data presented in Table 13 that the intensities of the analogous second-order lines in the chloroform and deuterochloroform spectra are similar to each other. The strongest are the lines interpreted as a combination of degenerate vibrations. The anharmonicity coefficients of the chloroform and deuterochloroform lines analogous in interpretation have the same signs and are of similar magnitude.

Table 13. Complete Second-Order Spectrum of Chloroform and Deuterochloroform

Interpret.	Symmetry	CDCl₃ Calc. freq., cm⁻¹	$-x_{ik}$, cm⁻¹	Raman obs. frequency cm⁻¹	I_0	ρ	Infrared obs. frequency, cm⁻¹	CHCl₃ Calc. freq., cm⁻¹	$-x_{ik}$, cm⁻¹	Raman obs. frequency cm⁻¹	I_0	ρ	Infrared exptl. frequency cm⁻¹
$2\nu_6$	A_1+E	520	0	520	0.4			528	3	521	0.5	d	—
$\nu_3+\nu_6$	E	625	6	619	0.4			630	1	629	4	d	
$2\nu_3$	A_1	730	4	—			722 m	732	4				724 m
$\nu_2+\nu_6$	E	909	0				909	933	3	930	0.3		932 m
$\nu_5+\nu_6$	A_1+A_2+E	996	6	990	0.2		989 w	1028	0				1028 w
$\nu_2+\nu_3$	A_1	1014	−4	1015	0.5		1023 w	1035	0	1035	1		
$\nu_3+\nu_5$	A_1+A_2+E	1101	3	1098	Band		1100 m	1130	13	1117	1		
$\nu_6+\nu_4$	E	1168	1	1167	0.7		1167 w	1483	3	—			1480 w
$\nu_3+\nu_4$	A_1	1272	−8	1280	0.1		1276 vw	1585	−6				1591 w
$2\nu_2$	A_1+E	1298	1	1297	2	p	1295 vw	1336	1	1334	2		
$\nu_2+\nu_5$	E	1385	5	1385	3	d	1380 w	1433	3				1430 m
$2\nu_5$	A_1+A_2+E	1472	9	1453	8	d	1465 m	1528	4	1519	12	0.7	—
$\nu_2+\nu_4$	A_1+E	1557	0				1557 vw	1888	2				1886 w
$\nu_4+\nu_5$	A_1+A_2+E	1664	16	1628	6	0.7	1635 w	1983	19	1964	1	0.7	1968 w
$2\nu_4$	A_1+E	1816	13	1790	5	d	1793 m	2438	19	2400	5	p	2400 s
$\nu_1+\nu_6$	E	2514	5	—			2509 vsw	3282	0				3282 vw
$\nu_1+\nu_3$	A_1	2619	0				2619 w	3384	−4				3388 vw
$\nu_1+\nu_2$	A_1	2903	1				2902 m	3687	1				3686 vw
$\nu_1+\nu_5$	E	2990	3				2907 m	3782	2				3780 w
$\nu_1+\nu_4$	E	3162	3				3159 s	4237	8				4229 s
$2\nu_1$	A_1	4508	34				4439 s	6036	57				5921 s

Note. p, polarized; d, depolarized; w, weak; vw, very weak; vsw, very strong and wide; s, strong; m, medium.

Table 14. Comparison of Experimental and Calculated Anharmonicity Coefficients (in cm⁻¹) for the Third-Order Lines in the Vibrational Spectra of Chloroform and Deuterochloroform

Interpretation	Combination of anharmonicity coeff.	CHCl₃ exptl. $-x_{ik}$	CHCl₃ calc.	CDCl₃ exptl.	CDCl₃ calc.
$3\nu_1$	x_{11}	58	57		
$3\nu_6$	x_{66}	1	3		
$2\nu_6+\nu_3$	$x_{66}+x_{36}$	16	4		
$\nu_3+\nu_2+\nu_4$	$x_{23}+x_{24}+x_{34}$	−3	−4		
$\nu_5+\nu_2-\nu_3$	$x_{25}-x_{35}-x_{23}$	4	−10		
$2\nu_4+\nu_6$	$x_{44}+x_{46}$	2	22		
$\nu_2+\nu_3+\nu_6$	$x_{23}+x_{26}+x_{36}$	−4	4		
$2\nu_4+\nu_2$	$x_{44}+x_{24}$	19	21		
$2\nu_3+\nu_5$	$x_{33}+x_{35}$	1	17		
$2\nu_5+\nu_2$	$x_{55}+x_{25}$	2	7		
$2\nu_5+\nu_2+\nu_6$	$x_{55}+x_{25}+x_{56}+x_{26}$	18	10		
$\nu_1+\nu_5+\nu_6$	$x_{15}+x_{16}+x_{56}$	15	2		
$2\nu_5+\nu_3$	$x_{55}+x_{35}$			15	12
$3\nu_1$	$3x_{11}$			32	34
$\nu_6+\nu_4+\nu_2$	$x_{46}+x_{26}+x_{24}$			3	1
$\nu_2+\nu_4+\nu_5$	$x_{24}+x_{25}+x_{45}$			−4	21
$2\nu_5+\nu_4$	$x_{55}+x_{45}$			20	25
$\nu_1+\nu_2-\nu_6$	$x_{12}-x_{16}-x_{26}$			−11	−4
$3\nu_2+\nu_5$	$x_{22}+x_{25}$			0	7
$\nu_1+\nu_3+\nu_6$	$x_{16}+x_{13}+x_{36}$			6	11

We note that a similar phenomenon occurs when comparing the anharmonicity coefficients of other molecules.

Table 14 shows combinations of the anharmonicity coefficients for third-order lines obtained experimentally; these are compared with calculated combinations obtained by reference to the anharmonicity coefficients of second-order lines. We see from the table that, of 20 combinations, agreement is completely satisfactory in six cases (the deviation lies within 1-3 cm^{-1}). In three cases there is a rather larger deviation (3-5 cm^{-1}), still within the limits of experimental error. In eleven cases the deviation is greater than that corresponding to possible experimental errors. These deviations are still difficult to explain, but they must of course be mentioned.

4. Anharmonicity of Some Characteristic Molecular Vibrations

The structure of the vibration—rotation bands of some of the simplest molecules has recently been studied and a certain amount of information has been obtained regarding the constants characterizing the coupling between vibrational and rotary motion as well as the anharmonicity coefficients [90]. This method of research is of great importance, but owing to the complexity of obtaining and analyzing the experimental material it is only applicable to very simple molecules. As indicated earlier, another method of studying anharmonicity lies in determining the overtones and composite frequencies in the Raman and infrared absorption spectra. In principle this method may be applied to complex molecules, although analysis of the experimental data also presents considerable difficulties owing to the indeterminacy of interpreting the spectra.

We shall now give some results of an investigation into the overtones of certain characteristic vibrations in Raman and infrared absorption spectra. We studied vibrations which to a first approximation could be considered as the characteristic vibrations of an isolated bond. A study of the overtones of such vibrations reduces the above-mentioned difficulties in the interpretation of the spectra. For characteristic vibrations we may in fact use the overtones in the spectra of several compounds containing the same characteristic bond. On the other hand, it is of quite independent interest to obtain the anharmonicity coefficients for a series of characteristic bonds.

The results of the measurements are shown in Table 15. The values of the anharmonicity coefficients were determined from the formulas (see [16] and [17]) $x_{11} = (\nu_1 - 2\nu)/2$ or $x_{11} = (\nu_2 - 3\nu)/6$, where ν is the fundamental frequency, ν_1 is the frequency of the first overtone, and ν_2 is the frequency of the second overtone. We see that the values of x_{11} determined from the first and second overtones practically coincide, which confirms the validity of the interpretation of the spectra.

The majority of the characteristics were studied in two different compounds. This enabled the resultant spectra to be interpreted with greater confidence. By way of example, Fig. 4 shows parts of the infrared spectra corresponding to the region of the first overtone of the SiH

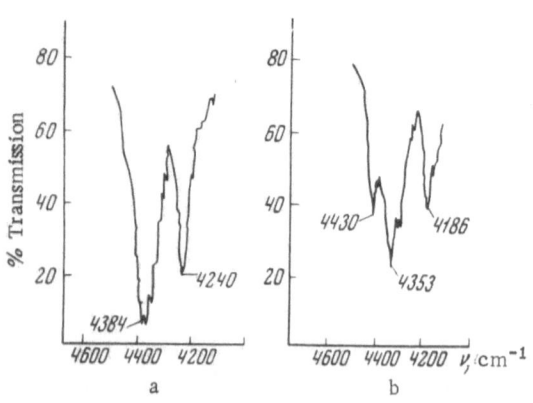

Fig. 4. Infrared absorption spectrum of: a) methyldichlorsilane; b) ethyldichlorsilane. LiF prism, cell thickness 1 mm.

Table 15. Anharmonicity Coefficients of the Characteristic Vibrations
of Various Molecules

Substance	Formula	Bond	Form of spectrum	$\Delta\nu$, cm^{-1}	$-x_{11}$, cm^{-1}
Chloroform. . . .	CHCl$_3$	C—H	IR	$\nu_1 = 5921$	57
			Ram	$\nu = 3018$	58
			IR	$\nu_2 = 8703$	
Deuterochloro-form	CDCl$_3$	C—D	IR	$\nu_1 = 4439$	33
			Ram	$\nu = 2253$	32
			IR	$\nu_2 = 6568$	
Methyldichlorsil-ane.	CH$_3$Cl$_2$SiH	Si—H	IR	$\nu_1 = 4384$	
			Ram	$\nu = 2221$	29
Ethyldichlorsilane	C$_2$H$_5$Cl$_2$SiH	Si—H	IR	$\nu_1 = 4353$	
			Ram	$\nu = 2210$	33
Acetone.	(CH$_3$)$_2$CO	C=O	IR	$\nu_1 = 3409$	
			IR	$\nu = 1711$	6
			IR	$\nu_2 = 5102$	6
Methylethylketone	C$_2$H$_5$CH$_3$CO	C=O	IR	$\nu_1 = 3414$	
			IR	$\nu = 1716$	9
			IR	$\nu_2 = 5100$	8
Diethylamine . .	(C$_2$H$_5$)$_2$NH	N—H	IR	$\nu_1 = 6480$	
			Ram	$\nu = 3311$	71
			IR	$\nu_2 = 9444$	81
Acetonitrile . . .	(CH$_3$CN)	C≡N	IR	$\nu_1 = 4520$	
			Ram	$\nu = 2246$	—14
			IR	$\nu_2 = 6870$	—22
Pentene-1	C—C—C—C=C	C=C	Ram	$\nu_1 = 3279$	
			Ram	$\nu = 1642$	3
Diallyl	C=C—C—C—C=C	C=C	Ram	$\nu_1 = 3278$	
			Ram	$\nu = 1642$	3
Pentadiene-1,3 .	C=C—C=C—C	C=C—C=C	Ram	$\nu_1 = 3301$	
			Ram	$\nu = 1655$	5
Diprotenyl	C—C=C—C=C—C	C=C—C=C	Ram	$\nu_1 = 3320$	
			Ram	$\nu = 1668$	8
Isobutadiene . . .	C=C—C=C ⎮ C	C=C—C=C	Ram	$\nu_1 = 3265$	
			Ram	$\nu = 1638$	6

Note. Ram, Raman spectrum; IR, infrared spectrum.

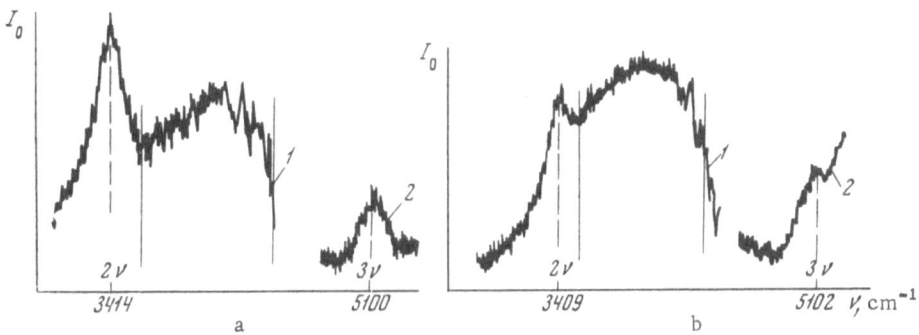

Fig. 5. First and second overtones in the infrared spectrum of: a) methyl-ethylketone; b) acetone. Cell thickness 1 mm (1) and 10 mm (2). The solid vertical lines show the region of absorption by water vapor.

vibration in methyl- and ethyldichlorsilanes. We see that together with the overtones bands of a different origin appear in this region. Thus, if we were to study just one compound, for example ethyldichlorsilane, the spectrum would clearly be interpreted in an ambiguous manner, since we could equally choose either the 4430 or the 4353 cm^{-1} band as the overtone of the SiH vibration. By studying two compounds at the same time, we can compare their spectra and choose the band frequencies nearest to each other. The nearest bands were in fact the 4353 cm^{-1} of ethyldichlorsilane and the 4384 cm^{-1} of methyldichlorsilane, and we interpreted these accordingly as the overtones of the SiH vibrations in these compounds.

The resultant data for the $C \equiv N$ bond should be considered as preliminary, since we only had one substance with a bond of this type, while the first overtone appeared in the spectrum as a "shoulder" only.

We see from Table 15 that, for the same characteristic vibration in different molecules, the anharmonicity coefficients are identical (within the limits of measuring error). We may conclude from this that the anharmonicity coefficients are also true characteristics of the characteristic vibrations of the type in question.

One notices a great difference between the anharmonicity coefficients of vibrations involving a hydrogen (or deuterium) atom and vibrations of the double bonds $C = C$ and $C = O$. The vibrations of the $C =$ bond in the presence of conjugation are in turn characterized by higher anharmonicity coefficients than those of an isolated $C = C$ bond.

It should be noted that the infrared bands corresponding to overtones of the characteristic vibrations in different compounds differ greatly from each other as regards structure and intensity. Figure 5 represents the parts of the acetone and methylethylketone spectra corresponding to the regions of the first and second overtones of the $C = O$ vibrations, obtained under identical experimental conditions.

CHAPTER IV

DISCUSSION OF RESULTS AND GENERAL CONCLUSIONS

1. General Conclusions Regarding the Anharmonicity Coefficients

In the modern version of the theory largely developed by M. V. Vol'kenshtein, a distinction is made between mechanical and electro-optical anharmonicity when considering overtones and composite frequencies. Mechanical anharmonicity means a deviation of the potential-energy function from the quadratic relationship. Electro-optical anharmonicity means a nonlinear dependence of the electrical moment or polarizability on the normal vibrational coordinates. The numerical values of the overtone frequencies are determined by the mechanical anharmonicity, and their electro-optical properties (intensity and polarization) by both the mechanical and electro-optical anharmonicities. In the presence of mechanical anharmonicity the concept of normal coordinates and normal vibrations loses its meaning. Nevertheless, for purposes of calculation one usually employs normal coordinates as a zero approximation, introducing anharmonic corrections as perturbations.

For the equilibrium configuration of the molecule, the potential-energy function has a minimum; around this minimum it may be expanded in series in powers of the natural coordinates:

$$U = U_{g=0} + \frac{1}{2} \sum_{ij} \left(\frac{\partial^2 U}{\partial g \, \partial g_i} \right)_{g=0} g_i g_j + \frac{1}{6} \sum_{ijk} \left(\frac{\partial^3 U}{\partial g_i \, \partial g \, \partial g_k} \right)_{g=0} g_i g_j g_k + \frac{1}{24} \sum_{ij..l} \left(\frac{\partial^4 U}{\partial g_i \, \partial g_j \, \partial g_k \, \partial g_l} \right)_{g=0} g_i g_j g_k g_l + \cdots \quad (20)$$

On transferring to normal coordinates Q_s, constituting linear combinations of the natural coordinates, the potential energy may be written in the form

$$U = U_{Q=0} + \frac{1}{2} \sum_{s} \left(\frac{\partial^2 U}{\partial Q_s^2} \right)_{Q=0} Q_s^2 + \frac{1}{6} \sum_{stu} \left(\frac{\partial^3 U}{\partial Q_s \, \partial Q_t \, \partial Q_u} \right)_{Q=0} Q_s Q_t Q_u +$$

$$+ \frac{1}{24} \sum_{stuv} \left(\frac{\partial^4 U}{\partial Q_s \, \partial Q_t \, \partial Q_u \, \partial Q_v} \right)_{Q=0} Q_s Q_t Q_u Q_v + \cdots \quad (21)$$

In solving the wave equation, the third and fourth terms of (21), due to the anharmonicity of the vibrations, are treated on the basis of perturbation theory. As a result of this the expression for the total energy takes the form

$$E = \sum_{s} h v_s \left(v_1 + \frac{f_s}{2} \right) + \sum_{st} h x_{st} \left(v_s + \frac{f_s}{2} \right) \left(v_t + \frac{f_t}{2} \right) + \Delta,$$

where f_s and f_t are the degrees of degeneracy of vibrations s and t, and x_{st} is the constant displacement of the level, which is expressed in terms of the anharmonicity constants

35

$$\left(\frac{\partial^3 U}{\partial Q_s\, \partial Q_t\, \partial Q_u}\right)_{Q=0} \quad \text{and} \quad \left(\frac{\partial^4 U}{\partial Q_s^2\, \partial Q_t^2}\right)_{Q=0}.$$

The x_{st} may be determined experimentally. Calculation gives the following formulas for x_{ss} and x_{st}:

$$x_{ss} = \frac{1}{2}\left[\nu_{v_s=2} - 2\nu_{v_s=1}\right] \tag{22}$$

or [see (16)]

$$x_{11} = \frac{\nu_\text{)} - 2\nu}{2},$$

$$x_{st} = \nu_{v_s=1,\, v_t=1} - \nu_{v_s=1} - \nu_{v_t=1}; \quad s \neq t$$

or [see (18)]

$$x_{12} = \nu_{12} - (\nu + \nu).$$

For determining the anharmonicity constants it is insufficient to know all the x_{st}, since the anharmonicity constant is always much larger than the constant displacements.

It is very difficult to calculate the anharmonicity for polyatomic molecules on the basis of the formulas for the total energy of the molecular vibrations. This can in fact only be done for the simplest molecules. Hence, for polyatomic molecules the anharmonicity has in practice to be considered semiempirically. Thus for hydrocarbons the anharmonicity is approximately taken into account by introducing "spectroscopic masses" when considering the valence vibrations of the C—H bonds, particularly when comparing these with the C—D bonds in the deutero-substituted compounds. By considering the C—H and C—D vibrations as vibrations of the H and D atoms, and taking the empirical value of the constant 4A as 0.080 in the formula $M' = M \cdot (1 + [4A/\sqrt{M}])$, where M is the reduced mass, we obtain the "spectroscopic mass" $M_H' = 1.088$, $M_D' = 2.128$ instead of the ordinary masses $M_H = 1.008$ and $M_D = 2.014$ (in atomic-mass units $M_{016}/16$). After the "spectroscopic masses" have been introduced, the vibrations are considered to be harmonic. This is really not a very good method. First, it is only to some extent justifiable for diatomic molecules. Secondly, the method only considers the anharmonicity of those vibrations involving hydrogen atoms, assuming that vibrations only arise from the motion of these. We found experimentally on studying characteristic vibrations that the anharmonicity coefficients of the vibrations of different bonds involving hydrogen atoms had quite different values.

If the vibrations were caused simply by the motion of the hydrogen atoms, we should expect identical x/ν values for all cases considered. However, experimental results (Table 16) show the contrary. It is therefore undesirable to use the method of spectroscopic masses in relation to the anharmonicity of the vibrations of complex molecules.

In our own investigation we obtained the anharmonicity coefficients of a large number of vibrations. It is of interest to carry out a general analysis of the results obtained in the experiments.

The following conclusions may be drawn from this analysis.

1. The anharmonicity coefficients of the vibrations are, generally speaking, fairly small. In the Raman spectra of cyclohexane, benzene, chloroform, and their deutero-derivatives we observed 118 overtones and composite frequencies (without allowing for the lines associated with Fermi resonance). We found that 73 lines had very small anharmonicity coefficients, between

Table 16. Comparison of the Anharmonicity Coefficients of the Vibrations
of Bonds Involving Hydrogen Atoms

Substance	Formula	Bond	$\Delta \nu$, cm^{-1}	$-x$, cm^{-1}	$-\dfrac{x}{\nu}$
Methyldichlorsilane .	CH$_3$Cl$_3$SiH	Si—H	$\nu_1 = 4384$ $\nu = 2221$	29	0,013
Chloroform	CHCl$_3$	C—H	$\nu_1 = 5921$ $\nu = 3018$ $\nu_2 = 8703$	57 58	0,019
Diethylamine	(CH$_3$CH$_2$)$_2$NH	N—H	$\nu_1 = 6480$ $\nu = 3311$ $\nu_2 = 9444$	71 81	0,022

Table 17. Anharmonicity Coefficients (in cm^{-1}) of the Overtones of the
Molecules Studied

Interpret.	$-x_{ik}$		Interpret.	$-x_{ik}$		Interpret.	$-x_{ik}$	
	C$_6$D$_{12}$	C$_6$H$_{12}$		C$_6$D$_6$	C$_6$H$_6$		CDCl$_3$	CHCl$_3$
$2\nu_1$	—1	—1	$2\nu_{20}$	—5	4	$2\nu_6$	0	3
$2\nu_7$	0	1	$2\nu_{18}$	—3	—2	$2\nu_3$	4	4
$2\nu_8$	—6	—2	$2\nu_4$	—3	—6	$2\nu_2$	2	1
$2\nu_{23}$	1	1	$2\nu_8$	2	1	$2\nu_5$	9	4
$2\nu_{18}$	0	0	$2\nu_{11}$	0	2	$2\nu_4$	13	19
$2\nu_9$	0,5	1	$2\nu_{19}$	4	2	$2\nu_1$	34	57
$2\nu_{10}$	0	5	$2\nu_2$	—	—1			
$2\nu_{25}$	21	—	$2\nu_6$	—	—1			
$2\nu_4$	5	—	$2\nu_7$	1	—			
$2\nu_{26}$	20	18	$2\nu_{17}$	0	—9			
$2\nu_{11}$	6	—3	$2\nu_9$	1	1			
			$2\nu_3$	—6	0			
			$2\nu_{13}$	1	5			
			$2\nu_{18}$	—	4			

0 and 6 cm^{-1}, 27 lines had anharmonicity coefficients between 6 and 11 cm^{-1}, and 18 lines exceeded
11 cm^{-1}.

2. The anharmonicity coefficients of the overtones were smaller than those of the composite frequencies. Table 17 shows the observed overtones in the molecules studied and the anharmonicity coefficients. We see from the table that the anharmonicity coefficients of the overtones lie mainly between 0 and 6 (the discrepancies in the case of $2\nu_{25}$ and $2\nu_{26}$ are due to Fermi resonance).

3. The anharmonicity coefficients of the vibrations of the molecules are as a rule of the same sign (negative). Of the 118 lines mentioned, 92 have a negative sign and 26 a positive sign. This result is of considerable importance from the point of view of estimating the influence of anharmonicity on the frequency of the vibrations. Each observed vibration frequency ν_i is in fact a sum of the form

$$\nu_i = \omega_i + 2x_{ii} + \frac{1}{2} \sum_{i \neq k} x_{ik}, \tag{23}$$

where ω_i is the "zero" frequency; x_{ii}, x_{ik} are the anharmonicity coefficients of the overtones and composite frequencies, respectively. In this formula the last term $\frac{1}{2}\sum_{i\neq k} x_{ik}$ constitutes the sum of the coefficients of the composite frequencies. The number of terms in this sum, corresponding to the number of fundamental frequencies, is large in polyatomic molecules. Since the greater part of the anharmonicity coefficients have a similar sign, the sum may make a considerable contribution to the observed frequency. The latter may differ considerably from the "zero" frequency, which in principle characterizes the molecular vibrations more accurately. For example, in the case of chloroform (where there are altogether six fundamental frequencies), the difference between the observed and "zero" frequencies lies between 12 and 120 cm^{-1}.

For polyatomic molecules we should clearly expect a still larger difference between the observed and "zero" frequencies. In this connection we would note that the calculations of molecular vibrations so far carried out are only approximate, since the theory of harmonic molecular vibrations on which the calculations are based presumes the use of the "zero" rather than the observed frequencies.

4. The overwhelming majority of observed lines are interpreted directly as the combination of fundamental frequencies identical in type of symmetry and form in both compounds. The anharmonicity coefficients of such lines are as a rule identical in sign and close in magnitude. We consider that this result is of considerable significance.

In interpreting second-order spectra, as indicated earlier, we encounter considerable difficulties arising as a result of the fact that the same lines may be interpreted as a number of different combinations of fundamental frequencies. In all such cases the case giving the lower anharmonicity coefficient is usually taken [6]. We have adhered to this principle in the present investigation, interpreting the second-order spectra of ordinary molecules and their deuterized derivatives independently of one another. The above result is therefore important in that it confirms the insignificant anharmonicity of the vibrations of the molecules studied. Thus, the basic principle employed in solving second-order spectra, the principle of minimum anharmonicity coefficients, has received a complete experimental confirmation.

5. The remaining point is the characteristic nature of the anharmonicity coefficients. This may be seen from the data presented in Table 15, which relate to a study of the characteristic vibrations. For the same characteristic vibration, the anharmonicity coefficients are identical (within measuring error) for all the molecules studied.

Thus, in complex molecules, the anharmonicity coefficients may, together with other parameters, be associated with specific characteristic structural elements of the molecules.

2. Electro-Optical Parameters
of Second-Order Lines

The measurement of the intensities and degrees of depolarization of second-order lines in Raman spectra is of great importance, mainly from the point of view of interpreting the spectra. As indicated repeatedly in the foregoing, when interpreting experimental results one often experiences difficulties in identifying the second-order lines. These difficulties are greatly alleviated if we know the degrees of depolarization and the intensities of the lines.

It is well known [5] that the intensities of second-order lines are affected by both the mechanical and the electro-optical anharmonicity. In the presence of both forms of anharmonicity the intensity of an overtone will be proportional to the square of the matrix element [4]

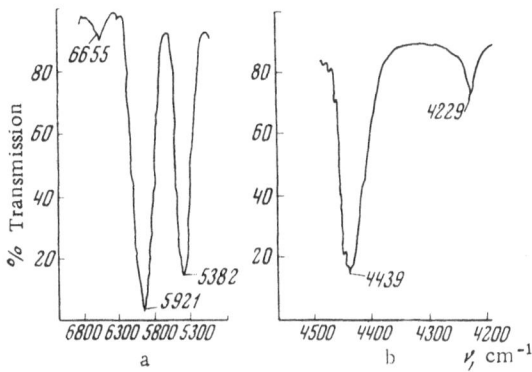

Fig. 6. Section of the infrared spectrum of: a) chloroform; b) deuterochloroform.

Table 18. Characteristics of the Absorption Bands of Chloroform and Deuterochloroform

Substance	Interpreta-tion	Obs. freq., cm^{-1}	Anharmonicity coeff., cm^{-1}
CHCl$_3$	$v_1 + 2v_4$	5382 m	30
	$2v_1$	5921 s	58
	$2v_1 + v_5$	6655 w	3
CDCl$_3$	$2v_1 - v_6$	4229 w	9
	$2v_1$	4439 s	34

Note. w, weak; s, strong; m, medium.

$$\left[\frac{\partial \alpha}{\partial Q_i}(Q_i)^{r_i}_{v_{i+2}} + \frac{1}{2}\frac{\partial^2 \alpha}{\partial Q_i^2}(Q_i^2)^{r_i}_{v_{i+2}} \right], \quad (24)$$

and the intensity of a composite frequency will be proportional to the square of

$$\left[\left(\frac{\partial \alpha}{\partial Q_i}Q_i + \frac{\partial \alpha}{\partial Q_k} \right)^{r_i v_k}_{v_{i+1}r_{k+1}} + \frac{\partial^2 \alpha}{\partial Q_i \partial Q_k}(Q_i Q_k)^{r_i v_k}_{v_{i+1}v_{k+1}} \right]. \quad (25)$$

The first term of each expression reflects the contribution of the mechanical and the second that of the electro-optical anharmonicity. Existing theory is inadequate for estimating the contribution of each term, and thus of the whole expression, to the intensity of the observed lines. Nor can anything be said as to when only mechanical or only electro-optical anharmonicity is to be expected.

In order to solve the problem as to the intensities of the overtones and composite frequencies we require a formula which would explicitly contain all the molecular parameters on which the quantities entering into the formula depended. Recently a problem of this kind has been solved by Gribov [15]. Assuming that the intensities of the overtones and composite frequencies are given by the quantities

$$\left(\frac{\partial^2 \bar{\mu}}{\partial Q_i^2} \right)_0 \sqrt{2}Q_{0i}^2 \quad \text{(transition } 0 \to 2)$$

and

$$\left(\frac{\partial^2 \bar{\mu}}{\partial Q_i \partial Q_j} \right)_0 Q_{0i}Q_{0j} \quad \text{(transition } 0 \to 1, \ 0 \to 1),$$

where Q_{0i} is the zero amplitude of the harmonic normal vibration, Gribov derived a general formula for the intensities of the second-order lines in the infrared spectra of polyatomic molecules. On the basis of this formula, Gribov drew a number of conclusions, which we cannot discuss at the moment owing to the inadequacy of the experimental data. In order to verify the conclusions derived from Gribov's formula to a certain extent, we shall need a great deal of experimental material on overtones and composite frequencies; at the moment, unfortunately, there is too little of this.

Gribov concluded that overtones and composite frequencies might appear in infrared spectra in the absence of both electro-optical and mechanical anharmonicity; the lines so appearing might be quite strong. This conclusion is drawn from the facts that, first, the formula is derived on the assumption of no mechanical anharmonicity and, secondly, that only one term of the formula contains elements associated with the electro-optical anharmonicity, so that in the absence of the latter the expression will depend on harmonic quantities only.

Clearly, in deriving the formula, the mechanical anharmonicity ought to have been considered; there seems to be no justification for its neglect. From this point of view it is interest-

Table 19. Intensity Distribution of Second-Order Lines

First group (0-4 cm^{-1})		Second group (4-8 cm^{-1})		Third group (over 8 cm^{-1})	
Intensity	No. of lines	Intensity	No. of lines	Intensity	No. of lines
Very weak and weak	30	Very weak and weak	14	Very weak and weak	6
Medium, strong, and very strong	19	Medium, strong, and very strong	16	Medium, strong, and very strong	13
Total	49	Total	30	Total	19

Table 20. Raman Spectrum of the Molecules Studied in the Neighborhood of the Vibrations of the C—H (C—D) Bond

C$_6$D$_{12}$				C$_6$H$_6$				C$_6$D$_6$				CHCl$_3$				CDCl$_3$			
Fundamental frequency, cm^{-1}	Overtones and composite frequency, cm^{-1}	I_0	$-x_{ik}$, cm^{-1}	Fundamental frequency, cm^{-1}	Overtones and composite frequency, cm^{-1}	I_0	p.	Fundamental frequency, cm^{-1}	Overtones and composite frequency, cm^{-1}	I_0	$-x_{ik}$, cm^{-1}	Fundamental frequency, cm^{-1}	Overtones and composite frequency, cm^{-1}	I_0	$-x_{ik}$, cm^{-1}	Fundamental frequency, cm^{-1}	Overtones and composite frequency, cm^{-1}	I_0	$-x_{ik}$, cm^{-1}
2082	1941	3.1	21		2923	8	8		2108	3	8	3018	3047	10	29	2254	2297	0.3	—4
2106	2062	10	6		2947	23	5	2265	2123	15	—6		3069	5	5		2339	1	20
2119	2230	18	5	3047	3164	7	4	2291	2417	6	—3								
2153	2313	27	20	3062	3184	20	8												
2170	2380	13	24		3208	4	4												
2194	2415	11	6																

ing to study the dependence of the second-order line intensities on the value of the anharmonicity coefficient. Figure 6 shows part of the infrared spectra of chloroform and deuterochloroform. The characteristics of the absorption bands are given in Table 18.

The considerable intensity of the 4439 and 5921 cm^{-1} bands of the C—D and C—H vibrations might be explained by the harmonic term in the Gribov formula, which in Gribov's opinion had a considerable value when the dipole moments of the bonds departed appreciably from the additivity law. For chloroform this could hardly be the case. The intensities observed might also be explained by electro-optical anharmonicity. However, then we should have to expect an absolutely identical intensity of these bands. The intensity of the 4439 cm^{-1} band in CDCl$_3$ is nevertheless much lower than that of the 5921 cm^{-1} band in CHCl$_3$, and the same applies in the anharmonicity coefficient.

In order to explain the relation between the intensity of the observed lines and the anharmonicity coefficients of the vibrations, we carried out a statistical analysis of the experimental material with respect to the infrared spectra of benzene, hexadeuterobenzene, chloroform, and deuterochloroform. The lines observed were divided arbitrarily into three groups.

The first group included lines with small anharmonicity coefficients between 0 and 4 cm^{-1}, the second those between 4 and 8 cm^{-1}, and the third those of 8 cm^{-1} and over. Altogether there were 98 bands (bands subject to Fermi resonance and uninterpreted bands were not included). The results are presented in Table 19.

We see from Table 19 that the majority of the weak lines have small anharmonicity coefficients, and that their number falls with increasing anharmonicity. Thus we may fairly assert that there is a direct dependence of the intensity of the overtones and composite frequencies on the value of the anharmonicity coefficient. Hence, in estimating the intensity of the overtones and composite frequencies the neglect of mechanical anharmonicity would appear quite unjustified.

We note that lines of a common origin observed in the Raman spectra of the ordinary molecule and its deutero-derivative have similar intensities.

Fermi resonance occurs quite often, particularly in the neighborhood of the vibrations of the C—H or C—D bonds. Table 20 shows all the lines subject to resonance in the C—H (C—D) neighborhood of the molecules studied. It is a typical feature that, the greater the number of the fundamental frequencies, the more the high-order lines are subject to resonance. The intensity of such lines becomes very great.

3. Use of Anharmonicity Coefficients for

Finding the Zero Vibration Frequencies

It is of great interest to use the anharmonicity coefficients in determining the "zero" frequencies of the vibrations ω_i, which in principle characterize the molecule more correctly than the experimental frequencies ν_i. We were able to solve this problem for comparatively simple molecules ($CHCl_3$ and $CDCl_3$), for which we had obtained the complete system of anharmonicity coefficients. Having the complete system of anharmonicity coefficients, obtained from the second-order lines, we were able to calculate the values of the "zero" frequencies. For this purpose we used the well-known formula obtained from the expression for the energies [25]:

$$\nu_i = \omega_i + x_{ii}(1 + d_i) + \frac{1}{2}\sum_{k \neq i} x_{ik}d_k + g_{ii}.$$

Here ν_i is the observed frequency, ω_i is the "zero" frequency, d_i is the degree of degeneracy, x_{ik} is the anharmonicity coefficient, and g_{ii} are the coefficients characterizing the relation between vibrations and rotations.

The values of the constants g_{ii} must be considered in calculating the zero frequencies of degenerate vibrations. However, experimental data for $CHCl_3$ and $CDCl_3$ showed that these constants might be neglected, since the splitting of the lines which should occur for large values of g_{ii} was hardly perceptible. In this case, all the quantities in the above formula for ω_i are known. The complete system of anharmonicity coefficients is given in Table 13.

The resultant "zero" frequencies ω_i shown in Table 21 were first of all used to prove the "rule of products." It is well known that the reliability of the interpretation of second-order lines depends on the accuracy of the values of the fundamental frequencies. For determining the fundamental frequencies, particularly those forbidden by the selection rules, one frequently uses the "rule of products." We therefore considered it necessary to discover the extent to which this rule was justified for experimental frequencies. The results given in Table 21 show that this rule is satisfied better for the zero frequencies than for the experimental values.

We see from Table 21 that the values of the "zero" frequencies in a number of cases differ considerably from the experimental values (the difference being between 12 and 120 cm^{-1}). Such a large difference, as noted earlier, occurs because the anharmonicity coefficients are, as a rule, of the same sign.

This result has a considerable significance from the point of view of the calculation of the force constants of the molecules from the vibrational frequencies. For such calculations it is in principle more correct to use the "zero" frequencies, since the theory of molecular vibrations

Table 21. Rule of Products for the Observed and Zero Frequencies, cm^{-1}

Type of vibrations	CHCl$_3$		CDCl$_3$		$\pi v^2(H)/\pi v^2(D)$		
					Frequencies		
	v_i	ω_i	v_i	ω_i	observed	zero	theoretical
A	3018	3140	2254	2333			
	669	679	649	659	1.91	1.99	2.01
	336	380	365	374			
E	1219	1300	903	963			
	764	793	736	780	2.00	1.96	1.95
	264	276	360	272			

is always based on the assumption of strict harmonicity of the vibrations. However, in practice one always uses the experimental frequency values, since there are no data regarding the anharmonicity coefficients. Thus, in determining the force field of molecules one introduces quite a large and practically uncontrollable error from the very beginning, and this subsequently affects the results of calculating the frequencies of complex molecules from the known force constants of simpler molecules.

It is to be hoped that a study of the anharmonicity of the molecular vibrations and the determination of the real force field of simple molecules by reference to the "zero" frequencies of the vibrations will bring substantial improvements in the accuracy of calculating the vibration frequencies of complex molecules.

CONCLUSION

The experimental results presented in the foregoing analysis show that an all-around investigation of the second-order spectra greatly broadens the possibilities of studying the structure of molecules. The number of frequencies in the second-order spectrum is considerably greater than the number of first-order lines, and the amount of information regarding the structure of molecules contained in second-order spectra is correspondingly greater. It is an important point that the second-order spectra contain the overtones of frequencies forbidden in both the Raman and infrared spectra. Quantitative data regarding the second-order lines in the vibrational spectra may be used for calculating quantities more broadly characterizing the molecular vibrations in question, namely, the second derivative of the polarizability and the dipole moment of the molecule with respect to the normal coordinates.

It is particularly interesting to use the anharmonicity coefficients for determining the "zero" frequencies of the vibrations, which in principle characterize the molecule more correctly than the experimental frequencies. We have been able to solve this problem for comparatively simple molecules ($CHCl_3$ and $CDCl_3$) and have found the complete system of anharmonicity coefficients for these. We have used the resultant zero frequencies to verify the "rule of products." The results showed that this rule was satisfied more accurately for the zero frequencies than for the experimental values.

The use of "zero frequencies" for calculating the force constants of the molecules clearly raises the accuracy of calculations relating to the vibration frequencies of polyatomic molecules.

By making an all-around study of the line parameters, we have established a number of laws in the second-order spectra as well as certain spectral singularities. The study and establishment of the laws governing the vibrations of certain characteristic structural elements are of particular interest.

In view of the fact that the theory of the anharmonicity of molecular vibrations has not yet been developed very far, the interpretation of the observed second-order lines has been mainly based on experimental data. This approach has of course been forced on us, and we should here like to emphasize the desirability of developing problems relating to the anharmonicity of molecular vibrations theoretically.

A method of allowing for the anharmonicity of the vibrations by introducing a more complex function than the ordinary quadratic one for the potential energy of the molecule has recently been proposed by a number of authors. This is especially true of Pliva and colleagues [91-93]. We hope that our experimental data relating to the anharmonicity coefficients may be used for constructing the potential function of the polyatomic molecules.

In conclusion, the author wishes to express his sincere thanks to his scientific director, Professor M. M. Sushchinskii, for constant interest and great help in this work, and also to Professor P. A. Bazhulin and G. V. Peregudov for valuable advice in discussing the results.

The author also wishes to thank his colleagues in the Optical Laboratory of the Physical Institute, whose kind cooperation has ensured the successful conclusion of this work.

LITERATURE CITED

1. M. V. Vol'kenshtein, M. A. El'yashevich, and B. I. Stepanov, Vibrations of Molecules, Vol. 1, Gostekhizdat, Moscow (1949).
2. V. D. Bogdanov and M. M. Sushchinskii, Izv. Akad. Nauk SSSR, Ser. Fiz., 22:1067 (1958).
3. W. H. Stockmayer, G. M. Kavanagh, and H. S. Mickely, J. Chem. Phys., 12:408 (1944).
4. H. L. Welsh, M. F. Crawford, and G. D. Scott, J. Chem. Phys., 16:97 (1948).
5. M. V. Vol'kenshtein, M. A. El'yashevich, and B. I. Stepanov, Vibrations of Molecules, Vol. 2, Gostekhizdat, Moscow (1949).
6. Yu. I. Naberukhin and M. M. Sushchinskii, Opt. i Spektroskopiya, 9:576 (1960).
7. S. Broderson and A. Langseth, Kgl. Danske Videnskab. Selskab, Mat.-Fys. Medd., 1:1 (1956).
8. L. S. Ornstein and J. J. Went, Proc. Am. Acad. Sci., 35:1024 (1932).
9. A. Carrelli and J. J. Went, Z. Phys., 76:236 (1932).
10. G. S. Landsberg and V. I. Malyshev, Dokl. Akad. Nauk SSSR, 3:265 (1936).
11. R. Ananthakrishnan, Proc. Indian Acad. Sci., 2:425 (1935).
12. Ya. S. Bobovich, Opt. i Spektroskopiya, 11:342 (1961).
13. B. Morzynska, Bull. Acad. Polon. Sci., 7:455 (1959).
14. K. Venkateswarlu and G. Thyagarajan, Z. Phys., 156:569 (1959).
15. L. A. Gribov, Opt. i Spektroskopiya, 13:594 (1962).
16. I. I. Kondilenko, V. E. Pogorelov, and V. L. Strizhevskii, Opt. i Spektroskopiya, 13:549 (1962).
17. V. A. Zubov, Tr. Fiz. Inst. Akad. Nauk SSSR, 30:3 (1964).
18. G. S. Landsberg, P. A. Bazhulin, and M. M. Sushchinskii, Basic Parameters of the Raman Spectra of Hydrocarbons, Izd. Akad. Nauk SSSR, Moscow (1956).
19. S. G. Rautian, Zh. Eksper. i Teor. Fiz., 27:625 (1954).
20. V. K. Prokof'ev, Photographic Methods of the Quantitative Spectral Analysis of Metals and Alloys, Gostekhizdat, Moscow (1951).
21. G. I. Distler, Kristallografiya, 1:218 (1956).
22. R. N. Jones, P. K. Faure, and W. Zaharias, Rev. Universelle Mines, 15:417 (1959).
23. A. N. Aleksandrov and V. A. Nikitin, Usp. Fiz. Nauk, 56:3 (1955).
24. R. Mecke and F. Oswald, Z. Phys., 130:445 (1951).
25. G. Herzberg, Vibrational and Rotational Spectra of Polyatomic Molecules [Russian translation], IL, Moscow (1949).
26. M. A. El'yashevich, Atomic and Molecular Spectroscopy, Fizmatgiz, Moscow (1962).
27. C. M. Beckett, K. S. Pitzer, and R. Spitzer, J. Am. Chem. Soc., 69:2488 (1947).
28. M. M. Sushchinskii, Tr. Fiz. Inst. Akad. Nauk SSSR, 12:54 (1960).
29. G. A. Aleksandrov, Opt. i Spektroskopiya, 3:202 (1957).
30. T. I. Kuznetsova and M. M. Sushchinskii, Optics and Spectroscopy, in collection: Molecular Spectroscopy (1963), p. 144.
31. A. Langseth and B. Bak, J. Chem. Phys., 8:403 (1940).

32.   R. W. Crowe and C. P. Smyth, J. Am. Chem. Soc., 73:5406 (1951).
33.   G. V. Peregudov, Dissertation, Minsk (1961).
34.   Z. M. Muldakhmetov and M. M. Sushchinskii, Optics and Spectroscopy, in collection:
      Molecular Spectroscopy (1963), p. 320.
35.   W. Gerlach, Sitzber. Bayer. Akad. Wiss., 39:112 (1932).
36.   I. Chedin, C. F. Hsuech, and T. I. Wu, J. Chem. Phys., 6:8 (1938).
37.   B. S. R. Rao, J. Chem. Phys., 6:343 (1938).
38.   A. Langseth and R. C. Lord, Kgl. Danske Videnskab. Selskab, Mat.-Fys. Medd., 16:6
      (1938).
39.   F. A. Miller and B. L. Crawford, J. Chem. Phys., 14:282 (1946).
40.   K. S. Pitzer and D. W. Scott, J. Am. Chem. Soc., 65:803 (1943).
41.   V. L. Broude and A. F. Prikhot'ko, Zh. Eksper. i Teor. Fiz., 22:60 (1952).
42.   V. L. Broude, V. S. Medvedev, and A. F. Prikhot'ko, Zh. Eksper. i Teor. Fiz., 21:665
      (1951).
43.   V. L. Broude, V. S. Medvedev, and A. F. Prikhot'ko, Opt. i Spektroskopiya, 2:317 (1957).
44.   A. Langseth and R. C. Lord, J. Chem. Phys., 6:203 (1938).
45.   C. R. Bailey, C. K. Ingold, H. G. Pool, and C. L. Wilson, J. Chem. Soc., 4:912 (1936).
46.   W. R. Angus, C. R. Bailey, C. K. Ingold, and C. L. Wilson, J. Chem. Soc., 14:222 (1946).
47.   R. D. Mair and D. F. Hornig, J. Chem. Phys., 17:1236 (1949).
48.   R. S. Halford and O. A. Schaffer, J. Chem. Phys., 14:141 (1946).
49.   G. Germain, J. Phys. Radium, 14:85 (1953).
50.   F. A. Miller, J. Chem. Phys., 24:996 (1956).
51.   N. I. Leonard and L. E. Sutton, J. Am. Chem. Soc., 70:1564 (1948).
52.   M. A. Kovner, Opt. i Spektroskopiya, 1:742 (1956).
53.   A. M. Bogomolov, Author's Abstract, Dissertation, Saratov Gos. Univ. (1963).
54.   A. M. Bogomolov, Opt. i Spektroskopiya, 4:311 (1960).
55.   I. P. Zietlow, F. F. Cleveland, and A. G. Meister, J. Chem. Phys., 18:1076 (1950).
56.   L. R. Nielson and N. E. Ward, J. Chem. Phys., 10:81 (1942).
57.   A. Dadien and K. W. F. Kohlrausch, Phys. Z., 31:514 (1930); Sitzber. Akad. Wiss. Wien,
      138:635 (1928).
58.   A. S. Ganesan and S. Venkateswaran, Indian J. Phys., 4:195 (1929).
59.   S. Bhagavantam, Proc. Roy. Soc. London, A127:300 (1930).
60.   S. Bhagavantam, Indian J. Phys., 5:35 (1930); 5:59 (1930).
61.   S. C. Sirkar, Indian J. Phys., 10:189 (1936).
62.   M. V. Rao, Proc. Indian Acad. Sci., 24A:510 (1946).
63.   W. M. Dabadghao, Indian J. Phys., 5:207 (1930).
64.   M. de Hemptinne, Ann. Soc. Sci. Bruxelles, 52:185 (1932).
65.   C. D. Cleeton and R. G. Dufford, Phys. Rev., 37:362 (1931).
66.   M. de Hemptinne and A. Peters. Bull. Acad. Roy. Belgique, 5:1107 (1931).
67.   R. W. Wood and D. H. Rank, Phys. Rev., 48:61 (1935).
68.   R. Truchet, Compt. Rend. Acad. Sci., 222:1997 (1936).
69.   O. Redlich and F. Pordes, Sitzber. Akad. Wiss. Wien, 116:145 (1936).
70.   I. T. Dhar, Indian J. Phys., 9:189 (1934).
71.   R. W. Wood, Phil. Mag., 6:729 (1928).
72.   R. M. Langer and W. F. Meggers, J. Res. Natl. Bur. Std., 4:711 (1930).
73.   P. Pringsrim and B. Rosen, Z. Phys., 50:741 (1928).
74.   M. de Hemptinne and I. Wonters, Nature, 138:864 (1936).
75.   D. H. Rank, J. Opt. Soc. Am., 37:798 (1947).
76.   L. Simons, Soc. Sci. Fennica, Commentationes Phys.-Math., 6:13 (1932).
77.   A. V. Rao, Z. Phys., 97:154 (1935).
78.   S. Venkateswaran, Phil. Mag., 15:263 (1933).

79. I. Cabannes and A. Ronsset, Ann. Phys., 19:229 (1933).
80. B. L. Crawford and W. Horwitz, J. Chem. Phys., 15:268 (1947).
81. C. Decins, J. Chem. Phys., 16:214 (1948).
82. M. P. Lisitsa and Yu. P. Tsyashchenko, Opt. i Spektroskopiya, 6:610 (1959).
83. E. Plyler and W. Benedict, J. Res. Natl. Bur. Std., 47:202 (1951).
84. I. Ellis, Phys. Rev., 23:48 (1924); 32:907 (1928).
85. C. Corin, Compt. Rend., 202:747 (1936).
86. T. Yeon, Compt. Rend., 206:1371 (1938).
87. A. Maione, Nuovo Cimento, 14:361 (1937).
88. I. R. Madigan, F. F. Cleveland, W. M. Boyer, and R. B. Bernstein, J. Chem. Phys., 18:1081 (1950).
89. A. Weisberger, E. Broskauer, J. Riddick, and E. Tuis, Organic Solvents [Russian translation], IL, Moscow (1958). [Organic Solvents (Techniques of Organic Chemistry), Vol. 7, 2nd edition, Interscience.]
90. H. Nielsen, Handbuch der Phys., 37/1:173 (1959).
91. I. Pliva, Coll. Czech. Chem. Commun., 23:777 (1958).
92. I. Pliva, Coll. Czech. Chem. Commun., 23:1839 (1958).
93. I. Pliva and Z. Chihla, Coll. Czech. Chem. Commun., 29:1232 (1963).

STUDY OF THE ROTATIONAL OSCILLATIONS
OF MOLECULES IN LIQUIDS
BY THE RAMAN METHOD*

U. A. ZIRNIT

*Dissertation in pursuit of the degree of Candidate of Physicomathematical Sciences. Defended April 19, 1965. Scientific director: Professor M. M. Sushchinskii.

INTRODUCTION

The rotational (torsional) oscillations of individual atomic groups in complex molecules appear in the low-frequency part of the vibrational spectra. Information regarding this spectral range is provided by the absorption spectra of the long-wave infrared radiation and the Raman spectra in the low-frequency range.

A study of the lines corresponding to the rotational oscillations of atomic groups is of great importance in understanding the forces acting between atoms not connected by chemical bonds in molecules. In addition to this, data relating to the rotational oscillations are required for the statistical calculation of the thermodynamic functions of molecules, while the exact values of the thermodynamic functions in turn are necessary for solving a wide variety of problems. It should be noted that the study of rotational oscillations in comparatively simple molecules may take on importance when considering the molecules of polymers.

Improvements to the values of the frequencies or the observation of new lines in the low-frequency range are important for the reliable interpretation of the whole vibration spectrum (including the overtones and composite frequencies). Data relating to ordinary deformation (strain) vibrations are of great importance in refining the force field of the molecule (we know that low frequencies are very sensitive to small changes in force constants).

Whereas research in the far-infrared part of the spectrum has been rapidly extended, despite great experimental difficulties, only one paper [1] has been devoted to the Raman spectra of liquids in the low-frequency range. There have been no papers at all on systematically studying rotational oscillations by examining the Raman scattering of light in the low-frequency region. The value of data obtained by reference to Raman spectra lies in the fact that the study of liquids in the far-infrared involves special difficulties as compared with the study of gases and solids.

The recording of low-frequency spectra in the infrared demands the creation of recording apparatus operating on entirely new principles. By photographically recording the Raman spectrum, we always obtain the whole spectrum, including the low-frequency region. The separation of the Raman spectrum into a low-frequency and an ordinary region is arbitrary. By the "remote low-frequency region" we shall understand that part of the spectrum extending to 400 cm^{-1}, i.e., the spectral range inaccessible to study by means of standard infrared apparatus. The lower limit of this region is set by the wing of the Rayleigh line, which prevents lines with frequencies below 80-100 cm^{-1} from being observed in liquids.

Remembering the great importance, in principle and in practice, of research into rotational oscillations, we set ourselves the problem of making a systematic study of low-frequency Raman spectra for a wide class of liquids, in some of which some indication of the appearance of rotational oscillations is to be expected.

CHAPTER I

REVIEW OF LITERATURE

1. Review of Experimental Data Relating
to the Low-Frequency Vibrational Spectra

In the low-frequency range we may find ordinary deformation (strain) vibrations, rotational oscillations of one part of the molecules relative to another (torsional oscillations), and intermolecular vibrations due to specific interactions between molecules, for example, the hydrogen bond. Frequencies characterizing the specific vibrations of molecules with four-membered rings have been observed very recently in the far-infrared. We shall confine ourselves to an analysis of the published data relating to the rotational oscillations. In parallel with results obtained by the Raman method, we have considered it desirable to consider data relating to long-wave infrared absorption spectra.

Very wide and strong lines were observed in the low-frequency part of the Raman spectra of the halogen derivatives of ethane even in the 30's. One of the earliest investigations was the study of 1,2-dichlorethane, in which a wide line was found near 125 cm^{-1} [2-4]. Interest in such compounds was increased because of their rotational isomerism, and this gave rise to a large number of papers on the Raman spectra of these compounds. Lines analogous to the 125 cm^{-1} line of 1,2-dichlorethane were also found in the extreme low-frequency part of the spectra of other halogen derivatives of ethane (70-150 cm^{-1}). These lines were interpreted as a manifestation of rotational oscillations around the C—C bond.

Other papers devoted to the long-wave infrared spectra of the halogen derivatives of ethane included [5] and [6]. The long-wave infrared absorption spectra of 1,2-dichlorethane CH_2Cl—CH_2Cl and 1,2-dibromethane CH_2Br—CH_2Br in the liquid and gaseous states were given in [5]. In addition to the infrared absorption spectra, the Raman spectra of these substances in the liquid state were also given. Unfortunately, the frequencies of the rotational oscillations observed in both infrared and Raman spectra could not be compared because they belonged to different steric isomers. We see from the infrared spectra of the substances in question that the frequency of the rotational oscillations in the liquid state are about 10 cm^{-1} greater than in the gas. In the far-infrared part of the spectrum, the spectrum of gaseous 1,1,1,2-tetrafluorethane CF_3—CH_2F was also studied in [6]. The authors ascribed the 120 cm^{-1} line to rotational oscillations.

The spectra of the halogen derivatives of butane were studied in [7] by the Raman method; however, no lines were found below 200 cm^{-1}.

Low frequencies were also studied by Green, Kynaston, and Gebbie [8], who obtained the low-frequency spectra of liquid substituted benzene products by the infrared and Raman methods. These authors ascribed the 330 cm^{-1} line of phenol to the rotational oscillations of the OH group. This line was only found in the infrared spectrum. The line at 130 cm^{-1} found in benzaldehyde had a considerable intensity in both the infrared and Raman spectra. This line was also attributed to rotational oscillations. In the same region a line was also observed in the long-wave infrared absorption spectrum of benzaldehyde in [9]. Rotational oscillations were observed in the 100 cm^{-1} region of the Raman spectra of diphenyl derivatives in [10]. The foregoing results constitute the limit (so far as we know) of available data regarding the appearance of rotational oscillations of heavy groups in the vibration spectra.

The rotational oscillations of the methyl group should have a higher frequency as compared with those of the heavy groups in view of its low moment of inertia. We see from experimental data [11-18] that in fact the frequency of the rotational oscillations of the methyl groups lies, in the majority of cases, between 150 and 250 cm^{-1}. Work on this subject has been performed fairly recently and has been concentrated on the far-infrared region of the rotational oscillations of comparatively simple molecules with only one or two methyl groups.

A great deal of fresh data has appeared in papers by Fataley and Miller [11-13]. These authors ascribe the following methyl group lines to rotational oscillations: the 150 cm^{-1} of

acetaldehyde $\overset{H}{\underset{O}{\diagdown}}C-CH_3$, the 243 cm^{-1} of chlorethane CH_2Cl-CH_3, the 222 cm^{-1} of 1,1-difluor-

ethane CHF_2-CH_3, the 109 cm^{-1} of acetone $CH_3-\overset{O}{\overset{\|}{C}}-CH_3$, the 242 cm^{-1} of dimethyl ether CH_3- $O-CH_3$, the 257 cm^{-1} of dimethylamine $CH_3-\underset{H}{\overset{|}{N}}-CH_3$, the 182 cm^{-1} of dimethyl sulfide CH_3-

$S-CH_3$, and the 269 cm^{-1} of trimethylamine $\overset{N}{\underset{CH_3\ \ CH_3\ CH_3}{\diagup|\diagdown}}$. Here, for the first time, the values of the potential barriers of the internal rotation of about 20 molecules were obtained by the method of long-wave infrared spectroscopy, including the following: for acetaldehyde, a potential barrier of $V_0 = 1180$ cal/mole, for chlorethane, 3580 cal/mole; 1,1-difluorethane, 3210 cal/mole; dimethyl ether, 2625 cal/mole; acetone, 830 cal/mole; dimethylamine, 3280 cal/mole; dimethyl sulfide, 2090 cal/mole; and trimethylamine, 4410 cal/mole. The interpretation of these lines as belonging to rotational oscillations is confirmed by comparing with existing data relating to the barriers for the internal rotation of molecules obtained from microwave spectra [19].

The majority of substances of this class have been studied in the gaseous state; however, in [11], the frequency of the rotational oscillations of acetaldehyde was also measured in the solid state. Since the frequencies of the rotational oscillations of the methyl group were the same in the gaseous and solid states, Fataley and Miller concluded that the rotational oscillations of the methyl groups depended little on the state of aggregation.

Proofs of the appearance of rotational oscillations of methyl groups in Raman spectra exist in two cases. The 274 cm^{-1} line of trimethylamine observed in the Raman spectra studied in [20] was attributed by Fataley and Miller [12] to the rotational oscillations of methyl groups after comparing with infrared absorption data. The 210 cm^{-1} frequency observed in the Raman spectrum of trans-butene-2 $CH_3-CH = CH-CH_3$ [21, 22] was also ascribed to the rotational motion of the methyl groups by Sverdlov on the basis of vibrational calculations [23].

We see from this review of the experimental study of low-frequency vibration spectra that there is very little experimental data in this field. This is particularly so for the low-frequency Raman spectra. Nevertheless, there are some quite clear cases in which the low-frequency Raman spectra have contained rotational oscillations of heavy atoms and methyl groups. This indicates that on improving the methods of studying low-frequency Raman spectra specific molecular vibrations may yet be found in substances for which they have not so far been observed.

2. Methods of Interpreting

Low-Frequency Spectra

In order to relate certain lines in the low-frequency part of the Raman spectrum to rotational oscillations we must consider suitable methods of interpreting low-frequency spectra.

Clearly, if we can assert that certain particular lines in the low-frequency region relate to oscillations of the deformation type, the remaining lines may be attributed to specific oscillations, for example, the rotational oscillations of atomic groups. However, general methods of calculating the normal vibrations of polyatomic molecules, as discussed in earlier monographs [24, 25], are insufficiently accurate to calculate the frequencies in the low-frequency part of the spectrum.

In view of this, all attempts at calculating frequencies in the low-frequency region are based on special methods of calculating simple models. All the models considered by various authors are based on the assumption that individual parts of the molecules may be regarded approximately as rigid. Clearly this method is inapplicable for certain classes of compounds.

The approximate calculation of deformation and valence vibrations for simple singly-substituted benzene derivatives between 150 and 550 cm^{-1} was carried out by Sechkarev [26, 27], who considered the benzene ring as rigid, basing this assumption on the fact that the vibrations of the benzene ring interacted very little with other molecular vibrations.

Approximate models have been widely used for calculating molecular vibration frequencies in the early stages of studying Raman spectra. This kind of calculation was often used in the well-known Kohlrausch monograph [28] and in a number of cases provided a qualitative explanation for the observed laws governing the behavior of the vibration frequencies on passing from one substance to another. However, such calculations often lead to unsatisfactory results, particularly for deformation oscillations lying in the low-frequency part of the spectrum. This state of affairs has been explained on the basis of later investigations in which the molecular vibrations were calculated with due allowance for all degrees of freedom. Thus, for example, the considerable interaction between the deformation vibrations of paraffins was noted in [24]. The results of a recent paper on calculating the normal vibrations of methyl-substituted cyclohexanes [29] shows a considerable interaction between the vibrations of the six-membered ring and the vibrations of the methyl groups. The substantial interaction between the deformation vibrations prevents us from separating the individual groups of atoms and using these groups as rigid.

The ascription of some of the lines in the low-frequency part of the spectrum to the rotational oscillations of atomic groups is a matter of considerable difficulty. In the ordinary methods of calculating the vibrations of molecules, the rotational oscillations have not been considered at all.

The fact that internal rotation around a single bond in the molecule is almost always retarded was first established by comparing the values of the thermodynamic quantities obtained by statistical calculations with the experimental values. Moreover, it is often found that, when the kinetic energy of internal rotation is lower than the work of overcoming the potential barrier, rotation is replaced by rotational oscillations (torsional oscillations) around the equilibrium position. The case of internal rotation may be seen most clearly by considering the case of the ethane molecule CH_3—CH_3. The equilibrium configuration of ethane corresponds to the trans-position of the C—H bonds. If we consider the molecule along the C—C bond, then the atoms in the C—H groups form two equalateral triangles rotated through 60° with respect to each other. The potential energy expressed as a function of the angle of rotation of one CH_3 group relative to the other will have the form of a sinusoidal function. The three identical minima on the resultant curve correspond to the three mutual dispositions of the CH_3 groups with $\varphi = 60, 180,$

and 300°. The minima are separated by maxima corresponding to the cis-position of the bonds. The difference in the heights of the maxima and minima determines the height V_0 of the potential barrier separating the minima from each other. The corresponding levels of the energy of rotational oscillations may also be found. Transitions between the two deepest levels v = 0 and v = 1 determine the fundamental frequency of the rotational vibrations. The question of retarded internal rotation has been treated in detail in a number of monographs [24, 30, 31] and reviews.

If the rotating group is a symmetric gyrostat, i.e., has C_{3v} symmetry similar to the methyl group, then the potential function is sinusoidal even in the case of an asymmetric core. The potential function of the rotational oscillations is in this case expressed by the simple formula

$$V(\varphi) = \frac{V_0}{2}(1 - \cos 3\varphi). \tag{1}$$

Here V_0 is the height of the potential barrier, and φ is the angle of relative rotation of the rotating parts. Later investigations of Fataley and Miller in the far-infrared part of the spectrum [13] showed that the deviations of the potential function representing the rotational oscillations of the methyl groups in the molecules were very slight. Absorption was observed corresponding to transitions not only from the zeroth to the first energy level of the rotational oscillations but also to transitions of the $1 \rightarrow 2$, $2 \rightarrow 3$, $3 \rightarrow 4$ types. As an example, we may cite the chlorethane molecule CH_2Cl-CH_3, for which 251.5, 235.5, 217, and 197 cm^{-1} were respectively observed. If we put the potential function in the form

$$V(\varphi) = \frac{V_0}{2}(1 - \cos 3\varphi) + \frac{V'}{2}(1 - \cos 6\varphi), \tag{1a}$$

then the term containing V' is no more than 3% of the first term. It should be emphasized that the second term does not change the height of the potential barrier, but only its shape. The observation of such series of absorption peaks in particular molecules may serve as an important criterion for ascribing lines to rotational oscillations.

In the general case, the maxima of the potential function of rotational oscillations may have different heights and the minima different depths, for example, the potential function of 1,2-dichlorethane. For this more general case Pitzer and Gwinn [32] introduced the following potential function:

$$V(\varphi) = \sum_{j=1}^{\infty} \frac{V_{0j}}{2}(1 - \cos jn\,\varphi). \tag{2}$$

Consideration of the motion corresponding to this more general form of potential function is extremely complicated. This general form of the potential function is associated with the general problem of rotational isomerism, which has constituted the theme of protracted investigations by many research workers.

In order to relate certain lines in the low-frequency region of the vibrational spectrum to rotational oscillations we use the following method [11-13]. From the values of those observed frequencies which, on the basis of some criterion or another, may be expected to be related to rotational oscillations, we calculate the height of the corresponding potential barrier of the retarded internal rotation. Then we compare the height of the potential barrier so calculated with the height of the barrier determined by other methods. If these values agree, we may assume that the assumed relation is correct.

Calculating the frequency of the rotational oscillations, knowing the height of the potential barrier and the geometrical parameters of the rotating group and the core of the molecule, is fairly simple [30, 31]. After substituting the potential energy of the rotational oscillations of the methyl group (1) into Schrödinger's equation for the one-dimensional rotator, the problem reduces to the solution of Mathieu's equation. On substituting the potential energy of the rotational oscillations of more general form (2) into Schrödinger's equation, we obtain Hill's equation. The latter case will not be considered in detail, since we shall not require it for the present investigation. The solution of the Mathieu equation presents no difficulties; the eigenvalues of this equation are tabulated for reasonably low barriers in [33]. Since the discovery of a relation between the value of the potential barrier and the frequency of the rotational oscillations is of great importance for our present work, we shall consider this question in more detail.

Knowing the frequency of the rotational oscillations ν and the coefficient F, depending on the reduced moment of inertia of the rotating groups, we may find the quantity

$$\Delta b = \frac{\nu}{2.25F}. \tag{3}$$

The quantity Δb is included in the tables of eigenvalues of the Mathieu equation; hence, if we know Δb we may determine the parameter s directly associated with the height of the barrier from the tables. In fact, the height of the potential barrier of the rotational oscillations of the methyl group V_3 is expressed in terms of the parameter s by the simple formula

$$V_3 = 2.25Fs. \tag{4}$$

The coefficient F is expressed in terms of the reduced moment of inertia of the rotational oscillation I_{red} in the following way:

$$F = \frac{h^2}{8\pi^2 I_{red}}, \tag{5}$$

where h is Planck's constant.

The fact that the potential functions of the rotational oscillations has several minima produces degeneracy of the energy levels of the rotational oscillations. Owing to the tunnel effect, splitting of the levels occurs. The splitting of the lowest levels is only slight; it diminishes rapidly with increasing height of the barrier. For the acetaldehyde molecule $COH-CH_3$ [11], which has a comparatively low barrier (1180 cal/mole), the splitting of the zero level is 0.09 cm^{-1}, that of the first level 2.2 cm^{-1}, and that of the second level 15.3 cm^{-1}; however, for the chlorethane molecule, CH_2Cl-CH_3, which has a high potential barrier (3800 cal/mole), the splitting of the first three levels is negligible, less than 0.1 cm^{-1}.

In the case of large values of V_3, the rotator executes rotational oscillations with a small angle of rotation φ. In this case, in the potential function we may expand cos 3φ in series, confining attention to terms in φ^2. The result is a harmonic-oscillation equation. The relation between the height of the potential barrier V_3 and the frequency of the rotational oscillations in this case takes the particularly simple form

$$\nu = \frac{3}{2\pi c}\sqrt{\frac{V_3}{2I_{red}}}. \tag{6}$$

We see from these formulas that for the calculations we must know the reduced moment of inertia (I_{red}) for the rotational oscillations. If we have a molecule consisting of any arbitrary core and a rotating group in the form of a symmetrical gyrostat, the expression for the reduced moment of inertia is that given by Crawford [34]:

$$I_{red} = I_\varphi \left(1 - I_\varphi \sum_g \frac{\lambda_g^2}{I_g}\right), \tag{7}$$

where g = 1, 2, 3; I_g are the principal moments of inertia of all the molecules; λ_g are the direction cosines of the angles formed by the axis of the gyrostat with the principal axes of inertia; and I_φ is the moment of inertia of the gyrostat relative to the axis of rotation. The accuracy of the Crawford formula has proved to be so high that a similar expression for the reduced moment of inertia is also used for the analysis of microwave spectra (see, for example, [33]).

In the presence of several symmetrical rotating groups, the question of the reduced moment of inertia of the rotational oscillations is considered in [12]. Here the reduced moment of inertia for molecules of a given type belonging to different symmetry groups is obtained with due allowance for the symmetry of the rotational oscillations. Thus, for example, for molecules with C_{2g} symmetry of the type of dimethyl ether CH_3-O-CH_3 there are rotational oscillations of symmetry A_2 and B_1. For rotational oscillations of type A_2, both CH_3 groups rotate in one direction (the line is allowed in the Raman spectrum but forbidden in the infrared spectrum). For rotational oscillations of the B_1 type, each CH_3 group rotates in its own direction (the corresponding line is allowed in the Raman spectrum and also in the infrared). Each type of symmetry of the rotational oscillations has its own reduced moment of inertia. It is clear that in this case there will be lines of different frequencies for rotational oscillations of different types of symmetry and the same potential barrier. The determination of reduced moments of inertia for molecules with C_S symmetry having two methyl groups and for molecules with C_{3V} symmetry having three methyl groups is also considered in [12]. It should be noted that the expressions used in [12] for finding reduced moments of inertia, and also the Crawford formula for the case in which the moment of inertia of the core is much greater than that of the rotating group, simply reduce to the equation $I_{red} = I_\varphi$.

For the rotational oscillations of heavier and asymmetric groups, the difficulties of assigning individual lines in the low-frequency part of the spectrum to this type of oscillations become much greater. As already discussed, in the case of the internal rotation of asymmetrical groups the potential function may have a complicated form. So far the literature has been confined to finding some average value of the potential barrier; however, even this finds direct practical use for thermodynamical calculations. The frequency of the rotational oscillations is associated with the height of the potential barrier in the manner described earlier. The reduced moment for the case in which both the rotating groups and the core constitute asymmetrical gyrostats may be expressed in the form of Pitzer's equation [35]. Owing to the complexity of this formula, we shall not give it here. The calculation of reduced moments by Pitzer's formula is very difficult.

The heights of the potential barriers of the rotational oscillations of the methyl groups have been calculated from the observed frequencies of the remote infrared spectrum by the method given in [11-13]. The good agreement between the resultant values and the height of the potential barriers of the same substances obtained from microwave measurements indicates that this description of the rotational oscillations, in which the methyl groups are regarded as rigid (i.e., in which the interaction between the rotational oscillations and other oscillations of the molecule is neglected) constitutes a good approximation.

We must also mention one further possible method of facilitating the ascription of some of the lines in the low-frequency part of the spectrum to rotational oscillations. Since, in ordinary methods of calculating the vibrations of molecules the rotational oscillations are not considered at all, the observed frequencies may be higher than the calculated values. By way of example, we may consider the trans-butene-2 molecule calculated by Sverdlov [23]. In such cases there is a possibility of ascribing "excess" lines in the low-frequency region to rotational oscillations. We note that the number of observed lines only exceeds the calculated number for molecules of high symmetry. For molecules of low symmetry the number of frequencies observed in the spectrum is as a rule smaller than the total number of frequencies allowed by the selection rules.

3. Presentation of the Problem

The foregoing review of papers relating to the Raman spectra of liquids at low frequencies shows that such papers are few and far between, while there have been no systematic studies of rotational oscillations whatsoever. Nevertheless, there are several clear cases in which the rotational oscillations of methyl groups and heavier atomic groups appear in Raman spectra. It should be noted that the direct observation of lines due to rotational oscillations in the spectrum greatly eases the determination of the thermodynamic functions of the molecules by statistical calculation.

In individual papers, attempts have been made to interpret lines in the low-frequency part of the spectrum by ascribing some of the lines to rotational oscillations or to some other specific type of oscillation (for example, the hydrogen bond). However, the results have never been systematized with respect to the interpretation of the lines in the low-frequency part of the spectrum.

In view of this we set ourselves the following problems.

1. To carry out systematic investigations into the rotational oscillations of a number of liquids of various classes by studying the Raman scattering of light at low frequencies. The most promising substances were paraffins and naphthenes.

2. To solve a number of methodical questions associated with the special aspects of working in the low-frequency part of the spectrum. This includes the choice of a recording method and of the spectral apparatus, as well as modifications of this apparatus for observations in the low-frequency part of the spectrum.

3. To develop methods of relating specific lines in the low-frequency part of the Raman spectrum of paraffins and naphthenes to rotational oscillations of methyl and heavier atomic groups, and so to improve methods of calculating potential barriers.

CHAPTER II

METHODS OF OBTAINING AND ANALYZING
LOW-FREQUENCY RAMAN SPECTRA

Two methods are now used for recording Raman spectra: photographic and photoelectric. The latter method, in the form in which it is usually employed for studying Raman spectra, is unsuitable for recording low frequencies. This is because of the very strong continuous background, which lies close to the exciting line and falls off gradually on moving away from it. In the present investigation we used the photographic method of recording spectra; this was better than the photoelectric method because long exposures might reveal weak lines not susceptible to photoelectric recording.

For photographing the low frequencies we used a two-prism Huet B-II spectrograph. The inverse linear dispersion of the apparatus in the 4358 Å region was 18 Å/mm. The spectral apparatus selected gave very little parasitic light-scattering inside the system, which is extremely important for recording low-frequency spectra.

The background in the low-frequency Raman range is due to a number of causes, the chief of which are as follows: 1) multiple reflection of the exciting light at optical components of the apparatus, and also the scattering of light in prisms and lenses; 2) insufficiently careful adjustment of the cuvette containing the sample; 3) the presence of interfering lines or continuous background in the spectrum of the source exciting the Raman scattering; 4) the development of a halo at the point at which an intense exciting line falls on the photographic plate; 5) the presence of a Rayleigh-line wing of a molecular nature. The first four factors are associated with the characteristics of the spectral apparatus, the source of exciting light, the illuminating system, and the photographic plate. If these elements are chosen to best effect, the continuous background in the low-frequency range may be greatly reduced. The fifth factor, the development of a continuous Rayleigh-line wing, extending to 200 cm^{-1} in some substances, cannot be even partly removed by these methods.

The penetration of source light reflected from the cuvette walls into the spectrograph makes it impossible to record the spectrum in the low-frequency region. However, if we calculate the condenser lens from the Rautian formula [36] and adjust the whole condenser by means of a hollow glass tube instead of the cuvette, this entirely prevents the light from the cuvette walls from falling into the spectrograph.

The Raman spectra were excited by the 4358 Å mercury line. The light source was a low-pressure mercury lamp. The spectra were photographed on high-sensitivity Kodak OaG plates. For taking these spectra, exposures of 6-20 h were required with a spectral-slit width of 5 cm^{-1} in the spectrograph. It should be noted that for reliable recording of the low-frequency part of the spectrum it was extremely important to choose the exposure time correctly.

The removal of the halo arising where the intense exciting line strikes the plate is of great importance when photographing low-frequency spectra. In the present investigation we

59

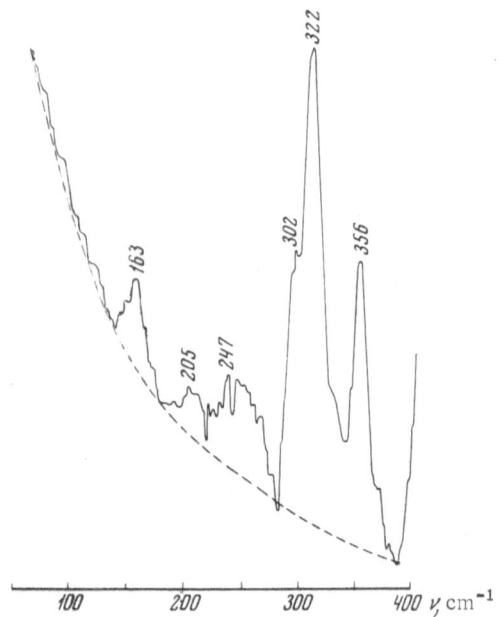

Fig. 1. Microphotogram of the Raman low-frequency spectrum of 1,1-dimethylcyclohexane.

made a small metal screen to remove the halo from the exciting line; this was placed immediately in front of the plate in the spectrograph. The screen could be moved by rotating a screw, thus varying the extent of the region near the exciting line so removed from the spectrum.

It should be mentioned that we also tried the difference method of photoelectrically recording Raman spectra, developed in the optical laboratory of the Physical Institute [37], for recording the low-frequency Raman spectra. However, we were unable to observe the desired Raman lines in the low-frequency part of the spectrum by this method. The chief reason for this was the fact that, when using the difference method, a large constant light signal fell on the photomultiplier near the exciting line. This caused the appearance of considerable light noise, interfering with the low-frequency lines, which in the majority of cases had a low intensity.

For our samples we were able to follow the spectrum as far as 80 or 90 cm^{-1} in favorable cases; however, these limits depended greatly on the nature of the substance and the intensity of the lines to be recorded.

In order to find the Raman frequencies we used a calibration graph based on the iron spectrum. We consider that the frequency-measuring error was no greater than 1-2 cm^{-1} for strong, narrow lines. For weak, well-spread lines and lines at the edge of the Rayleigh-line wing, it was hard to determine the frequency. The error in determining the frequencies of such lines reached an estimated 5-8 cm^{-1}.

The determination of intensities at the line maxima was carried out in an MF-2 microphotometer with a recording attachment. The intensities and frequencies here recorded were obtained by averaging over several plates for each substance. A typical microphotogram of the Raman spectrum at low frequencies is given in Fig. 1. We see from this graph that the continuous background increases in the direction of the exciting line. This background greatly interferes with the photometry of the lines. It should be noted that, for the simultaneous photometric recording of both the lines being studied and the comparison lines lying in another part of the spectrum, the exposure time has to be selected carefully for each particular substance. The background line is drawn rather arbitrarily on the microphotogram in our present case. We considered it most natural to draw the line in the manner shown in Fig. 1. For lines lying in the background, the errors in determining the intensities were very large owing to background fluctuations. As shown in [38], when working within the range of ordinary photometric densities, the average value of the background fluctuations is approximately proportional to the intensity of the background itself. In view of this, the intensities of the lines at the very edge of the Rayleigh-line wing must be regarded as very rough. The intensities are reduced to a common scale [38] in which the 802 cm^{-1} line of cyclohexane has a maximum intensity of 250 units.

In the majority of cases, our sample substances were specially synthesized in order to study their Raman spectra (these spectra were published in [38]). Before taking the spectra, the samples were vacuum-distilled by the usual method [39].

It should be noted that, although individual aspects of the method of investigation employed were by no means new, the results nevertheless show that our method of recording the low-frequency Raman spectrum was fairly efficient and provided new information regarding this little-studied spectral region in which the frequencies of the rotational oscillations of atomic groups fall.

CHAPTER III

APPEARANCE OF THE ROTIONAL OSCILLATIONS OF METHYL GROUPS IN LOW-FREQUENCY RAMAN SPECTRA

1. Experimental Results on the Low-Frequency Raman Spectra of Methyl-Substituted Cyclohexanes

For studying the rotational oscillations of the methyl groups in the low-frequency Raman spectrum we considered that methyl-substituted cyclohexanes constituted a good type of sample. We see, for example, from the structural formula of methylcyclohexane

$$CH_2 \diagdown CH_2-CH_3 \diagup CH-CH_3 \diagdown CH_2-CH_2 \diagup$$

that, apart from the rotational oscillations of the CH_3 group, only deformation oscillations can appear in the spectrum. We can hardly expect substantial intermolecular interactions in these substances. The latter factor facilitates analysis of the results obtained.

For present purposes a very important fact was that, even in the course of our preliminary investigations, we observed an "excess" line in the low-frequency part of the spectrum of trans-1,4-dimethylcyclohexane

$$CH_3-CH \diagup CH_2-CH_2 \diagdown CH-CH_3 \diagdown CH_2-CH_2 \diagup$$

i.e., there were more lines in this region than would be allowed by the selection rules if we exclude the rotational oscillations of the methyl groups. This was discovered by analyzing the results of a complete calculation of the vibrations of trans-1,4-dimethylcyclohexane [29] and comparing with experimental data. Naturally, we thought that the excess line should be ascribed to the rotational oscillations of the methyl groups. An analogous conclusion follows from [23].

Trans-1,4-dimethylcyclohexane has a center of symmetry; hence, the number of allowed lines in its Raman spectrum is smaller than in the other methylcyclohexanes. In the spectra of the other methylcyclohexanes some of the allowed lines fail to occur. However, we may well expect that the rotational oscillations of the methyl groups will also appear in the low-frequency part of the spectrum of other methyl-substituted cyclohexanes. We studied the low-frequency Raman spectra of nine methyl-substituted cyclohexanes [40]. The structural formulas of the substances studied were as follows:

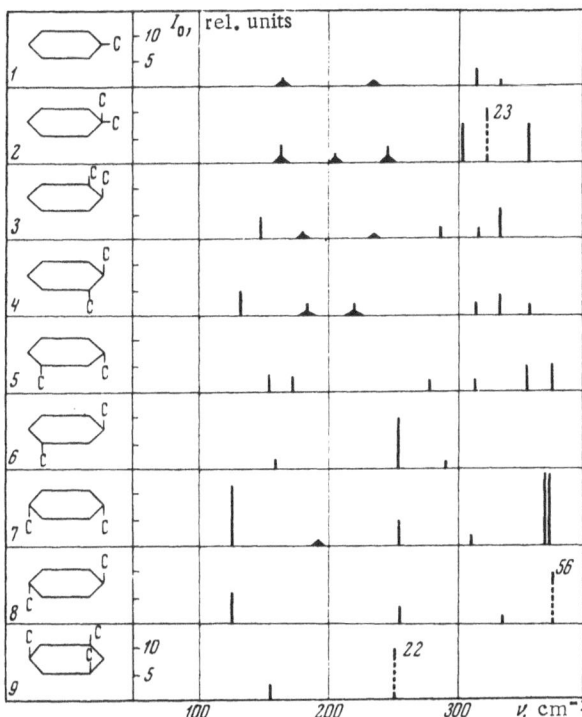

Fig. 2. Schematic representation of the low-frequency Raman spectra of methyl-substituted cyclohexanes. 1) Methylcyclohexane; 2) 1,1-dimethylcyclohexane; 3) cis-1,2-dimethylcyclohexane; 4) trans-1,2-dimethylcyclohexane; 5) cis-1,3-dimethylcyclohexane; 6) trans-1,3-dimethylcyclohexane; 7) cis-1,4-dimethylcyclohexane; 8) trans-1,4-dimethylcyclohexane; 9) cis-1,3,5-trimethylcyclohexane.

methylcyclohexane

1,1-dimethylcyclohexane

cis- and trans-1,2-dimethylcyclohexane

cis- and trans-1,3-dimethylcyclohexane

cis- and trans-1,4-dimethylcyclohexane

$$CH_3-CH \begin{matrix} CH_2-CH_2 \\ \\ CH_2-CH_2 \end{matrix} CH-CH_3,$$

cis-1,3,5-trimethylcyclohexane

$$CH_2 \begin{matrix} CH_3 \\ CH-CH_2 \\ \\ CH-CH_2 \\ CH_3 \end{matrix} CH-CH_3.$$

We found a large number of lines in the low-frequency spectra of the methyl-substituted cyclohexanes. The substances studied had a comparatively low-intensity Rayleigh-line wing; hence, in the low-frequency range we were able to record weak Raman lines starting from 100 cm⁻¹. The lines observed in the low-frequency region had various widths and intensities; however, the majority of the lines were weak, their intensities equalling a few units on the common scale.

We consider it appropriate to give all possible factual data regarding the frequencies and intensities in the low-frequency Raman spectra of our methylcyclohexanes, as well as of other substances studied in the following chapter. This is first necessary because it is undesirable to consider the rotational oscillations separately from other oscillations appearing in this part of the spectrum. Secondly, remembering that the low-frequency part of the spectrum has hardly been studied at all by other authors, this material may be of independent interest in connection with the purposes indicated in the Introduction.

Table 1

Substance	Published data [38]		Results of our measurements		Substance	Published data [38]		Results of our measurements	
	ν, cm⁻¹	I_0	ν, cm⁻¹	I_0		ν, cm⁻¹	I_0	ν, cm⁻¹	I_0
Methylcyclohexane	—	—	165 w	1	cis-1,3-Dimethylcyclo-hexane	—	—	155	3
	—	—	237 w	0.2		—	—	173	3
	312 w	0	313	3		—	—	278	2
	340 w	0	333	1		314	0	313	2
1,1-Dimethylcyclo-hexane	—	—	163 w	3		355	5	355	5
	—	—	205 w	1		375	5	375	5
	—	—	247 w	3	trans-1,3-Dimethyl-cyclohexane	—	—	161	2
	302	4	303 dif	7		254	7	257	8
	322	23	323	23		—	—	292	1
	356	7	356	7	cis-1,4-Dimethylcyclo-hexane	125	5	127	12
cis-1,2-Dimethyl-cyclohexane	—	—	148	4		—	—	193 w	1
	—	—	181 w	1		258	5	256	5
	—	—	236 w	0.5		310	3	311	2
	286	0	287	2		373 d	17	372 d	14
	315	0	317	2	trans-1,4-Dimethyl-cyclohexane	127 *	6	127	6
	333	6	334	6		254	0	258	3
trans-1,2-Dimethyl-cyclohexane	—	—	133	5		335	0	337	1
	—	—	185-222	2		376	65	376	56
	—	—	315	2	cis-1,3,5-Trimethyl-cyclohexane	—	—	156	3
	—	—	334	4		254	22	253	22
	—	—	357	2					

*From [29].

Note. w, wide band; d, double; dif, diffuse.

Literature devoted to the study of methyl-substituted cyclohexanes is restricted to a few papers. This is probably because of the difficulty of synthesizing and purifying the substances in question.

For comparison with the results obtained, we chose the data presented in [38] as the fullest and most reliable.

In all we observed over 40 lines in the low-frequency part of the Raman spectra of methyl-substituted cyclohexanes; half of these, including the lowest frequencies, had not been mentioned before. The resultant frequency and intensity values at the line maxima are given in Table 1. Figure 2 shows schematic representations of the spectra recorded.

The interpretation of the observed lines, particularly the attribution of some of the lines to rotational oscillations, is difficult without a special theoretical analysis. We therefore calculated the frequencies of the rotational oscillations of the methyl groups in the substances studied.

2. Referring the Lines in the Raman Spectra of Methyl-Substituted Cyclohexanes to Rotational Oscillations

There is some difficulty in relating the lines in the low-frequency Raman spectra of methyl-substituted cyclohexanes to rotational oscillations owing to the large number of lines in the spectra of these substances. This means that deformation vibrations occur together with rotational oscillations. Whereas, for molecules containing heavier groups, lines lying in the very low-frequency part of the spectrum are usually ascribed to rotational oscillations, in the case of the rotational oscillations of the methyl groups the corresponding lines may lie over a comparatively wide spectral range, starting from the wing of the Rayleigh line and extending to 300 cm^{-1}. This is because the moment of inertia of the methyl group is quite small. Considering the experimental material, we see that it is heard to separate a specific group of lines in the substances in question by reference to any distinguishing feature.

As already mentioned in the review of literature, if we know the height of the potential barrier of the rotational oscillations it is fairly easy to find the corresponding frequency. By comparing the calculated values of the frequencies with those found by experiment, we may identify some frequencies with rotational oscillations. In view of this we attempted a theoretical calculation of the potential barriers and corresponding frequencies of the rotational oscillations of methyl-substituted cyclohexanes.

a) Theoretical Calculation of the Heights of Potential Barriers

Since reliable experimental values of the barriers only exist for a comparatively small number of molecules, methods of calculating the potential barriers theoretically are of great interest.

The physical nature of the retarding potential is still not completely understood. The heights of the internal-rotation barriers are small in comparison with other forms of energy in the molecule; hence, their theoretical calculation presents an extremely difficult problem. Attempts at a strict quantum-mechanical calculation of the barriers have only been made for the simplest molecule with internal rotation, ethane, and the results of these calculations cannot be regarded as at all successful [41-43].

There have been some attempts at using the theory of the electrostatic interaction of bonds in order to explain the potential barrier of internal rotation [44-46]. It was found, however, that calculations based on this principle gave values too low for the heights of the barriers. In a number of papers there has been the idea of replacing the interaction of chemically uncoupled atoms and atomic groups by the interaction of isoelectronic or free atoms and molecules, estimated experimentally [47-49]. These papers have also not fully explained the origin and magnitude of the potential barrier.

Apart from papers in which the authors have attempted to find the heights of the potential barriers from general physical considerations, there are others in which simple semi-empirical methods have been proposed for calculating the potential barriers. For our practical purpose (identifying the frequencies in the low-frequency part of the spectrum), it is these papers which are of greatest interest from our own point of view. The calculation of the barriers by these methods is comparatively simple, and comparison with experimental values is satisfactory for the majority of cases considered.

The basic assumption of the semi-empirical method of calculating the barriers is that the height of the potential barrier equals the difference between the energies of repulsion between the two rotating parts of the molecule in the positions of greatest and least repulsion, respectively. For example, in the case of ethane this will be the difference between the repulsive energies of the trans and cis forms. There are several methods for the semi-empirical calculation of the barriers [50-51]. Aston and colleagues [50], using the experimental values of the barriers

for ethane and tetramethylmethane, $CH_3-\overset{\overset{\displaystyle CH_3}{|}}{\underset{\underset{\displaystyle CH_3}{|}}{C}}-CH_3$, and considering that the barrier in each case

was due to the repulsion of the hydrogen atoms, obtained the following value for the interaction potential of the methyl groups (in cal/mole)

$$V = \sum_{i,j} \frac{4.99 \cdot 10^5}{r_{ij}^5}. \tag{8}$$

Here r_{ij} is the distance in Å between the two hydrogen atoms of the two rotating groups (i is the index of the atoms in the first group and j that of the atoms in the second). Calculation by this formula agrees quite well with experiment for a number of molecules with methyl groups. Pentin and Tatevskii [52] proposed the following more general formula for the interaction potential:

$$V = \sum_{i,j} \left(\frac{k_{ij}}{r_{ij}^n} + \frac{h_{ij}}{r_{ij}^m} \right). \tag{9}$$

In this expression the first term describes the forces of repulsion and the second term other interactions.

The second method was proposed by Magnasco in [51]. The main idea of Magnasco was no different from that expressed in [50]. The difference was simply that the interaction potential of atoms not joined by bonds was represented as a Morse function and the interaction of all the atoms was taken into account. The method proposed was used in order to calculate the potential barriers of many molecules, and fair agreement with experiment was achieved. It should be mentioned that the behavior of the Morse function in the range $r > 3r_0$, where r_0 corresponds to an extremum, differs little from a function of the form r^{-5} (see, for example, [53]). This shows that there is no very great difference between the two methods.

Table 2

Substance	Structural formula	V, cal/mole			Literature ref.	
		Experiment	Calculation		for V	for geometric parameter
			without $C \cdots H$ interac.	with $C \cdots H$ interac.		
Dimethylsilane.....	$CH_3-SiH_2-CH_3$	1647	1560	1930	[54]	[54]
Dimethyl ether.....	CH_3-O-CH_3	2720	1760	3660	[55]	[55]
Dimethyl sulfide...	CH_3-S-CH_3	2132	2290	3770	[56]	[56]
Propane..............	$CH_3-CH_2-CH_3$	3400	3090	4160	[57]	[58]
Trimethylamine...	$N-CH_3$ $CH_3 \diagup \diagdown CH_3$	4400	3850	7680	[59]	[59]
Isobutane............	$CH-CH_3$ $CH_3 \diagup \diagdown CH_3$ $CH_3 \diagup \diagdown CH_3$	3620	3440	5590	[57]	[60]

We used the semi-empirical Magnasco method for calculating the potential barriers; however, we modified the method considerably in order to extend it to more complex molecules. In methyl-substituted cyclohexanes there are two types of interactions playing a part in such calculations: the interactions of chemically uncoupled $H \cdots H$ and $C \cdots H$ pairs of atoms. Magnasco, in considering molecules containing $H \cdots H$ and $C \cdots H$ interactions but not interactions between other uncoupled atoms, unjustifiably (as we think) neglected the interactions between other neighboring hydrogen atoms. In order to check the validity of this, we made a calculation of the potential barriers using Magnasco's repulsive potential for a number of molecules in which only $H \cdots H$ and $C \cdots H$ interactions occurred (see structural formulas in Table 2). We selected molecules having reliably determined values of the barriers, obtained from microwave data (except for propane and isobutane). Exact structural data (bond lengths and angles) were also available for the molecules chosen. This latter point is extremely important in view of the sharp variation in potential values with varying interatomic distance. We have emphasized this circumstance because the calculated value of the barrier is given by the difference between two large numbers: the energies of repulsion in the positions with the largest and smallest repulsions between the groups. For obtaining reliable results, a high accuracy of the calculations is demanded, and this we ensured.

Our calculation showed that the use of the inter-hydrogen-atom repulsion function given in [51]

$$V = 932.2 \exp(-1.944r) - 1951 \exp(-2 \cdot 1.944r) \tag{10}$$

for molecules of dimethylsilane, dimethyl ether, dimethyl sulfide, propane, trimethylamine, and isobutane, without allowing for the interaction of other atoms, in the majority of cases gave satisfactory agreement with experiment. The results obtained by calculating the potential barriers of the above-mentioned substances are shown together with the experimental barrier heights (obtained from the literature) in Table 2. For comparison, the table also shows the heights of the potential barriers which we calculated with due allowance for the $C \cdots H$ interaction. We see that allowance for these interactions leads to too large values of the calculated barriers. It should be noted that in the case of dimethyl ether the calculated values deviate considerably from experiment. It is not impossible that this is due to special features of the bonds formed by the oxygen atom. Since there were no oxygen atoms in our own substances, we did not take this disagreement too seriously. In calculating the repulsive energies, we considered that the other groups were in their equilibrium positions. Of course, in this method of calculation we are not considering the force interactions of the rotational oscillations of different methyl

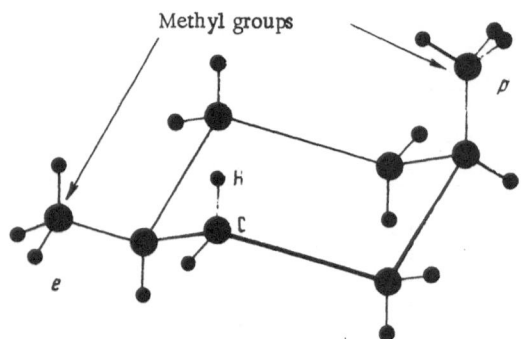

Fig. 3. Arrangement of the cis-1,4-dim-
ethylcyclohexane molecule.

groups, which in principle should take place [30].
The authors of [54-56] came to the conclusion
that such interactions were small. A calculation
of the potential barriers, using the repulsive function (10) of chemically uncoupled hydrogen atoms,
was carried out for substances in the gaseous
state. However, as indicated in [11], the state of
aggregation of the substance had little effect on
the frequency of the rotational oscillations of the
methyl groups. In view of this we considered that
it was possible to extend this method of calculation
to liquid hydrocarbons also.

We note that the function expressing the repulsion of uncoupled hydrogen atoms is only
known for a limited range of r values. For small r it is only known up to those minimum distances between the hydrogen atoms of the methyl group and the core which exist in dimethylsilane, dimethyl ether, and certain other molecules which we studied. If the hydrogen atoms
come very close together, we are really no longer justified in using the Morse function or the
r^{-5} law.

It should be noted that the values of the frequencies are not very sensitive to changes in
the heights of the potential barriers. This may be seen from the relation between the height of
the potential barrier V_3 and the frequency of the rotational oscillations ν in the case of a large
V_3 (case of the harmonic approximation):

$$\nu \sim \sqrt{V_3}.$$

For calculating the heights of the potential barriers, we have to start from the structural
parameters of the molecules. Since the methyl-substituted cyclohexanes have quite a complicated structure, we must consider this in a little more detail.

b) Structure of the Molecules of
Methyl-Substituted Cyclohexanes

By structural parameters we mean the bond lengths, the angles between the bonds, and
the mutual arrangement of the atoms, i.e., the configuration of the molecule.

The individual configurations of methyl-substituted cyclohexanes are considered in reviews [61, 62]. In the "armchair" configuration of cyclohexane the C—H bonds may be of two
types: parallel to the axis of symmetry of the ring and roughly perpendicular to this axis. The
first of these are called polar and the others equatorial (abbreviated to p and e). The methyl
group replacing the hydrogen atom may find itself in the p or e position, according to which
particular hydrogen atom it replaces. The positions of the methyl groups may be seen from
Fig. 3, which illustrates the cis-1,4-dimethylcyclohexane molecule (for methyl-substituted cyclohexane molecules the significance of cis and trans is rather arbitrary). One methyl group
in this molecule is equatorial (e) and the other polar (p). Energetically the e configuration is
more favorable for the methyl group owing to the closer positioning of the hydrogen atoms of
the core in the p configuration. In view of this the single methyl group in methylcyclohexane
lies in the e position only. However, in dimethylcyclohexanes polar groups may occur as well.
In 1,1-dimethylcyclohexane, cis-1,2-dimethylcyclohexane, and trans-1,3-dimethylcyclohexane
there are methyl groups in both e and p positions. Trans-1,2-dimethylcyclohexane and cis-1,3-
dimethylcyclohexane have two different configurations: 1) both methyl groups in the e position;

2) both methyl groups in the p position. Our trans-1,4-dimethylcyclohexane and cis-1,3,5-trimethylcyclohexane molecules also have all the methyl groups in the e configuration. We note that all the configurations mentioned really exist in nature.

For complex molecules of the substituted-cyclohexane type, the bond lengths and valence angles are not known to great accuracy. Hence, for the calculation, the C—C bond lengths were taken as equal to 1.54 Å, the C—H as 1.09 Å, and all the angles as tetrahedral.

c) General Characteristics of the Potential Barriers of Methyl Groups in Various Positions

The calculation of the potential barriers by the method described shows that, in the case of a methyl group in the polar position, the corresponding barrier is sixfold, and much higher than the barriers of the methyl groups in the equatorial positions. Since, in the case of polar groups the hydrogen atoms come very close together, the values of the potential function used are not very reliable and cannot give good results. Evidently, the corresponding frequencies of the rotational oscillations should lie above the region studied. In our investigation these frequencies are not considered; this eliminates from consideration the configurations of trans-1,2- and cis-1,3-dimethylcyclohexane, which have methyl groups in the polar positions. In addition to this, in the case of trans-1,2-dimethylcyclohexane with two equatorial methyl groups, there is also a close approach between the hydrogen atoms. This is not so obvious as in the case of the polar groups; however, calculation of the H···H distance confirms this conclusion.

The determination of the frequency of the rotational oscillations from the known value of the barrier, using Mathieu's equation, was considered in Chapter I. We used this method for calculating the frequencies of the rotational oscillations of the methylcyclohexanes on the basis of our own calculated potential-barrier heights. We verified that, in view of the small moment of inertia of the methyl group (which is attached to a much heavier core), the reduced moment of inertia used in the calculations could be replaced by the moment of inertia of the methyl group relative to the C—C axis. The use of the harmonic approximation gave high values of the frequencies as compared with the Mathieu equation.

d) Number of Rotational-Oscillation Lines in the Low-Frequency Range

Considering all possible configurations of our methyl-substituted cyclohexanes and the symmetry of the molecules, we may determine the number of rotational-oscillation lines in the part of the spectrum under examination from general considerations regarding the symmetry of the molecular vibrations.

For methyl-substituted cyclohexanes having one methyl group in the e position, naturally, only one line of the rotational oscillations falls within the part of the spectrum studied. In the case of cis-1,3- and trans-1,4-dimethylcyclohexane and cis-1,3,5-trimethylcyclohexane, there are several methyl groups in the e position. Cis-1,3-dimethylcyclohexane belongs to the point symmetry group C_S; hence, rotational oscillations of both A' and A" types of symmetry should be allowed in the spectrum. However, owing to the heavy core and the weakness of the interaction between the rotational oscillations and other vibrations, these lines should coincide. In trans-1,4-dimethylcyclohexane and cis-1,3,5-trimethylcyclohexane, the selection rules only allow lines corresponding to rotational oscillations of the B_g and E symmetry types, respectively, i.e., one line for each substance.

The calculated barrier heights of the methyl groups and frequency values are given in Table 3 together with experimental results.

Table 3

Substance	Possible positions of CH_3 groups	V, cal/mole (calc.)	v, cm^{-1} calc.	v, cm^{-1} exptl.
Methylcyclohexane	e	4000	240	237
1,1-Dimethylcyclohexane	ep	4800	265	247
cis-1,2-Dimethylcyclohexane . .	ep	2400	190	181
cis-1,3-Dimethylcyclohexane . .	ee	4000	240	278
trans-1,3-Dimethylcyclohexane .	ep	4000	240	257
cis-1,4-Dimethylcyclohexane . .	ep	3700	230	256
trans-1,4-Dimethylcyclohexane .	ee	4000	240	258
cis-1,3,5-Trimethylcyclohexane.	eee	4000	240	253

e) Comparison of the Calculated Frequencies of the Rotational Oscillations with the Experimental Values

As already noted, the frequencies of the rotational oscillations are approximately proportional to the square roots of the heights of the potential barriers. In order to discover what error inaccuracy in the determination of the potential-barrier height introduced into the frequency values for our case, we carried out the corresponding calculations for methylcyclohexane. We found that a change of 1000 cal/mole in the height of the potential barrier of methylcyclohexane changed the frequency of the rotational oscillation of the methyl group by 30 cm^{-1}. According to our estimates, the accuracy of determining the barriers was a little better than this, and the error in determining the frequencies averaged no worse than ±25 cm^{-1}. We see from Table 3 that the calculated and experimental values of the frequencies are in satisfactory agreement. In the case of cis-1,2-dimethylcyclohexane, either 181 or 236 cm^{-1} may correspond to the rotational oscillation, since the calculated frequency value equals 190 cm^{-1}.

A calculation of the barriers for methylcyclohexane, cis-1,3-, trans-1,4-dimethylcyclohexane, and cis-1,3,5-trimethylcyclohexane gives identical barrier heights owing to the fact that the methyl groups lie a long way from each other and their hydrogen atoms do not interact. Thus, the frequencies found for the rotational oscillations of these substances should lie in roughly the same region even if the repulsion function used is not altogether correct. We see from Table 1 that, within the limits of error indicated this corresponds to experiment for frequencies in the region of 240 cm^{-1} (see Table 3). The corresponding frequencies are 237, 278, 258, and 253 cm^{-1}.

A complete calculation of the oscillations of trans-1,4-dimethylcyclohexane, given in [29] without allowing for the rotational oscillations, shows that in the low-frequency range this compound can only have three lines. According to calculation these lines have frequencies of 185, 304, and 323 cm^{-1}. The intense line at 376 cm^{-1} may be ascribed to the calculated frequency of 323 cm^{-1}. The frequency of 185 cm^{-1} may be attributed either to the observed 127 cm^{-1} line or to the line at 258 cm^{-1} (see Table 1). Considering the calculated value of the rotational-oscillation frequency (240 cm^{-1}) the experimental lines should be assigned as follows. The 258 cm^{-1} line should be regarded as belonging to the rotational oscillations, and the 127 cm^{-1} line as belonging to the ordinary deformation vibrations of this molecule.

Thus, the calculation of the rotational-oscillation frequency values enables us to interpret a number of lines in the low-frequency spectra of methyl-substituted cyclohexanes quite unequivocally. The remaining lines not considered in this region may evidently be referred to the deformation vibrations of these molecules. The lines corresponding (according to our calculations) to the rotational oscillations of the methyl groups are mostly weak and wide. However there are also many weak and wide ones among the other lines, so that these lines can only be attributed to rotational oscillations after calculating the frequencies.

APPEARANCE OF THE ROTATIONAL OSCILLATIONS OF ETHYL AND HEAVIER GROUPS IN THE LOW-FREQUENCY RAMAN SPECTRA

1. Experimental Results on the Low-Frequency Raman Spectra of Methyl-Substituted Butanes

In addition to the rotational oscillations of the methyl groups, we were also interested in the appearance of the rotational oscillations of ethyl and heavier groups in the low-frequency Raman spectra. From this point of view the most convenient class of substances was that comprising the methyl-substituted butanes ($CH_3-CH_2-CH_2-CH_3$).

Interest in molecules of the methyl-substituted butane type was evoked by the great analogy between the structure of the molecules of these substances and the halogen derivatives of ethane. It is well known that the low-frequency part of the Raman spectra of the halogen derivatives of ethane contains lines generally accepted as being due to rotational oscillations. The literature contains no indications regarding the appearance of rotational oscillations in the Raman spectra of methyl-substituted butanes. However, owing to the similarity in the molecular structure of these two groups of substances, we confidently expected the appearance of rotational oscillations around the central C—C bond in the low-frequency spectrum of methyl-substituted butanes.

We studied the low-frequency Raman spectra of four branched paraffins of the methyl-substituted butane type [63]:

$$isopentane \quad CH_3-\overset{\overset{\textstyle CH_3}{|}}{CH}-CH_2-CH_3,$$

$$2,2\text{-dimethylbutane} \quad CH_3-\overset{\overset{\textstyle CH_3}{|}}{\underset{\underset{\textstyle CH_3}{|}}{C}}-CH_2-CH_3,$$

$$2,3\text{-dimethylbutane} \quad CH_3-\overset{\overset{\textstyle CH_3}{|}}{CH}-\!\!\!-\overset{\overset{\textstyle CH_3}{|}}{CH}-CH_3,$$

$$2,2,3\text{-trimethylbutane} \quad CH_3-\overset{\overset{\textstyle CH_3}{|}}{\underset{\underset{\textstyle CH_3}{|}}{C}}-\!\!\!-\overset{\overset{\textstyle CH_3}{|}}{CH}-CH_3.$$

In order to compare these molecules more clearly with the molecules of the ethane halogen derivatives, the structures of our four molecules are shown alongside that of 1,2-dichlorethane in Fig. 4.

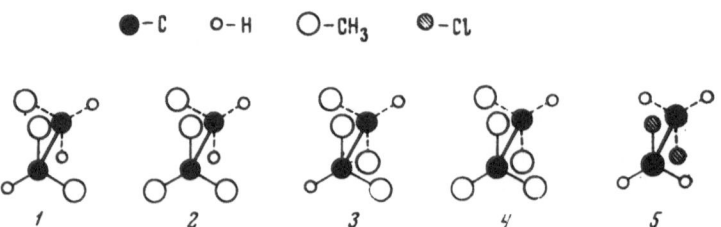

Fig. 4. Structure of the molecules of methyl-substituted butanes and 1,2-dichlorethane. 1) Isopentane; 2) 2,2-dimethylbutane; 3) 2,3-dimethylbutane; 4) 2,2,3-trimethylbutane; 5) 1,2-dichlorethane.

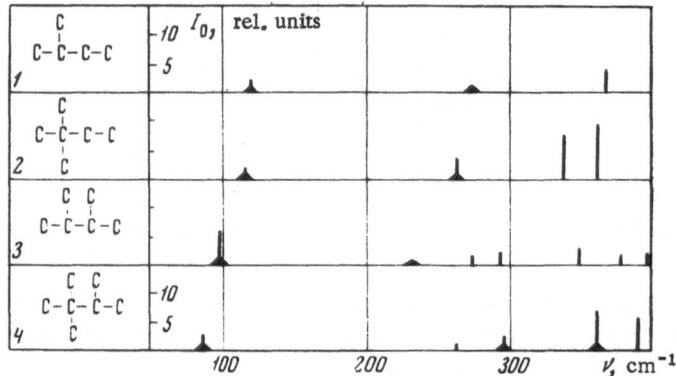

Fig. 5. Schematic representations of the low-frequency Raman spectra of methyl-substituted butanes. 1) Isopentane; 2) 2,2-dimethylbutane; 3) 2,3-dimethylbutane; 4) 2,2,3-trimethylbutane.

We see from the structural formulas of the substances studied that these contain many methyl groups capable of experiencing rotational oscillations. The appearance of these in the low-frequency spectrum is by no means impossible; however, it is very hard to take proper account of rotational oscillations of this kind at the present stage of the investigations for the class of compounds in question, owing to the complexity of the internal motions in these molecules. However, by basing our considerations on the results of our earlier study of the rotational oscillations of the methyl groups in methyl-substituted cyclohexanes, we may expect that the corresponding lines should lie in a frequency range no lower than 200 cm^{-1} and that these will therefore not constitute interference when studying the rotational oscillations of the ethyl and heavier groups.

The low-frequency Raman spectra of methyl-substituted butanes, according to our observations, contain a considerable number of lines, mainly weak ones. In the extreme low-frequency part of the spectrum (around 100 cm^{-1}) all the substances studied show wide lines. The wing of the Rayleigh line is of low intensity in all cases.

The fullest information regarding the Raman spectra of isopentane, 2,2- and 2,3-dimethylbutane, and 2,2,3-trimethylbutane is given in an earlier monograph [38].

The resultant values of the frequencies and intensities at the line maxima are given, together with corresponding published data, in Table 4. Figure 5 gives the schematic representations of the spectra recorded.

Table 4

Substance	Published data		Lit. cited	Our own measurements	
	ν, cm^{-1}	I_0		ν, cm^{-1}	I_0
Isopentane	—	—	—	120w	2
	271 w	1	[38]	274w	1
	367	6	[38]	368	4
2,2-Dimethylbutane	—	—	—	116w	2
	261 w	3	[38]	263w	4
	276 w	0	[38]	—	—
	340	9	[38]	338	8
	362	12	[38]	362	10
2,3-Dimethylbutane	96	—	[66]	98w	6
	220	—	[66]	231w	0.5
	267	1	[64]	274	1
	293	2	[64]	294	2
	346	3	[64]	349	3
	379	2	[64]	378	2
	395	2	[64]	397	2
2,2,3-Trimethylbutane	—	—	—	86w	3
	261	0	[38]	262	1
	297 w	2	[38]	296w	2
	361 w	9	[38]	361	7
	392	10	[38]	390	6

Note. w, wide.

Assignment of Lines to Rotational Oscillations

The 120 cm^{-1} line of isopentane, the 116 cm^{-1} line of 2,2-dimethylbutane, the 98 cm^{-1} line of 2,3-dimethylbutane, and the 86 cm^{-1} line of 2,2,3-trimethylbutane may be explained as being due to rotational oscillations around the central C—C bond. In favor of this assumption, we have the already mentioned analogy between the structure of the molecules under consideration and the halogen derivatives of ethane. In addition to this, these lines lie in the very low-frequency region, well removed from the remaining lines due to the deformation vibrations of the molecule. These lines are very wide (20-30 cm^{-1}), which is a characteristic of the lines corresponding to the rotational oscillations of the ethane halogen derivatives. For comparison we obtained the spectrum of 1,2-dichlorethane under the same conditions and observed that the 125 cm^{-1} line, already known to be due to rotational oscillations, had the same width.

An additional confirmation of the relation between the lines in the 100 cm^{-1} region and the rotational oscillations comes from a comparison between experimental data and the results of a detailed calculation of the frequencies of 2,3-dimethylbutane given in [64]. It follows from this calculation that there should be five lines in the low-frequency Raman spectrum of 2,3-dimethylbutane. All these lines are observed experimentally (Table 5). Thus, the 98 cm^{-1} line found in the spectrum cannot be compared with any of the calculated frequencies (the calculation of the frequencies was carried out in [64] without considering the coordinates characterizing the retarded rotation of the atomic groups). Hence, this frequency undoubtedly belongs to the rotational oscillations.

The interpretation of the lines in question may be obtained to a first approximation if we represent the molecules as consisting of two rigid sections executing rotational oscillations one around the other. This assumption is a coarse simplification of the problem, since it fails to

Table 5

Data of [64]				
Experimental			Calculated	
ν, cm^{-1}	I_0	ρ	trans-isomer ν, cm^{-1}	isomer ν, cm^{-1}
98	6	—	—	—
231	0.5	—	—	194 (B) *
274	1	—	—	201 (A)
294	2	0.32	270 (A_g)	—
349	3	0.72	—	347 (B)
378	2	0.18	—	306 (A)

*Symbols in brackets indicate the symmetry of the oscillation.

consider the interaction of the rotational oscillations with the other vibrations of the molecule. However, we may expect that these interactions will not be very strong in view of the considerable difference between the frequencies of the lines under consideration and those of the other lines in the low-frequency part of the spectrum of the molecules in question.

In considering approximate methods of calculating the potential barriers from structural data, we have already noted that the use of the simple form of the potential function for the repulsion of chemically uncoupled atoms is inapplicable for the case in which the hydrogen atoms come close together. A close approach between the hydrogen atoms occurs in the present case; hence, the approximate method of calculation must not be used. Of course, in order to find the frequencies of the rotational oscillations, we could use the values of the potential barriers taken from published data. However, in the present case such data are either absent altogether or quite unreliable. In view of this we considered it more natural to proceed in the following way: To find the internal-rotation potential barrier from the frequency values which for various other reasons might be expected to belong to rotational oscillations. Then, by comparing the height of the barrier with experimental values of potential barriers taken from the literature (even though imperfect), to find whether the frequency in question could in fact justifiably be ascribed to rotational oscillations. If there were no experimental data for the potential barrier of the substance in question, we should simply have to compare the value of the potential barrier with the region in which the potential barriers of other substances containing heavy rotating groups fell. (The majority of the potential barriers of heavy groups lie between 3000 and 5000 cal/mole [24, 30].) At first glance it may seem that this method of considering lines in the very low-frequency part of the Raman spectrum is unreliable. However, it must not be forgotten that the height of the potential barrier is proportional to the square of the frequency of the rotational oscillations. If we assume that the height of the potential barrier of the rotational oscillations of 2,2-dimethylbutane around the central C—C bond varies from 3000 to 5000 cal/mole, the corresponding frequency of the rotational oscillations varies over comparatively narrow limits: from 95 to 120 cm^{-1}.

We checked that in the case of the rotational oscillations of ethyl and heavier groups the calculated values of the heights of the potential barriers were the same for the harmonic approximation as for the more precise method of calculation. We therefore used formula (6) for calculating the potential barriers from the observed frequencies of the rotational oscillations.

Regarding the calculations of the reduced moment of inertia we should note that we are concerned with the case of asymmetric rotating groups. The reduced moment of inertia for this case is given by Pitzer's formula. The calculation of reduced moments of inertia by Pitzer's formula is very troublesome. In view of this we proposed a simplified formula enabling us to calculate the reduced moments of inertia for molecules consisting of two asymmetric groups in a simpler way:

$$I_{\text{red}} = \frac{I_1 I_2}{I_1 + I_2}, \tag{11}$$

where I_1 and I_2 are the moments of inertia of each group relative to an axis parallel to the axis of rotation and passing through the center of mass of the corresponding group. The formula

Table 6

Substance	ν, cm^{-1}	I_{red}, amu·A^2	V, cal/mole
Isopentane	120	16.5	4500
2,2-Dimethylbutane ...	116	18.6	4700
2,3-Dimethylbutane ...	98	31.8	5800
2,2,3-Trimethylbutane .	86	40.4	5600

was verified for our branched paraffins, i.e., the resultant values of the reduced moment of inertia were compared with the corresponding values calculated by the Crawford or Pitzer formulas. The verification was carried out for the following substances: isopentane, 2,2- and 2,3-dimethylbutane in the trans- and gosh-positions, 2,2,3-trimethylbutane, and 2,2-dimethylpentane. For the latter substance the asymmetry was extremely great. In all cases the deviation was no greater than 5%, which was quite adequate for our present aims. The time spent in using our formula was about 10% less.

We also note that in the retarded rotation of the asymmetric group all three barriers may be of different heights. In this case our calculation gives the height of the potential barrier corresponding to the rotational oscillations of the rotational isomer predominating in the given substance in the liquid state. In the case of 2,2-dimethylbutane and 2,2,3-trimethylbutane, all three barriers are of the same height because the rotating groups constitute symmetrical gyrostats. The computing formula is more accurate for these substances. Clearly, the form of the barriers of isopentane and 2,3-dimethylbutane cannot be refined at the present stage of the investigation.

Table 6 shows the calculated reduced moments of inertia and the potential barriers corresponding to the observed frequencies at the very low-frequency part of the spectrum for all the molecules considered. We see from the table that the values of the barriers fall in the region of the ordinary potential barriers indicated in the literature. There are no more precise indications of the barriers corresponding to internal rotation around the C—C axis for the substances under consideration.

If we suppose that any of the other recorded lines (apart from those mentioned) are due to rotational oscillations, then the calculation of potential barriers gives values exceeding 25,000 cal/mole. We see that these values greatly exceed those given in the literature.

Thus the calculation confirms our interpretation of the very low-frequency lines as lines due to rotational oscillations. The rest of our observed lines clearly belong to deformation vibrations of the molecule. In principle, the appearance of rotational oscillations of the methyl groups also is not to be excluded.

2. Experimental Results on the Low-Frequency Raman Spectra of Methyl-Substituted Pentanes

Methyl-substituted pentanes are interesting because they contain atomic groups analogous to the atomic groups in methyl-substituted butanes (considered in the previous section). In view of this we may also expect rotational oscillations of ethyl and heavier groups to appear in the spectra of the methyl-substituted pentanes, i.e., lines analogous to those observed in the low-frequency part of the Raman spectrum of methyl-substituted butanes. It is also not impossible that there will be rotational oscillations of the methyl groups; however, for reasons analogous to those mentioned in the case of methyl-substituted butanes, we shall not consider these. We note that, as in the case of methyl-substituted butanes, there are here no specific intermolecular interactions (for example, a hydrogen bond); this facilitates the treatment of the lines obtained.

We studied the low-frequency Raman spectra of eight methyl-substituted pentanes and n-pentane [65]. The substances studied had the following structural formulas:

n-pentane $CH_3—CH_2—CH_2—CH_2—CH_3$,

2-methylpentane $CH_3—\overset{\displaystyle CH_3}{\overset{|}{CH}}—CH_2—CH_2—CH_3$,

3-methylpentane $CH_3—CH_2—\overset{\displaystyle CH_3}{\overset{|}{CH}}—CH_2—CH_3$,

2,2-dimethylpentane $CH_3—\overset{\displaystyle CH_3}{\overset{|}{CH}}—CH_2—CH_2—CH_3$,

2,3-dimethylpentane $CH_3—\overset{CH_3}{\overset{|}{CH}}—\overset{CH_3}{\overset{|}{CH}}—CH_2—CH_2$,

3,3-dimethylpentane $CH_3—CH_2—\overset{\displaystyle CH_3}{\underset{\displaystyle CH_3}{\overset{|}{\underset{|}{C}}}}—CH_2—CH_3$,

2,4-dimethylpentane $CH_3—\overset{CH_3}{\overset{|}{CH}}—CH_2—\overset{CH_3}{\overset{|}{CH}}—CH_3$,

2,2,3-trimethylpentane $CH_3—CH_2—\overset{CH_3}{\overset{|}{CH}}—\overset{CH_3}{\underset{CH_3}{\overset{|}{\underset{|}{C}}}}—CH_3$,

2,2,4-trimethylpentane $CH_3—\overset{CH_3}{\underset{CH_3}{\overset{|}{\underset{|}{C}}}}—CH_2—\overset{CH_3}{\overset{|}{CH}}—CH_3$.

According to our observations, the low-frequency Raman spectra of the methyl-substituted pentanes and n-pentane contained an average of five to six lines. The majority of the substances had wide lines in the very low-frequency part of the spectrum, from 95 to 120 cm^{-1}. Like the methyl-substituted butanes, all the methyl-substituted pentanes studied had a weak Rayleigh line wing, which promoted the revelation of the weak lines in the very low-frequency part of the spectrum.

The resultant frequencies and intensities at the line maxima are compared with published data in Table 7. Figure 6 presents the recorded spectra schematically.

Assignment of Lines to Rotational Oscillations

The 131 cm^{-1} line of n-pentane, the 121 cm^{-1} of 2,2-dimethylpentane, the 104 cm^{-1} of 2,3-dimethylpentane, the 108 cm^{-1} of 3,3-dimethylpentane, the 123 cm^{-1} of 2,4-dimethylpentane, and the 96 cm^{-1} of 2,2,3-trimethylpentane may reasonably be explained as being lines due to rotational oscillations around one of the two central C—C bonds. (We found no lines in the region of 100 cm^{-1} in the spectra of 2-methylpentane and 2,2,4-trimethylpentane. These may have been

Table 7

Substance	Published data [38]		Our measurements	
	ν, cm^{-1}	I_0	ν, cm^{-1}	I_0
n-Pentane	—	—	131 w	4
	—	—	179 w	1
	—	—	270 w	1
	338 d	6	336	6
	364	0	—	—
	377	0	—	—
	401	25	400	25
2-Methylpentane	182 [66]	—	190 w	1
	—	—	247 w	0.5
	325	13	324 w	10
	387 w,d	3	381 w	1
3-Methylpentane	—	—	217 w	2
	—	—	260 w	2
	314	0	314	2
	388	12	388	12
2,2-Dimethylpentane	—	—	121 w	4
	—	—	192	2
	—	—	262	0.5
	320	12	320	11
	338	18	342	17
	346			
2,3-Dimethylpentane	—	—	104 w	3
	—	—	220 d	1
	—	—	303	2
	320 w	2	323	2
	337 w	2	342	2
	374 w	1	379	2
3,3-Dimethylpentane	—	—	108 w	5
	243 w	3	239 w	3
	295	1	294	1
	347	8	346	7
	374	9	378	9
2,4-Dimethylpentane	—	—	123 w	3
	179 [67]	—	183	4
	246	4	246	1
	303	17	305	20
	356 w	1	351	1
	388 w	0	—	—
2,2,3-Trimethylpent-ane	—	—	96 w *	—
	—	—	215 w	0.5
	—	—	303 w	2
2,2,3-Trimethylpent-ane	336	3	332	5
	350	6	347	7
	374 w	1	369	3
	393	4	392	4
2,2,4-Trimethylpent-ane	192	1	196	1
	—	—	257 w	2
	295 w	7	293 w	7
	312 w	7	311 w	7
	353 w	1	354 w	2
	373	0	379	1

*Not photometered owing to the strong background.

Note.   w, wide; d, double.

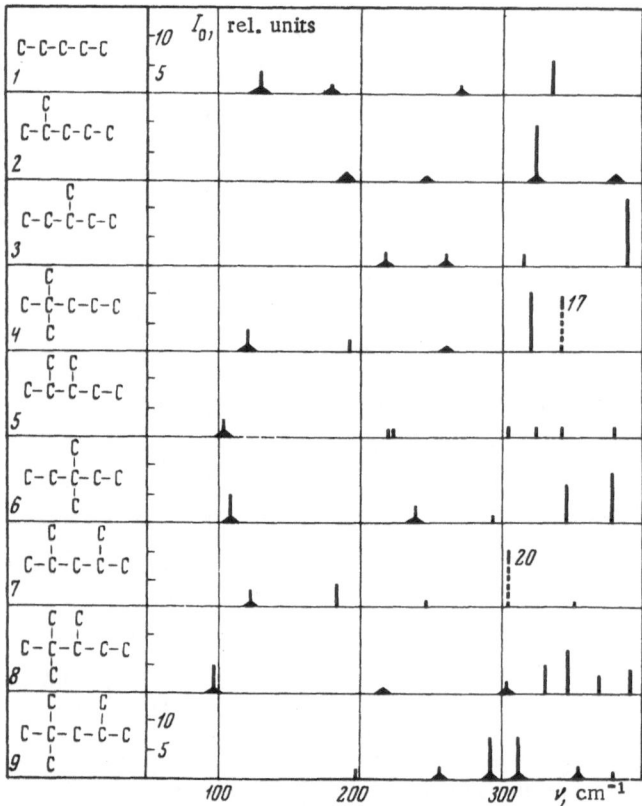

Fig. 6. Schematic representation of the low-frequency Raman spectra of n-pentane and methyl-substituted pentanes. 1) n-Pentane; 2) 2-methylpentane; 3) 3-methylpentane; 4) 2,2-dimethylpentane; 5) 2,3-dimethylpentane; 6) 3,3-dimethylpentane; 7) 2,4-dimethylpentane; 8) 2,2,3-trimethylpentane; 9) 2,2,4-trimethylpentane.

invisible against the background owing to their low intensities.) The arguments in favor of this supposition are roughly the same as those employed in connection with methyl-substituted butanes: the considerable width of the lines (20-30 cm^{-1}), and the fact that they lie at the extreme low-frequency end of the spectral range studied, in a region where lines corresponding to the rotational oscillations of heavy groups usually occur (for example, the lines of the rotational oscillations of 1,2-dichlorethane).

In order to verify our assumption, we made a calculation of the potential barriers of all the molecules in which lines were observed in the very low-frequency part of the spectrum, considering that these lines corresponded to rotational oscillations around the central bonds. For this purpose we used a simple model, in principle analogous to that used in calculating the potential barriers of methyl-substituted butanes. According to this model the molecules in question consist of two parts executing rotational oscillations with respect to each other. The difference in relation to the methyl-substituted butanes is in this case simply that the model of the molecule may be selected in two ways. For example, the molecule of 2,2-dimethylpentane may be considered as consisting of two parts in the following ways: 1) $CH_3C(CH_3)_2$—$CH_2CH_2CH_3$ or, 2) $CH_3C(CH_3)_2CH_2$—CH_2CH_3 (the dash indicates the bond relative to which the rotational oscillations are being considered). We do not know which of these rotational oscillations appears in the spectrum. We appreciate that questions may well be raised as to the legitimacy of separately

Table 8

Substance	v, cm^{-1}	I_{red}, amu · Å^2	V, cal/mole
n-Pentane	131	12.8	4200
		16.3	5300
2,2-Dimethylpentane	121	18.6	5100
		23.6	6500
		39.4	10900
2,3-Dimethylpentane	104	19.0	3900
		19.4	4000
		19.9	4100
		34.6	7000
		41.2	8400
		43.7	8900
3,3-Dimethylpentane	108	19.2	4200
		19.9	4400
2,4-Dimethylpentane	123	32.8	9300
		39.5	11000
2,2,3-Trimethylpentane	96	20.0	3500
		45.1	7800
		56.8	9900
		62.2	10800

considering the rotational oscillations about two central bonds. However, remembering that the angle between these bonds differs little from a straight line, we may readily assert from general mechanical considerations that the rotational oscillations relative to each bond will be almost independent.

It should also be noted that in methyl-substituted pentanes there may be steric isomerism (there are no reliable data on this point for our samples). In view of this, we calculated the potential barriers for all the steric isomers of the molecules under consideration and both arrangements of the central bonds.

The reduced moments of inertia were calculated by means of the approximate formula (11). It is clear that the reduced moments of inertia should be different for different steric isomers.

Using the observed values of frequencies and the calculated reduced moments of inertia, we may find the corresponding potential barriers from the simple formula of the harmonic approximation (6).

Table 8 shows the reduced moments of inertia and the potential barriers corresponding to the lines in the very low-frequency part of the spectrum. We see from the table that in all the molecules considered there are barrier values corresponding to the observed frequencies, falling in the same region as those of the methyl-substituted butanes. This result appears quite natural if we consider the presence of similar rotating groups in both cases, and it confirms that the frequencies in the 95-120 cm^{-1} range belong to rotational oscillations around the central bonds. An exception occurs for 2,4-dimethylpentane, which has a higher barrier. The greater height of the barrier in this case may be due to the arrangement of the methyl groups in the 2 and 4 positions, which differs from other methyl-substituted pentanes. We consider that the 123 cm^{-1} line of 2,4-dimethylpentane may also be ascribed to rotational oscillations.

We may thus assert that rotational oscillations of the ethyl and heavier groups appear in the very low-frequency region (about 100 cm^{-1}) of the low-frequency Raman spectra of both methyl-substituted butanes and methyl-substituted pentanes. The corresponding lines are wide (20-30 cm^{-1}) and of low intensity: a few units on the common scale.

3. Experimental Results on the Low-Frequency Raman Spectra of Ethyl-Substituted Naphthenes

Our promising results relating to the rotational oscillations of ethyl and heavier groups in branched paraffins inspired us to turn to another class of substances containing analogous atomic groups: the ethyl-substituted naphthenes. We studied the low-frequency Raman spectra of four ethyl-substituted naphthenes, the structural formulas of which were as follows:

$$\text{ethylcyclopropane} \quad \begin{array}{c} CH_2 \\ | \quad \diagdown \\ \quad \quad CH{-}CH_2{-}CH_3, \\ | \quad \diagup \\ CH_2 \end{array}$$

$$\text{ethylcyclobutane} \quad CH_2 \diagup\diagdown \begin{array}{c} CH_2 \\ \\ CH{-}CH_2{-}CH_3, \\ CH_2 \end{array}$$

$$\text{ethylcyclopentane} \quad CH_2 \begin{array}{c} \diagup CH_2 \diagdown \\ \quad\quad CH_2 \\ \diagdown \quad | \\ CH_2{-}CH{-}CH_2{-}CH_3, \end{array}$$

$$\text{ethylcyclohexane} \quad CH_2 \begin{array}{c} \diagup CH_2{-}CH_2 \diagdown \\ \quad\quad\quad CH{-}CH_2{-}CH_3. \\ \diagdown CH_2{-}CH_2 \diagup \end{array}$$

In these molecules the ethyl group may experience rotational oscillations with respect to the core. We might expect that, as in methyl-substituted butanes and pentanes, vibrations of this type would appear in the Raman spectrum.

The low-frequency spectra of ethylcyclopropane, ethylcyclobutane, and ethylcyclopentane obtained in our investigations had a small number of lines; in the case of ethylcyclohexane there was a considerable number of lines in the low-frequency region. The lines with the lowest frequencies were all wide. The wing of the Rayleigh line was not very strong, although stronger than in the paraffins.

Reliable data regarding the frequencies in the Raman spectra of our ethyl-substituted naphthenes only occur in [38, 39].

The resultant frequencies and intensities at the line maxima are given in Table 9, together with the corresponding published data. Figure 7 gives a schematic representation of the spectra recorded.

Assignment of the Lines to Rotational Oscillations

By analogy with the analysis of the spectra of the methyl-substituted butanes and pentanes, we started in this case also from the idea that the molecules of the ethyl-substituted naphthenes could be considered as consisting of two rigid parts executing rotational oscillations relative to one another. The choice of model in this case is unambiguous: a rigid core (ring) and a rigid ethyl group executing rotational oscillations about the C—C bond. The lowest-frequency lines

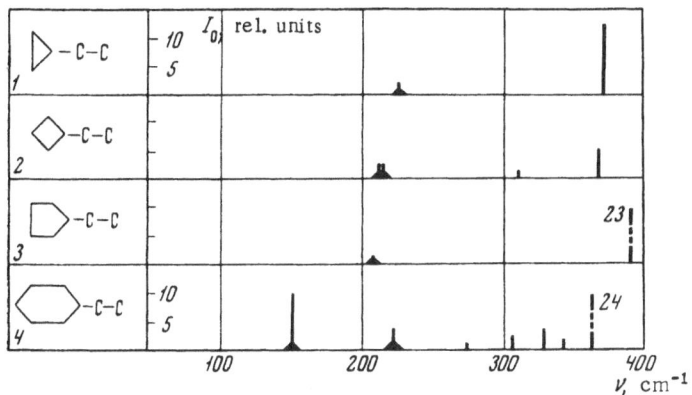

Fig. 7. Schematic representations of the low-frequency Raman spectra of the ethyl-substituted naphthenes. 1) Ethylcyclopropane; 2) ethylcyclobutane; 3) ethylcyclopentane; 4) ethylcyclohexane.

Table 9

Substance	Published data [38, 39]		Our measurements	
	ν, cm^{-1}	I_0	ν, cm^{-1}	I_0
Ethylcyclopropane	—	—	226 w	2
	290	2	—	—
	317	2	—	—
	331	2	—	—
	347	2	—	—
	370	13	371	13
Ethylcyclobutane	—	—	213 w	2,5
	—	—	312	1
	—	—	368	5
Ethylcyclopentane	213 w	0	209 w	1
	391	23	391	23
Ethylcyclohexane	—	—	154 w	10
	236	0	224 w	3,5
	—	—	275	1
	—	—	305	2
	330	0	330	4
	—	—	343 s	2
	364	24	364	24

Note. s, sharp; w, wide.

Table 10

Substance	ν, cm^{-1}	$I_{red}, amu \cdot \overset{\circ}{A}^2$	$V, cal/mole$
Ethylcyclopropane · · ·	226	13,9	13,400
Ethylcyclobutane. · · ·	213	16,9	14,300
Ethylcyclopentane · · ·	209	18,2	15,100
Ethylcyclohexane · · ·	154	18,9	8,500
	224		17,900

are close in width to the lines of the rotational oscillations in the methyl-substituted butanes and pentanes, but have frequencies considerably greater than in the case of the paraffin molecules. Assuming nevertheless that these lines related to the rotational oscillations of the ethyl groups, we calculated the heights of the potential barriers by the method used for the paraffin molecules.

The calculated values of the barriers are shown together with the reduced moments of inertia in Table 10. We see from the table that the heights of the barriers are in the present case two or three times those of the potential barriers found for the methyl-substituted butanes and pentanes.

There are no published data regarding the potential barriers of our present group of materials. Unfortunately, this means that we cannot give an unambiguous interpretation of the observed lines in this case. Two possibilities may clearly exist.

In the first case, all the observed very low-frequency lines may belong to deformation vibrations; the lines belonging to the rotational oscillations are too weak and lie in a very low-frequency part of the spectrum inaccessible to observation. If we attempt to calculate the frequencies of the rotational oscillations for these substances, assuming that the barrier lies between 4000 and 5000 cal/mole, then we find that the frequencies should lie in the region of 110 to 130 cm^{-1}. Since the wing of the Rayleigh line in this group of compounds is stronger than in the paraffins, it is quite likely that we shall be unable to record any of the rotational oscillations owing to their low intensities.

In the second case, the potential barriers are far higher than in the methyl-substituted butanes and pentanes, and our observed lines may belong to rotational oscillations. Although we cannot exclude the possibility of this explanation, it would seem more natural (having regard to a certain similarity between the structures of the ethyl-substituted naphthenes and the paraffins, for example, between ethylcyclohexane and isopentane) to expect that the barriers corresponding to the two cases would not differ too severely.

CONCLUSION

We see from our data relating to the low-frequency Raman spectra of different classes of substances that in almost all the molecules studied we have been able to establish the presence of the rotational oscillations of individual atomic groups in the spectra. The importance of studying rotational oscillations was mentioned earlier in [24], a monograph which appeared in 1949, and in which it was pointed out that rotational oscillations were by no means forbidden either in the infrared or Raman spectra, although their intensities would be low. From our point of view, the low intensity of these lines, confirmed by our own observations, and the inconvenient part of the spectrum in which they appear are responsible for the absence of any systematic studies of rotational oscillations by the Raman method. However, our investigations have shown that, by taking proper precautions and choosing the optimum conditions for photographing the Raman spectra, frequencies in the region of 100 cm^{-1} or even lower may be reliably recorded.

A detailed consideration of existing published data regarding the interpretation of low-frequency spectra has revealed methods which in more or less altered form are suitable for interpreting the spectra of our samples. Our research has demonstrated the possibility of using approximate calculations in order to ascribe individual lines to rotational oscillations. In the approximate calculations the molecules are regarded as consisting of two rigid sections joined by a chemical bond and executing rotational oscillations with respect to the latter.

We have shown the possibility of using a semi-empirical computing method for obtaining the potential barriers of the methyl groups in methyl-substituted cyclohexanes. The main principle of the method is to allow for repulsion between chemically uncoupled hydrogen atoms.

The ascription of frequencies to rotational oscillations involves calculating the heights of the potential barriers of the rotational oscillations. The resultant heights of the potential barriers are of independent interest, in view of their possible use in thermodynamic calculations. Despite the fact that our calculations have enabled us to systematize some experimental material, we must emphasize the importance of developing better methods of theoretically studying low-frequency oscillations of various types. Since the frequencies of the rotational oscillations may differ for different rotational isomers, it is in principle possible to decide which isomer we are dealing with by studying the frequency of the rotational oscillations.

In studying rotational oscillations further, it may be of considerable value to investigate deutero-substituted substances. This is so in particular for the rotational oscillations of methyl groups. To a first approximation the replacement of the hydrogen in a methyl group by deuterium should reduce the corresponding rotational-oscillation frequency by a factor of 1.4.

According to our observations, the lines of the rotational oscillations are very weak (a few units on the cyclohexane scale). The only exception is the 253 cm^{-1} line of cis-1,3,5-tri-methylcyclohexane, which has an intensity of 22 units. It is not impossible that this may involve the superposition of a line of different origin.

The lines of the rotational oscillations are very wide (with rare exceptions in the spectra of the methyl-substituted cyclohexanes). Particularly wide lines correspond to the rotational oscillations of the ethyl and heavier groups of methyl-substituted pentanes. The width of these lines according to visual observation is 20-30 cm^{-1} or more. We have considered several possible reasons for the wide lines of rotational oscillations in these substances, including the tunnel effect and also anharmonicity due to the sinusoidal form of the potential function of the rotational oscillations. We have established that the tunnel effect produces a splitting of the corresponding energy levels of no more than a small fraction of a cm^{-1} for the rotational oscillations under consideration, and thus contributes nothing to the width of the line. In order to consider the contribution of anharmonicity to the line width of the rotational oscillations, we must find the higher energy levels of the rotational oscillations of the molecules in question. We found that the frequency of the transition from the zeroth to the first energy level of the rotational oscillations differed from the frequency of the $1 \rightarrow 2$ transition by about 1-2 cm^{-1}, and hence could not be the cause of the wide rotational-oscillation lines. The great width of the rotational-oscillation lines in the substances studied probably arises from quite different causes, and the discovery of these demands a special investigation.

The author wishes to thank his scientific director, Professor M. M. Sushchinskii, for great help in the work and a number of valuable comments.

LITERATURE CITED

1. P. D. Simova, Dokl. Akad. Nauk SSSR, 69:27 (1949).
2. K. W. F. Kohlrausch and P. Ipsilanti, Z. Phys. Chem., (B)29:274 (1935).
3. S. Mizushima, Y. Morino, and S. Noziri, Sci. Papers Inst. Phys. Chem. Res., Tokyo, 29:111 (1936).
4. R. Ananthakrishnan, Proc. Indian Acad. Sci., A5:285 (1937).
5. I. Ichishima, H. Kamyama, T. Shimnouchi, and S. Mizushima, J. Chem. Phys., 29:1190 (1958).
6. A. Danti and J. L. Wood, J. Chem. Phys., 30:582 (1959).
7. L. Kahovec and K. W. F. Kohlrausch, Z. Phys. Chem., (B)48:7 (1940).
8. J. H. S. Green, W. Kynaston, and H. A. Gebbie, Nature, 195:595 (1962).
9. P. Delorme, Comptes Rend. Acad. Sci., 256:3272 (1963).
10. A. Kortüm and R. Maier, Z. Phys. Chem., N. F., 7:207 (1956).
11. W. G. Fataley and F. A. Miller, Spectrochim. Acta, 17:856 (1961).
12. W. G. Fataley and F. A. Miller, Spectrochim. Acta, 18:977 (1962).
13. W. G. Fataley and F. A. Miller, Spectrochim. Acta, 19:611 (1963).
14. K. D. Möller, Comptes Rend. Acad. Sci., 251:686 (1960).
15. A. Hadni, J. Phys. Radium, 15:375 (1954).
16. E. Catalano and K. S. Pitzer, J. Phys. Chem., 62:873 (1958).
17. A. Hadni, Comptes Rend. Acad. Sci., 239:349 (1954).
18. R. L. Redington, J. Mol. Spectr., 9:469 (1962).
19. C. C. Lin and J. O. Swalen, Rev. Mod. Phys., 31:841 (1959).
20. K. W. F. Kohlrausch and J. Wagner, Z. Phys. Chem., (B)52:185 (1942).
21. J. Kilpatrick and K. S. Pitzer, J. Res. Natl. Bur. Stds., 38:191 (1947).
22. C. Richards and J. Nilsen, J. Opt. Soc. Am., 40:442 (1950).
23. L. M. Sverdlov, Opt. i Spektroskopiya, 1:752 (1956).
24. M. V. Vol'kenshtein, M. A. El'yashevich, and B. I. Stepanov, Vibrations of Molecules, Vols. 1 and 2, Gostekhizdat, Moscow (1949).
25. E. Wilson, J. Decius, and P. Cross, Theory of the Vibrational Spectra of Molecules [Russian translation], IL, Moscow (1960).
26. A. V. Sechkarev, Izv. Vyssh. Uchebn. Zaved., Fizika, No. 3, p. 88 (1961).
27. A. V. Sechkarev, Izv. Vyssh. Uchebn. Zaved., Fizika, No. 5, p. 68 (1961).
28. K. W. F. Kohlrausch, Ramanspectren (Vol. 9, Part 6 of Hand- und Jahrbuch der chemischen Physik, A. Eucken, ed.), Edwards, Ann Arbor, Michigan.
29. T. I. Kuznetsova and M. M. Sushchinskii, "Optics and Spectroscopy," in: Molecular Spectroscopy (1963), p. 144.
30. M. V. Vol'kenshtein, Configuration Statistics of Polymer Chains, Izd. AN SSSR, Moscow (1959).
31. I. N. Godnev, Calculation of Thermodynamic Functions from Molecular Data, Gostekhizdat, Moscow (1956).

32.   K. S. Pitzer and W. Gwinn, J. Chem. Phys., 9:485 (1941).
33.   D. R. Herschbach, J. Chem. Phys., 31:91 (1959).
34.   B. L. Crawford, J. Chem. Phys., 8:273 (1940).
35.   K. S. Pitzer, J. Chem. Phys., 14:239 (1946).
36.   S. G. Rautian, Zh. Eksper. i Teor. Fiz., 27:625 (1954).
37.   V. A. Zubov, G. G. Petrash, and M. M. Sushchinskii, Opt. i Spektroskopiya, 6:827 (1959).
38.   G. S. Landsberg, P. A. Bazhulin, and M. M. Sushchinskii, Basic Parameters of the Raman
       Spectra of Hydrocarbons, Izd. AN SSSR, Moscow (1956).
39.   G. S. Landsberg, B. A. Kazanskii, P. A. Bazhulin, T. F. Bulanova, A. L. Liberman, E. A.
       Mikhailova, A. F. Platé, Kh. E. Sterin, M. M. Sushchinskii, G. A. Tarasova, and S. A.
       Ukholin, Determination of the Individual Hydrocarbon Composition of Direct-Distillation
       Gasolines by a Combined Method, Izd. AN SSSR, Moscow (1959).
40.   U. A. Zirnit and M. M. Sushchinskii, Opt. i Spektroskopiya, 15:190 (1963).
41.   G. M. Harris and F. E. Harris, J. Chem. Phys., 31:1450 (1959).
42.   M. Karplus, J. Chem. Phys., 33:316 (1960).
43.   W. L. Clinton, J. Chem. Phys., 33:632 (1960).
44.   E. Lassetre and L. Dean, J. Chem. Phys., 17:317 (1949).
45.   L. J. Oosterhoff, Disc. Faraday Soc., 10:79 (1951).
46.   Au Chin-Tang, J. Chin. Chem. Soc., 18:2 (1951).
47.   E. A. Mason and M. M. Kreevoy, J. Am. Chem. Soc., 77:5808 (1955).
48.   N. P. Borisova and N. V. Vol'kenshtein, Zh. Strukt. Khim., 2:469 (1961).
49.   J. van Dranen, J. Chem. Phys., 20:1982 (1952).
50.   J. G. Aston, S. Isserow, G. J. Szasz, and R. M. Kennedy, J. Chem. Phys., 12:336 (1944).
51.   V. Magnasco, Nuovo Cimento, 24:425 (1962).
52.   Yu. A. Pentin and V. M. Tatevskii, Vestn. Mosk. Univ., No. 3, p. 631 (1955).
53.   H. J. Bernstein, J. Chem. Phys., 17:262 (1949).
54.   L. Pierce, J. Chem. Phys., 34:498 (1961).
55.   P. H. Kasai and R. J. Myers, J. Chem. Phys., 30:1096 (1959).
56.   L. Pierce and M. Hayashi, J. Chem. Phys., 35:479 (1961).
57.   K. S. Pitzer, Disc. Faraday Soc., 10:66 (1951).
58.   D. R. Lide, J. Chem. Phys., 33:1514 (1960).
59.   D. R. Lide and D. E. Mann, J. Chem. Phys., 28:572 (1958).
60.   D. R. Lide, J. Chem. Phys., 33:1519 (1960).
61.   I. N. Nazarov and L. D. Bergel'son, Usp. Khimii, 26:3 (1957).
62.   W. Klein and P. de la Mar, Advances in Stereochemistry, Collection [Russian translation],
       IL, Moscow (1961).
63.   U. A. Zirnit and M. M. Sushchinskii, "Optics and Spectroscopy," in: Molecular
       Spectroscopy (1963), p. 153.
64.   R. I. Podlovchenko, L. M. Sverdlov, and M. M. Sushchinskii, Opt. i Spektroskopiya,
       6:146 (1959).
65.   U. A. Zirnit and M. M. Sushchinskii, Opt. i Spektroskopiya, 16:902 (1964).
66.   F. F. Cleveland and P. Porcelli, J. Chem. Phys., 18:1459 (1950).
67.   G. B. Bonino and R. M. Ansidei, Proc. Indian Acad. Sci., A8:405 (1938).

STUDY OF THE OPTICAL AND ELECTRICAL PROPERTIES OF CERTAIN FOURTH-GROUP METALS*

A. I. GOLOVASHKIN

* Abbreviated text of a dissertation defended on June 21, 1965, at the P. N. Lebedev Physical Institute of the Academy of Sciences of the USSR.

INTRODUCTION

Considerable advances have been made in recent years in the optical study of metals. This has occurred because of the development of theory and experimental technology in the actual optics of metals and also because of the general successes of the electron theory and the whole of solid-state physics.

The theory of the anomalous skin effect has been developed [1-19] and this has given a good explanation for the optical properties of metals at low temperatures [5, 20-21]. Great progress has been made in experimental technology; many methods have been proposed for measuring optical constants, measuring systems complementary to one another have been devised, the accuracy and reliability of optical measurements have been greatly improved by a number of experimental procedures (multiple reflections from the metals under examination, modulation of the light beam with subsequent signal amplification, complex research into the properties of metals, the use of sensitive receivers, extremely pure metals, good procedures for finishing surfaces, etc.), and computing techniques have been greatly improved [12, 22-64].

In addition to this, the success of the Harrison scheme [65-67], which gives a very simple method of constructing the Fermi surfaces of metals, restores the value of the idea of "free" electrons and gives great significance to the optical study of metals in which this idea is largely used [12, 78].

Optical studies of metals are divided into two groups. The first group includes research in the infrared and part of the visible parts of the spectrum. In this range the optical properties of metals are determined by "free" electrons. Work of this kind provides information on the Fermi surfaces (the characteristics obtained relating to the main bulk of the conductivity electrons) and enables us to refine the Harrison system, determine the microcharacteristics of metals, and verify various theories of metal optics. The advantages of optical methods over other methods also capable of revealing a number of the electron characteristics of metals (cyclotron resonance, van Alphen—de Haas effect, measurement of magnetoresistance, the electron specific heat of metals, etc.) lies in the very wide energy range involved and in the possibility of varying the temperature of the metal over wide limits.

The second group involves research in the field of quantum transitions. The visible and ultraviolet parts of the spectrum are usually involved here. For some metals quantum transitions start even in the near-infrared. Research in this field enables us to establish the band structure of metals, to verify energy-band calculations, and to check the quantum theories of optical transitions [79, 80].

This paper will be mainly concerned with the first group.

Despite the fact that polyvalent metals constitute a very large class, in the optical respect they have until recently been very little studied. Reliable data only exist for metals of the first group in the periodic table: gold, silver, and copper. This is due both to experimental difficulties and to difficulties in the theoretical analysis of the results associated with polyvalent metals, since it is by no means known in advance how applicable the concepts developed for "good" metals are to the former. In contrast to monovalent metals, these have a complicated Fermi surface; frequently electrons of different shells emerge on the Fermi surface, while in metals of even groups the metallic properties are solely due to the overlapping of bands. This explains the interest recently evoked in polyvalent metals [81-97].

In this paper we shall be concerned with the properties of metals belonging to the fourth group, tin and lead. Apart from the facts already mentioned, interest in these arises from the fact that they are superconductors. The determination of their microcharacteristics (concentration of conductivity electrons, electron–phonon collision frequency, velocity of electrons on the Fermi surface) is thus of particular interest in the theory of superconductivity [9]. In addition to this, these metals offer an excellent opportunity of studying electron/phonon interactions in pure form. The losses associated with electron/phonon interaction are in these metals much greater (at all temperatures) than the losses associated with collisions between the electrons and lattice defects, impurities, and the surface.

We shall make a complex study of the optical and electrical properties of tin and lead. We shall study: 1) the optical properties in the spectral range 0.9-12 μ at temperatures of 293, 78, and 4.2°K; 2) the static conductivity and its temperature dependence between 4.2 and 293°K; 3) the density; 4) the Hall field; and, 5) the temperature corresponding to the transition into the superconducting state. We shall also determine the Debye temperature.

The optical properties of tin and lead have been studied in the infrared part of the spectrum over a wide range of wavelengths; this is very important, since only so shall we be able to secure a reliable determination of the microcharacteristics of the metal [98]. For tin and lead the region of quantum transitions starts in the near-infrared; hence measurements over a wide range are desirable from the point of view of separating the effects of "free" electrons and electrons participating in quantum transitions on the optical properties. This in turn enables us to obtain additional information regarding the band structure.

We shall thus indeed be making a complex study of the properties of tin and lead. As indicated earlier [92], only in this way may we derive the microcharacteristics of the metal and make a full comparison between theory and experiment. Such investigations give us a good idea of the quality of the samples.

The optical properties of tin and lead have been studied over a wide range of temperatures, including the temperature of liquid helium. Only measurements at such low temperatures, for which $T \ll \Theta$, are capable of allowing a reliable verification of the theory of the skin effect and indicating the part played by electron/phonon interaction and its temperature dependence (here, Θ is the Debye temperature, T is the absolute temperature). In addition to this, the measurement of the optical constants at low temperatures enables us to determine the microcharacteristics more reliably, to study their temperature dependence, and to elucidate such fundamental questions as the character of the reflection of electrons from the surface of the metal and the part played by electron/electron interaction.

These investigations enable us: 1) to demonstrate the applicability of the skin-effect theory to metals of the fourth group; 2) to determine the most important microcharacteristics of these metals; 3) to observe the purely quantum effect associated with the interaction of an electron, a photon, and phonons, resulting in the large residual absorption of light; 4) to observe the temperature dependence of the concentration of conduction electrons in tin and lead; and, 5) to estimate some parameters of the band structure of tin.

THEORETICAL STUDY OF THE OPTICAL PROPERTIES OF METALS

1. Classical Theory of the Skin Effect

It is well known that the optical properties of metals may be described by a complex refractive index $n' = n - i\varkappa$ or surface impedance Z [12, 44]. The relation between the quantities and the microscopic characteristics of the metals is given by the skin-effect theory.

At the present time the most satisfactory microtheory is that based on the kinetic equation for the electron-distribution function in the metal. In the linear approximation this equation has the form

$$i\omega f + \mathbf{v}\,\mathrm{grad_r}\,f + e\mathbf{E}\,\mathrm{grad_p}\,f_0 = -\frac{f}{\tau}. \tag{I.1}$$

Here f_0 is the equilibrium Fermi distribution, f is a small increment added to f_0, reflecting the effect of the field $\mathbf{E}$; $\mathbf{p}$ is the momentum; $\mathbf{v}$ is the velocity of the electrons; e is the charge on an electron; and ω is the angular frequency of the light. The time dependence is expressed in the form of $\mathbf{E}(t) = \mathbf{E}\cdot e^{i\omega t}$. The collision integral is written as f/τ, where τ is the relaxation time of the electrons. The question as to the conditions under which we may introduce the relaxation time τ is discussed in [12, 78, 99–101]. The introduction of τ is strictly justifiable for a number of practically important cases, in particular, low and high temperatures.

The simultaneous solution of the kinetic equation and Maxwell's equations

$$\left.\begin{array}{l} \mathrm{rot}\,\mathbf{H} = \dfrac{4\pi}{c}\mathbf{j} + \dfrac{i\omega}{c}\mathbf{E}, \\[2mm] \mathrm{rot}\,\mathbf{E} = -\dfrac{i\omega}{c}\mathbf{H} \end{array}\right\} \tag{I.2}$$

enables us to find the distribution of fields and currents under the simultaneous action of an electromagnetic wave and electron collisions. In formula (I.2), $\mathbf{H}$ is the magnetic field intensity and c is the velocity of light in vacuum. This in principle offers the possibility of determining the surface impedance of the metal or the associated complex dielectric constant $\varepsilon' = (n')^2$.

The simplest solution is obtained in the case of the normal skin effect, when we may neglect the term $\mathbf{v}\,\mathrm{grad_r}\,f$ of Eq. (I.1). Thus, $f = -\dfrac{e}{\nu + i\omega}\mathbf{E}\,\mathrm{grad_p}\,f_0$. Here $\nu = 1/\tau$ is the frequency of electron collisions. If we consider the surface of the metal as coinciding with the z = 0 plane and lay the x axis along the field $\mathbf{E}$, we obtain an expression for the total current density

$$j = \left(-\frac{2e^2}{h^3}\int\frac{v_x}{\nu + i\omega}\frac{\partial f_0}{\partial p_x}\,d\mathbf{p}\right)E. \tag{I.3}$$

Here h is Planck's constant, $d\mathbf{p} = dp_x dp_y dp_z$, the integration is carried out over the whole of momentum space. It is assumed that the metal is isotropic and that the wave is incident normally to the surface. Taking account of the delta-function nature of the function $\partial f_0 / \partial p_x$, we obtain ε' from (I.3):

$$\varepsilon' = 1 - \frac{4\pi e^2}{m\omega(\omega - i\nu)}\left(\frac{2m}{3h^3}\int v\,dS\right). \tag{I.4}$$

The integral is taken over the Fermi surface. The quantity in round brackets plays the part of the conductivity-electron concentration

$$N = \frac{2m}{3h^3}\int v\,dS = \frac{2}{3}mv^2\left(\frac{dY}{dW}\right)_F, \tag{I.5}$$

where $(dY/dW)_F$ is the density of states at the Fermi surface.

Separating the real and imaginary parts of $\varepsilon' = \varepsilon - (4\pi\sigma/\omega)i$, we obtain

$$\left.\begin{aligned}\varepsilon &= n^2 - \varkappa^2 = 1 - \frac{4\pi e^2 N}{m(\omega^2 + \nu^2)}, \\ \sigma &= \frac{n\varkappa\omega}{2\pi} = \frac{e^2 N\nu}{m(\omega^2 + \nu^2)}.\end{aligned}\right\} \tag{I.6}$$

The formulas relate to the contribution of the "free" electrons and enable us to find N and ν from measurements of n and $\varkappa$.

The conditions of the normal skin effect may be obtained by comparing the real and imaginary parts of the rejected and residual terms of Eq. (I.1):

$$\left.\begin{aligned}l &\ll \delta, \\ v &\ll \frac{c}{n}.\end{aligned}\right\} \tag{I.7}$$

Here $l = v\tau$ is the free path of the electrons, and δ is the depth of the skin layer.

2. Anomalous Skin Effect

In the case of the anomalous skin effect, when the term $\mathbf{v}\,\mathrm{grad}_{\mathbf{r}}f$ is not small, Eq. (I.1) must be solved in general form. This leads to a nonlocal relation between the current and the field and to a complicated distribution of fields and currents in the metal. We note that in the case of the anomalous skin effect n and $\varkappa$ signify the effective quantities defined in terms of the surface impedance, which retains its significance. The relation between the optical constants and the microcharacteristics of the metal becomes very complicated in the case of the anomalous skin effect. An important aspect of this is that an additional surface absorption, keeping its value right down to absolute zero temperature, then appears in the metals.

An experimental study of the optical properties of metals in the infrared part of the spectrum has shown that, although in the majority of cases the skin effect is abnormal, it has only a slightly anomalous character [5, 45, 88-90, 92, 102-104). The term $\mathbf{v}\,\mathrm{grad}_{\mathbf{r}}f$ is not in fact so small as to be negligible, but small enough to allow the system of Eqs. (I.1) and (I.2) to be solved by the method of successive approximations. The expression for the field obtained in the first approximation, $\mathbf{E} = \mathbf{E}_0\,e^{-i\frac{\omega}{c}\sqrt{\bar{\varepsilon}'}z}$, where $\mathbf{E}_0$ is the field on the surface of the metal, is substituted into (I.1). The resultant equation is solved with due allowance for the diffusion character of the reflection of electrons at the metal boundary [5, 12, 78, 105]. The final formulas relating the microcharacteristics to the optical constants of the metals were obtained for a weakly anomalous skin effect by Motulevich [19]. For the case of a spherical Fermi surface these expressions may be put in the form:

$$N = \frac{m\omega^2}{4\pi e^3} \frac{(n^2 + \varkappa^2)^2}{\varkappa^2 - n^2} \frac{1}{1 - \beta_1} = \frac{0.1115 \cdot 10^{22}}{\lambda^2} \frac{(n^2 + \varkappa^2)^2}{\varkappa^2 - n^2} \frac{1}{1 - \beta_1},$$

$$\nu/N = \frac{4\pi e^2}{m\omega} \frac{2n\varkappa}{(n^2 + \varkappa^2)^2} (1 - \beta_2) = 1.690 \cdot 10^{-6} \lambda \frac{2n\varkappa}{(n^2 + \varkappa^2)^2} (1 - \beta_2),$$

$$\beta_1 = \frac{3}{8} \frac{v}{c} \varkappa \cdot \frac{\sqrt{1 + \dfrac{n^2}{\varkappa^2}}}{\sqrt{1 + \dfrac{\nu^2}{\omega^2}}} \left[\cos(\varphi_1 - \varphi_2) + \frac{2n\varkappa}{\varkappa^2 - n^2} \sin(\varphi_1 - \varphi_2) \right],$$

$$\beta_2 = \frac{3}{8} \frac{v}{c} \varkappa \cdot \frac{\sqrt{1 + \dfrac{n^2}{\varkappa^2}}}{\sqrt{1 + \dfrac{\nu^2}{\omega^2}}} \left[\cos(\varphi_1 - \varphi_2) - \frac{\varkappa^2 - n^2}{2n\varkappa} \sin(\varphi_1 - \varphi_2) \right].$$

$$(I.8)$$

Here, $\tan \varphi_1 = n/\varkappa$, $\tan \varphi_2 = \omega/\nu$. The wavelength of the light λ is in microns. The formulas are obtained to an accuracy of the first order in the small parameter $\gamma = \frac{3}{8} \frac{v}{c} \frac{n - i\varkappa}{1 - i\frac{\nu}{\omega}}$.

In the case of the weakly-anomalous skin effect, the measurement of n and $\varkappa$ over a wide frequency range enables us to determine N, ν, and v.

3. Frequency of Electron Collisions

In the linear approximation of the classical theory of the skin effect

$$\nu = \nu_{cl}^{ef} + \nu_{cl}^{ee} + \nu^{ed}. \qquad (I.9)$$

Here ν_{cl}^{ef} is the electron−phonon collision frequency; ν_{cl}^{ee} is the interelectron collision frequency, and ν^{ed} is the frequency of collisions of electrons with impurities or defects. The statistical conductivity is $\sigma_{st} = e^2 N / m\nu$.

The frequency ν_{cl}^{ef} characterizes the interaction of the electrons with phonons [96, 106]

$$\nu_{cl}^{ef} = \nu_\Theta \frac{T}{\Theta} \Phi(T),$$

$$\Phi(T) = 4 \left(\frac{T}{\Theta} \right)^4 \int_0^{\Theta/T} \frac{x^5 \, dx}{(e^x - 1)(1 - e^{-x})}. \qquad (I.10)$$

For $T \gg \Theta$, $\nu_{cl}^{ef} = \nu_\Theta T/\theta$; for $T \ll \Theta$, $\nu_{cl}^{ef} \sim (T/\Theta)^5$, except for the lowest temperatures, for which $\nu_{cl}^{ef} \approx 0$. Thus, for T = 4.2°K and metals with $\Theta \approx 100$-300°K, ν_{cl}^{ef} is 10^{-6} to 10^{-7} of the frequency of electron−phonon collisions at room temperature.

The classical frequency of interelectron collisions [107], $\nu_{cl}^{ee} \sim T^2$, only has an effect at fairly low temperatures.

The frequency of electron collisions with impurities or defects ν^{ed} is regarded as independent of temperature and leads to a finite absorption of light at the very lowest temperatures. In addition to this, the overall collision frequency ν contains a term associated with interaction between the electrons and high-frequency lattice vibrations. So far, however, no estimate of ν^{ed} has been made for metals.

4. Quantum Kinetic Equation

The theory under consideration was based on the classical kinetic equation (I.1). However, at high frequencies, for which $\hbar\omega > kT$, the interaction between the electrons and the electromagnetic field is of a quantum nature, and the question of the applicability of this equation arises. Even at room temperature, the inequality $\hbar\omega > kT$ means $\lambda < 40\,\mu$. At lower temperatures, quantum effects must be considered over the whole infrared region. The quantum kinetic equation must also be used. The derivation of this equation is considered in [108-110]. A quantum kinetic equation for electrons in metals was obtained by Gurzhi [110]. In order to obtain the collision integral, it was considered that the interaction of photons with electrons took place by the generation or absorption of phonons. In deriving the equation a number of stringent but acceptable assumptions were made (only acoustic phonons were considered; the interaction of the electrons with the phonons and photons was considered weak, so that the theory of perturbations could be used). The collision integral and the term containing the field thus acquired increments which vanished in the classical limit $\hbar\omega \rightarrow 0$. The quantum equation [110] is used for studying the optical properties of metals in the infrared part of the spectrum in [100, 111, 112]. In addition to this, the effect of interelectron collisions on the optical properties of metals is considered in [111, 112] on the basis of the quantum kinetic equation.

The general case of the anomalous skin effect was considered in [110] from the quantum point of view and expressions were obtained for the impedance and absorption in the infrared region in the form of series. These expressions became much simpler on satisfying the conditions

$$e^{\hbar\omega/kT} \gg 1 \text{ and } e^{\hbar\omega/kT} \gg e^{\Theta/T}. \tag{I.11}$$

Comparison of the results obtained on the basis of the quantum kinetic equation with the results of classical theory [1, 4] reveals the following:

1. To a first approximation, the volume and surface losses simply add, as in the classical case.

2. The expressions for the optical characteristics (impedance, absorption, dielectric constant) are the same in both theories to an accuracy of correction terms of the first order. The interaction between the electrons and the phonons is nevertheless characterized by a different frequency ν^{ef}. In the quantum case (in the presence of a high-energy light quantum),

$$\nu^{ef} = \nu_\Theta \frac{T}{\Theta} \varphi(T),$$

$$\varphi(T) = \frac{2}{\alpha}\left(\frac{T}{\Theta}\right)^4 \int_0^{\Theta/T} x^4 \left(\frac{2\alpha}{e^x - 1} + \frac{x - \alpha}{e^{x-\alpha} - 1} - \frac{x + \alpha}{e^{x+\alpha} - 1}\right) dx, \tag{I.12}$$

where $\alpha = \hbar\omega/kT$. In the case (I.11), we obtain $\varphi(T)$ in the following simpler form:

$$\varphi(T) = 2\left(\frac{T}{\Theta}\right)^4 \int_0^{\Theta/T} x^4 \operatorname{cth} \frac{x}{2} dx. \tag{I.13}$$

Conditions (I.11) are not too severe and even at room temperature these are satisfied over a wide range of the spectrum.

3. The second and subsequent correction terms obtained by Gurzhi [100] differ from the corresponding terms found by Dingle [4, 6]. This is because, even in the classical case, under the conditions of the anomalous skin effect, the replacement of the collision integral by an expression of the form f/τ is (strictly speaking) not always valid. However, the fact that the expressions in the two cases agree to an accuracy of first-order correction terms indicates that the approximation is a good one for all T.

The form of the function $\varphi(T)$ was calculated by Gurzhi [112]. For $T \gg \Theta$, $\varphi(T) = 1$ and $\nu^{ef} = \nu^{ef}_{cl} = \nu_{\Theta}(T/\Theta)$. For $T \ll \Theta$, $\varphi = 2\Theta/5T$ and $\nu^{ef} = \frac{2}{5}\nu_{\Theta}$ (here, $\nu^{ef}_{cl} \sim T^5$). Thus, in the presence of a large light quantum the electron—phonon collision frequency depends relatively weakly on temperature. In contrast to the classical electron collision frequency, ν^{ef} remains finite right down to absolute zero temperature. At a fairly low temperature the ratio ν^{ef}/ν^{ef}_{cl} may reach several orders. This theoretical result has not been proved experimentally.

In addition to these results, a quantum consideration of the question of interelectron collisions leads [111] to the expression

$$\nu^{ee} = \nu^{ee}\left[1 + \left(\frac{\hbar\omega}{2\pi k T}\right)^2\right].$$

The theory of the skin effect has also been developed with due allowance for inclined incidence [1, 4, 17, 113], anisotropy [10, 11, 13, 16, 114], interelectron collisions and Fermi-fluid effects [14, 15, 107, 111, 115-118], and interband transitions [119]. In studying the optical properties of metals in the region of quantum transitions or in the case of several groups of electrons, the question as to how the contributions from these various states or groups may be separated arises [119-125].

CHAPTER II

MEASURING METHOD

1. Method of Measuring Optical Constants

1. For measuring the optical constants of tin and lead we used the polarization method. The main idea of this method is illustrated in Fig. 1. In the figure, S_1 is the entrance slit of the polarization system (in particular cases the light source may appear in place of slit S_1). The slit S_1 is placed at the focus of the objective O_1. In all forms of the system, even in the short-wave part of the spectrum, we used mirror optics. In this case, O_1 and O_2 are spherical mirrors; P and A are the polarizing elements. The first of these is usually called the polarizer, and the second the analyzer. Between these are the samples L. The exit slit S_2 lies at the focus of objective O_2 (in particular cases there may be a light receiver in place of slit S_2).

A parallel beam of natural light falls on the polarizer P. Light polarized in a specific plane falls on the samples L and experiences multiple reflection (in Fig. 1 we only show two reflections for the sake of simplicity). The angle of incidence of the light on the samples remains the same for all m reflections. The light reflected from the samples is analyzed in the analyzer. The measurements are made in monochromatic light.

This method may be analyzed on the basis of an expression giving the intensity of the light passing through the polarizer—samples—analyzer system as a function of the position of the polarizer and analyzer [29]:

$$I = \frac{I_0}{2}\left(r_p^m \cos^2 \psi_P \cos^2 \psi_A + r_s^m \sin^2 \psi_P \sin^2 \psi_A + \frac{1}{2}\sqrt{r_p^m r_s^m}\,\sin 2\psi_P \sin 2\psi_A \cos \Delta_m \right). \qquad \text{(II.1)}$$

Here I_0 is the intensity of the light falling on the polarizer (polarizer and analyzer considered ideal), r_p and r_s are the reflection coefficients for the p and s components of the light $(\sqrt{r_p/r_s} = \log \rho$; ρ is the azimuth corresponding to single reflection); ψ_P and ψ_A are the azimuths of the polarizer and analyzer. We assume an m-fold reflection of light from the sample; $\Delta_m = m\Delta$, Δ is the phase difference between the p and s components of the light corresponding to single reflection.

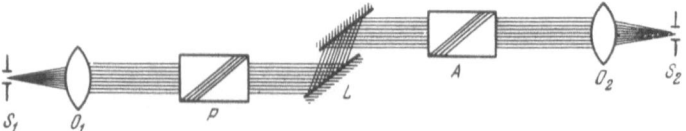

Fig. 1. Basic arrangement of the polarization method.

By determining Δ and ρ we may find n and $\varkappa$. On condition that $\sin^2 \varphi / |\varepsilon'| \ll 1$, which is usually satisfied adequately in metals,

$$
\left.
\begin{aligned}
n &= \frac{\sin \varphi \tan \varphi \cos 2\rho}{1 + \sin 2\rho \cos \Delta}, \\
\varkappa &= \frac{\sin \varphi \tan \varphi \sin 2\rho \sin \Delta}{1 + \sin 2\rho \cos \Delta}.
\end{aligned}
\right\}
\qquad \text{(II.2)}
$$

Here φ is the angle of incidence of the light on the sample.

2. We tried a number of different forms of the polarization method. Most of all we used the method based on a rotating polarizer. In this method the polarizer is rotated at a frequency Ω. We see from expression (II.1) that the intensity of the light passing through the polarization system depends on the azimuths of the polarizer and analyzer in the same way. From this point of view it does not matter which of the polarizing elements rotates. In the first forms of our system [41, 45] we rotated the analyzer. It was found that at low temperatures this led to the appearance of a parasitic signal which had to be compensated. We therefore used the arrangement with a rotating polarizer for subsequent experiments at temperatures differing from the room value. For uniform rotation of the polarizer, the azimuth of the latter varies with time in accordance with $\psi_P = \Omega t$. Expression (II.1) may then be converted to the form

$$
I = I_{01} + I_1 \cos(2\Omega t + \gamma), \qquad \text{(II.3)}
$$

where

$$
I_{01} = \frac{I_0}{4} \left(r_p^m \cos^2 \psi_A + r_s^m \sin^2 \psi_A \right),
$$

$$
I_1 = \frac{I_0}{4} \sqrt{(r_p^m \cos^2 \psi_A - r_s^m \sin^2 \psi_A)^2 + r_p^m r_s^m \sin^2 2\psi_A \cos^2 \Delta_m},
$$

$$
\tan \gamma = \sqrt{r_p^m r_s^m} \sin 2\psi_A \cos \Delta_m / (r_p^m \cos^2 \psi_A - r_s^m \sin^2 \psi_A).
$$

Here I_{01} and I_1 are independent of time. The light flux is modulated at a frequency of 2Ω. The light receiver is connected to the input of a resonance amplifier by means of which the alternating signal of frequency 2Ω is amplified. The amplitude of the signal is determined by the value of I_1. The signal vanishes for $I_1 = 0$. This only occurs if

$$
\left.
\begin{aligned}
\cos \Delta_m &= 0, \\
\tan^2 \psi_A &= \left(\frac{r_p}{r_s} \right)^m.
\end{aligned}
\right\}
\qquad \text{(II.4)}
$$

The way in which I_1 (and hence the signal) depends on the analyzer azimuth ψ_A for various values of $\cos \Delta_m$ is shown in Fig. 2. If $\cos \Delta_m \neq 0$, the signal does not vanish for any value of ψ_A. Condition (II.4) may be selected by simultaneously varying λ and ψ_A.

By measuring the azimuth of the analyzer for which the signal vanishes, and knowing the angle of incidence φ and the phase difference Δ_m, we may use formula (II.2) to find n and $\varkappa$.

3. We also used a variation on the polarization method in which linear polarization of the light reflected from the sample was achieved. In this case the light was modulated by an additional rotating interrupter. (For measurements at low temperature the position of the interrupter was not immaterial because of the possible appearance of a parasitic low-temperature signal. This signal is absent if the interrupter is placed in front of the polarizer.) The modulation frequency was 2Ω, so that the same receiving and amplifying system as before could be used.

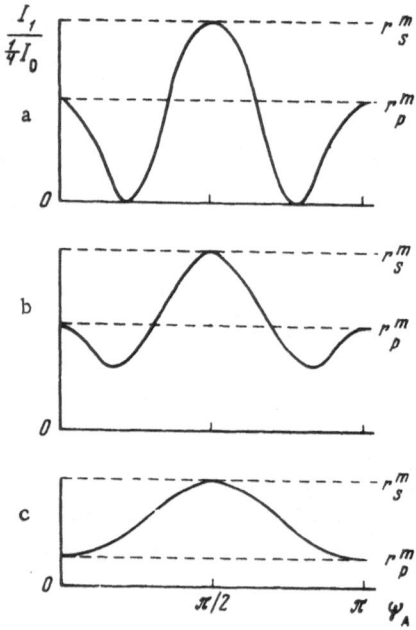

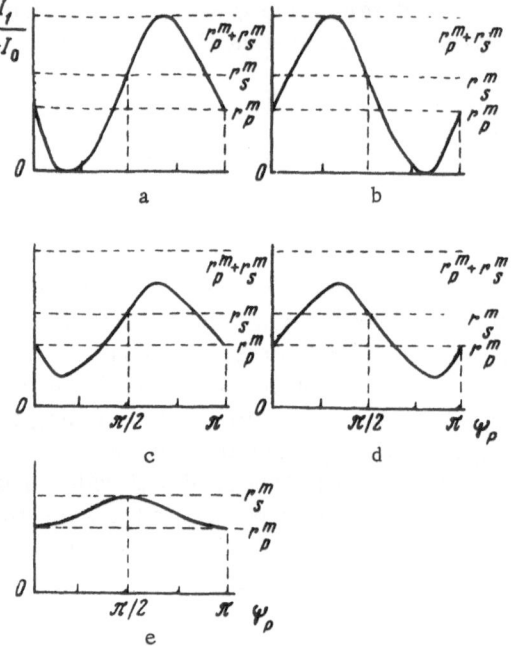

Fig. 2. Signal as a function of analyzer azimuth in the arrangement with a rotating polarizer for various values of: a) $\cos \Delta_m = 0$; b) $\cos^2 \Delta_m < \frac{1}{2}(1 + r_p^m/r_s^m)$; c) $\cos^2 \Delta_m \geq \frac{1}{2}(1 + r_p^m/r_s^m)$.

Fig. 3. Signal as a function of the polarizer azimuth in the system incorporating an interrupter for various values of $\cos \Delta_m$: a) $\cos \Delta_m = -1$; b) $\cos \Delta_m = +1$; c) $-1 < \cos \Delta_m < 0$; d) $0 < \cos \Delta_m < 1$; e) $\cos \Delta_m = 0$.

In this method the analyzer azimuth was made equal to 45°. Then the intensity of the light passing through the polarizer system will be equal to

$$I = \frac{I_0}{4}\left(r_p^m \cos^2 \psi_P + r_s^m \sin^2 \psi_P + \sqrt{r_p^m r_s^m}\, \sin 2\psi_P \cos \Delta_m\right). \qquad (\text{II.5})$$

The signal vanishes on condition that

$$\left.\begin{aligned}\cos \Delta_m &= -1,\\ \tan \psi_P &= \sqrt{\frac{r_p^m}{r_s^m}}\end{aligned}\right\} \qquad (\text{II.6})$$

or on condition that

$$\left.\begin{aligned}\cos \Delta_m &= 1,\\ \tan \psi_P &= -\sqrt{\frac{r_p^m}{r_s^m}}\end{aligned}\right\}. \qquad (\text{II.6a})$$

We see from expression (II.1) that in this method it is quite immaterial which of the polarizing elements has its amplitude made equal to 45°. In this case, when the polarizer amplitude is made equal to 45°, we obtain expressions analogous to (II.5) and (II.6) with ψ_P in place of ψ_A. The two methods are equivalent even at low temperatures. In our own case it was more convenient to fix the azimuth of the analyzer.

In the method under consideration, the vanishing of the signal corresponds to zero light flux. The light flux I (or the signal proportional to this) is shown as a function of the polarizer azimuth ψ_P for various values of $\cos \Delta_m$ in Fig. 3.

By measuring the polarizer azimuth at which the signal vanishes and knowing φ and Δ_m, we may use formula (II.2) to find n and $\varkappa$.

Both in the first and in the second variants of the polarization method, a specific, preassigned polarization of the light reflected from the metal was achieved. This polarization may be obtained by varying either the angle of incidence of the light on the samples φ or the wavelength of the light λ. In the first case the wavelength is kept fixed and in the second a fixed angle of incidence is used.

The arrangement used in our present investigation was based on a fixed angle of incidence. This was because this arrangement was simpler and easily applicable to measuring optical constants at various temperatures.

4. In addition to methods in which a preassigned polarization of the reflected light was achieved, we also used methods based on the analysis of the full relationship between the azimuth of the polarizing element and the intensity.

For the case in which the flux was modulated with a rotating polarizer, the dependence of the variable part of the intensity I_1 on ψ_A is very substantial. This relationship is shown in Fig. 2b for $\cos \Delta_m \neq 0$. The relationship is periodic with a period of π. The maximum values of I_1 are reached for $\psi_A = 0$ and 90° and are respectively equal to $(I_0/4) r_p^m$ and $(I_0/4) r_s^m$. The minimum values of I_1 occur at azimuths for which

$$\tan^2 \psi_A = \frac{r_p^m}{r_s^m} \frac{1 - \dfrac{2r_s^m}{r_p^m + r_s^m} \cos^2 \Delta_m}{1 - \dfrac{2r_p^m}{r_p^m + r_s^m} \cos^2 \Delta_m}. \tag{II.7}$$

By measuring the ratio of the signals at the maxima, equal to r_p^m / r_s^m, and the analyzer azimuth, we may use formulas (II.7) and (II.2) to find the optical constants. Measurements may be made for values of λ such that $\cos^2 \Delta_m < (r_p^m + r_s^m)/2r_s^m$.

For the case in which the flux is modulated with an interrupter, the intensity I is shown as a function of polarizer azimuth in Fig. 3. This relationship also has a period of π. In this case, for determining the optical constants, it is convenient to use the values of I at $\psi_P = 0$ and 90°, respectively equal to $(I_0/4)r_p^m$ and $(I_0/4)r_s^m$. The azimuth of the polarizer for which the intensity of the light reaches a minimum may be found from the relation

$$\psi_P = -\frac{1}{2} \tan^{-1} \left(\frac{2\sqrt{r_p^m r_s^m}}{r_s^m - r_p^m} \cos \Delta_m \right). \tag{II.8}$$

By measuring the ratio of the signals at $\psi_P = 0°$ and $\psi_P = 90°$, equal to r_p^m / r_s^m, and the azimuth of the polarizer for which the signal reaches a minimum, we may find n and $\varkappa$ from formulas (II.8) and (II.2).

The two latter forms of the polarization method enable us to measure the optical constants of metals for a fixed value of the angle of incidence and an arbitrary wavelength of the light.

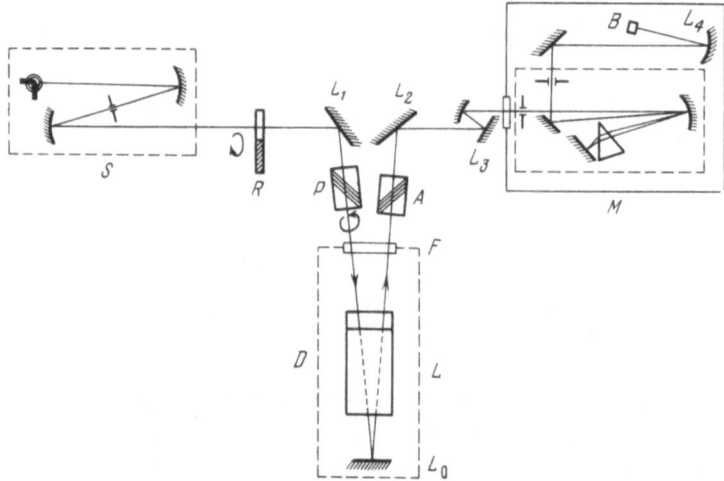

Fig. 4. General arrangement of the apparatus for measuring
the optical constants of metals. S) Illuminating system; R) interrupter; L_0, L_1, L_2) plane mirrors; P) polarizer; A) analyzer;
F) rocksalt plate; L) samples; D) low-temperature part of the
apparatus; L_3) mirror system focusing the beam on the monochromator slit M; L_4) spherical mirror; B) germanium bolom-
eter. The illuminating system S and monochromator M are
shown turned through 90°.

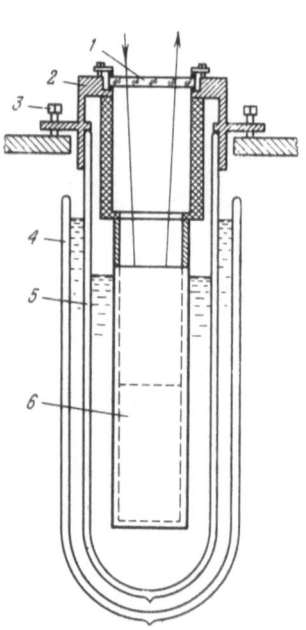

Fig. 5. Low-temperature part of the apparatus. 1) Salt window sealed with rubber gaskets; 2) sealed end; 3) adjusting
screws; 4) nitrogen Dewar; 5) helium Dewar; 6) holder containing mirrors under examination (samples).

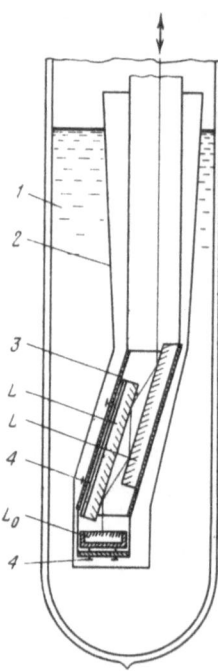

Fig. 6. Arrangement of sample mirrors
used in the measurement of the optical constants. 1) Helium; 2) casing; 3) holder; 4)
adjusting screws; L) sample mirrors; L_0)
mirror reflecting the beam in the reverse
direction.

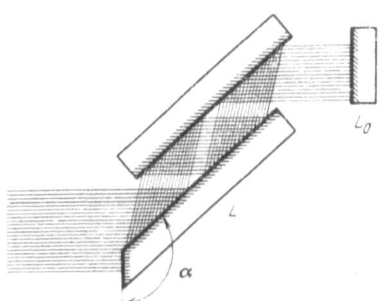

Fig. 7. Path of the rays inside the sample system. L) Sample mirrors; L_0) mirror reflecting the beam in the reverse direction, $\alpha = 180° - \varphi$ (φ is the angle of incidence of the light).

2. Experimental System for Measuring Optical Constants

We designed an experimental apparatus suitable for measuring optical constants of metals at both room and low temperatures. The main idea of the system is illustrated in Fig. 4. The low-temperature system, including a sealed end, a holder for the samples under examination, and Dewars containing cooling liquids, is shown separately in Figs. 5 and 6. The signal from the receiver created by the modulation of the light beam with a rotating polarizer or interrupter was amplified by means of a low-frequency, narrow-band amplifier [126]. For a pass band of 1 cps the minimum detectable power was $4 \cdot 10^{-10}$ W.

The path of the light beam inside the system of samples is shown in Fig. 7. The samples form two plane mirrors L placed parallel to each other. The angle of incidence of the light is fixed. After undergoing fourfold reflection from the samples, the light falls on the plane mirror L_0 and is reflected in the opposite direction, after which it again undergoes a fourfold reflection from the sample mirrors L. The mirror L_0 also serves to separate the incident and reflected beams by a small angle (~4°). The beams are separated in a plane perpendicular to the plane of incidence.

The sample mirrors and mirror L_0 (Figs. 5 and 6) were held in a special holder designed to make mirrors L parallel and also to set mirror L_0 perpendicular to the light beam. The mirrors placed in the holder were fixed with springs. The holder and mirrors were suspended from the sealed end by means of stainless steel and Plexiglas plates. This kind of suspension minimized the leakage of heat from the "hot" parts of the apparatus to the holder when working at low temperatures. The mirrors were placed in a casing and surrounded by helium or nitrogen vapor. The level of cooling liquid was much higher than the mirrors, which were at a uniform temperature practically coinciding with that of the liquid. The whole system, including the sealed end, the holder with sample mirrors, and the Dewars, was adjusted by means of three adjusting screws on the common base of the apparatus. The position of the system relative to the other parts of the apparatus was strictly fixed, and this was verified by special experiments.

The polarizer and analyzer were formed by piles of eight free-selenium films [127]. The angle of incidence of the light on the pile was 67.5°. For the apertures used in our experiments this kind of polarizer passed less than $\frac{1}{500}$ of the unwanted light component over a wide range of wavelengths. This corresponds to a degree of polarization better than 99.8%. The transmission of such polarizers was 40%. In the short-wave range glass polarizers were also used. The analyzer and polarizer were arranged vertically in order to avoid other elements introducing polarization.

The IKS-12 monochromator was calibrated by reference to absorption lines [128-130]. In addition to ordinary adjustment [131], carried out in visible light, the arrangement of the sample mirrors had to be checked so as to ensure that they should be parallel to one another and should receive the light at the specified angle of incidence. The mirrors were adjusted to the parallel state by means of the arrangement shown in Fig. 8.

By making the images of the source coincide in the focal plane of the objective L^m the mirrors could be made parallel to an accuracy better than 10″. The samples were set to the required angle of incidence with the help of a special face making an angle of α with the working face (see Fig. 7). When this special face was set perpendicular to the incident light beam, the

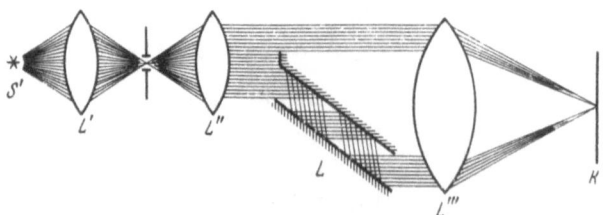

Fig. 8. Optical system for making the sample mirrors parallel. S') Source (glow lamp); L', L", L''') objectives; L) working mirror; K) screen.

angle of incidence on the working faces was $\varphi = \pi - \alpha$. The angle α was measured with a GSP-5 goniometer to an accuracy of better than 5". In practice, the angle of incidence was set to an accuracy of a few minutes, which led to an error of under 0.5% in the determination of the optical constants.

In parallel with the optical-constant measurements made at room temperature in the apparatus just described, we also measured these constants for the same samples on the four-mirror system described by Motulevich and Shubin [34]. In these measurements we used a fixed light wavelength λ and determined the angle of incidence φ and azimuth ρ leading to circular polarization of the light reflected from four identical samples. The advantage of this system is that the optical constants of the metal may be obtained very rapidly for any preassigned wavelength of the light.

Agreement between the results of the measurements on the two sets of apparatus served as a reliable criterion for ensuring correct adjustment and the absence of systematic errors.

3. Measurement of the Real and Imaginary

Parts of the Complex Refractive Indices n and $\varkappa$

At the beginning of this chapter, we described the theory of the polarization method used for the measurement of optical constants. In this section we describe the measuring process. The experiment was conducted in the following sequence. With the same mirrors we first measured the optical constants at room temperature, then at the temperature of liquid helium, then again at room temperature, and finally at liquid-nitrogen temperature. The initial adjustment was very well maintained on subsequent measurements. During the measurements at room temperature the sample mirrors were kept in a helium atmosphere.

In the first method the analyzer azimuth was defined as half the angle between the two positions at which the signal vanished. For any λ at which the signal vanishes, the phase difference Δ_m equals an odd number of $\pi/2$. The determination of this number for each such λ is usually quite easy and may be carried out over the whole set of measurements with a given angle of incidence φ. A graph of $\Delta_m(\omega)$ greatly assists in this determination. In the part of the spectrum in which the optical properties of the metal are determined by the conduction electrons, the relationship $\Delta_m(\omega)$ is almost linear and the signal vanishes at points quite uniformly distributed along the scale of frequencies. The measured values of the amplitude ψ_A, phase difference Δ_m, and wavelength λ, and also the angle of incidence of the light φ preassigned on adjustment of the apparatus, enable us to find the optical constants n and $\varkappa$ for a given λ. In order to determine n and $\varkappa$, we must know the phase difference Δ and azimuth ρ corresponding to a single reflection. These values are obtained from the relations $\Delta = \Delta_m/m$ and $\tan^m \rho = \tan \rho_m$. Here ρ_m means the measured value of the analyzer azimuth, while n and $\varkappa$ may be found from formulas (II.2). These formulas were obtained for the condition $\sin^2\varphi/|\varepsilon'| \ll 1$; otherwise one must use the exact formulas, which are rather cumbersome. For tin and lead we may use the approximate formulas for all the measured λ, i.e., formulas incorporating a correction term of the first order in $\sin^2\varphi/|\varepsilon'|$. Here, n and $\varkappa$ are found from the formulas:

$$
\left.
\begin{aligned}
n &= n_0\left[1 + \frac{\sin^2\varphi}{2\,(n_0^2 + \varkappa_0^2)}\right], \\
\varkappa &= \varkappa_0\left[1 - \frac{\sin^2\varphi}{2\,(n_0^2 + \varkappa_0^2)}\right], \\
n_0 &= \frac{\sin\varphi\,\tan\varphi\,\cos 2\rho}{1 + \sin 2\rho\,\cos\Delta}, \\
\varkappa_0 &= \frac{\sin\varphi\,\tan\varphi\,\sin 2\rho\,\sin\Delta}{1 + \sin 2\rho\,\cos\Delta}.
\end{aligned}
\right\}
\tag{II.9}
$$

The maximum error when using these formulas instead of the exact ones was a few hundredths of a percent. Usually, in the range 0.8 to 1.2 μ values of n and $\varkappa$ could be obtained by this method for four or five wavelengths.

In measurements made by the second method, in which linear polarization of the reflected light is achieved, the analyzer was set at an azimuth of 45° and the polarizer served to find the azimuth at which the signal became equal to zero. This amplitude ρ_m and the corresponding wavelength were determined analogously to the previous method. The difference was simply that the second position of the azimuth of the polarizer for which the signal vanished occurred at $\psi_A = -45°$. The wavelengths for which Δ_m was a multiple of π lay between the wavelengths for which the signal vanished in the previous method. For a given angle of incidence the first such λ was about twice as short as the first λ of the previous method. Hence, the use of this method for very long waves was troublesome, since it demanded too large angles of incidence. The determination of Δ_m by this method was simpler, since, for one particular value of ψ_A a change of $\triangle_m$ in π caused the polarization of the light to pass into the neighboring quadrant. The values of n and $\varkappa$ were determined by the same formulas as in the first case. Three or four values of wavelength were obtained by this method. Measurements based on both methods were carried out in a single experiment. Thus, for a given φ, seven to nine values of λ were obtained and the corresponding n and $\varkappa$ determined.

These methods of measuring optical constants have a number of advantages. First of all, they are direct. The parameters measured are directly related to the optical constants. Secondly, the parameters are determined by means of a nul experiment rather than by the measurement of intensities. Hence, the stability of the source and the linearity of the receiver-amplifier system play a secondary part. Thirdly, the conditions for the vanishing of the signal are unique. There is no confusion or double meaning. The use of these methods, together with multiple reflections, which increase the effective changes in azimuth and phase differences, enables us to work with small angles of incidence and still secure a high accuracy in measuring the optical constants. It should be noted that the use of the first method enables us to make measurements at very small angles of incidence and to secure great accuracy.

The use of methods based on analyzing the intensity as a function of azimuth is very limited. In view of the fact that in these methods one has to measure the relative intensity, strict requirements are imposed on the stability of the source and the linearity of the receiver—amplifier system. These methods were only used for tin at low temperatures in order to discover the general relationship between the optical constants and λ. The azimuth corresponding to the minimum signal was determined much as in the other variants of the method. The use of the two forms of the method was called for because the method with the rotating polarizer was unsuitable near the λ values for which Δ_m was close to a whole number of π, while the method with the interrupter was unsuitable near the λ values for which Δ_m was close to an odd number of $\pi/2$. The linearity of the receiver—amplifier system was verified by special experiments. The correctness of the values obtained by the latter methods was verified by comparing the measurements made by these and the fundamental methods for the same wavelength.

4. Possible Systematic Errors

The polarization method enables us to achieve extremely high accuracy when studying the optical properties of metals. However, there are a number of possible sources of systematic errors which may seriously distort the results. Some of these errors are considered in [28, 29, 132]. Errors arise most frequently from the partial polarization of light by various elements in the optical system and from the displacement of the beam by the rotating polarizer. Below we consider sources of error which might in principle introduce distortions into the results for the optical system employed.

a) Parasitic Polarization of Light by Elements
of the Optical System

In the form of optical system used (Fig. 4), only the polarization introduced by elements in front of the polarizer (illuminating system S, mirror L_1) are dangerous. Polarization introduced by elements of the illuminating system (source, slit, mirrors) for the apertures [29], slits [133], and angles of incidence used led to a negligible error in the optical constants.

The worst polarization came from the mirror L_1, situated at an angle of 45° to the incident beam. The effect of this polarization was allowed for by a method analogous to that given in [29]. For partial polarization of the incident beam, the quantity I_0 in (II.1) ceases to be independent of the azimuth of the polarizer ψ_p and should be replaced by $I_0[1 + 2\alpha \cos(2\psi_p + \gamma)]$. Here, α is the degree of polarization, which is assumed small. For our system the initial phase $\gamma = 0$. Under these conditions, in contrast to (II.4), the signal vanishes at

$$\left.\begin{array}{l} \cos \Delta_m = 0, \\[2mm] \tan^2 \psi_A = \dfrac{r_p^m}{r_s^m}(1 + 4\alpha). \end{array}\right\} \qquad (\text{II}.10)$$

Comparison of (II.10) with (II.4) shows that the partial polarization of the light by the mirror L_1 only produces an error in measuring the azimuth and has no effect on the measurement of wavelength. If we express the degree of polarization α in terms of the optical characteristics of the mirror L_1, the second of the conditions (II.10) will have the form

$$\tan^2 \psi_A = \frac{r_p^m}{r_s^m}\frac{r_s'}{r_p'}.$$

Here, r_p' and r_s' are the reflection coefficients of the mirror L_1 for p and s polarization, respectively. For determining r_p' and r_s' we used the values of the optical constants of aluminum obtained in one of our own papers [88] and also in [31, 93]. The resultant value of α in the range 1.7-12 μ was no greater than 0.4%, reaching 1% in the neighborhood of 0.9 μ. The correction for the polarization of the light by the mirror L_1 chiefly affected n and had little effect on $\varkappa$. The value of the correction to n was greatest for lead at helium temperature, where it equalled 1-2%, rising to 3% at $\lambda \sim 1.5 \mu$. For lead at other temperatures, and tin at all temperatures, this correction was under 1% over the whole spectral range.

We note that the polarization of the light by the mirror L_1 and the illuminating system has no effect at all on the measurements in the method with the rotating interrupter, since, in this method, there is no periodic variation in the azimuth of the polarizer.

Special attention must be given to the additional elements situated between the polarizer and analyzer, since the phase difference and change in azimuth which these introduce reflect most directly on the results of the measurements. In the arrangement used such elements include the mirror L_0 and the rocksalt window. The beam falls at an angle of 2° on the mirror L_0. Calculation shows that the phase difference and change in azimuth thus introduced are negligible. The light also falls on the rocksalt window almost normally. However, this window might have an effect in the presence of artificial anisotropy. Special experiments were therefore made to check the effect of the window. These experiments failed to reveal any such influence.

b) Quality of the Polarizers

The question as to the quality of the polarizers is considered in [29]. The most widespread source of errors is the displacement of the beam by the rotating polarizer. However, when using free-selenium films a few microns thick as polarizer, this displacement is very slight. Various defects in the rotating polarizers may introduce an additional modulation of the light flux. However, the frequency of this modulation is half that of the signal, and by using a narrow-band amplifier the effect of such modulation may be eliminated. The degree of polarization produced by the polarizers is very important. We therefore paid special attention to the quality of the polarizers. For the polarizers used in the present investigation these parasitic effects were negligible.

c) Finite Signal/Noise Ratio

Conn and Eaton [132] indicated yet another possible source of systematic errors. If there is an error of either sign in determining the angle of incidence (or for our apparatus, the wavelength of the light) corresponding to zero signal, an error of one particular sign occurs in the determination of azimuth. The resultant azimuth values are therefore always too low. This error is associated with the finite nature of the signal/noise ratio and can thus only be substantial in the long-wave region. An estimate of the value of this error under our experimental conditions shows that in the range $\lambda \approx 8\text{-}12\ \mu$ in the worst case we have an error of ~0.1% in n and ~0.02% in $\varkappa$.

d) Illumination of the Samples with Light
of Wide Spectral Composition

In our apparatus the samples were illuminated with light over a wavelength range of ~0.7-15 μ. A certain proportion of the electrons in the samples were continually in the excited state, and in principle this might affect the optical constants. Estimates show that the number of excited electrons under our conditions was quite low (~10^7 per cm³). However, special experiments were made to check the influence of this effect. The optical constants of the metal in the range ~2-12 μ were measured both under ordinary conditions and with the use of filters cutting off the short-wave radiation. As filters we used germanium plates and MgO filters of various densities. A dense MgO filter in particular reduced the intensity of the light in the short-wave range by several hundreds of times. No influence of the effect in question was observed.

The absence of any influence of the illumination of the samples by light of a wide spectral composition on the optical constants was also confirmed by the results of measurements with the second apparatus. Parallel measurements of the optical constants at room temperature in the apparatus described in [34], in which light of a narrow spectral range fell on the sample, gave the same results for both tin and lead.

e) Low-Temperature Parasitic Signal

In measuring the optical constants of metals at low temperatures, there may be yet another source of systematic errors. If the modulation of the light beam is effected by a rotating analyzer or interrupter between the sample mirrors and receiver, then the parasitic signal mentioned earlier [41] arises as a result of the flow of radiation emerging from bodies at room temperature and then polarized by the sample mirrors; this is particularly important in the long-wave region and appears even with an ideal polarizer. By putting a monochromator between the analyzer and receiver, we may reduce this unwanted signal, but then we shall have to consider the polarization of the light by the monochromator. The parasitic low-temperature signal may be completely eliminated by placing the interrupter between the source and sample mirrors or using a system with a rotating polarizer situated in front of these mirrors. This latter arrangement was in fact used in our apparatus.

Consideration of the widespread sources of systematic errors shows that in our apparatus these were either completely removed or reduced to extremely small values. The absence of additional sources of systematic errors was verified by making parallel measurements of the optical constants of lead and tin at room temperature in two types of apparatus. The results obtained on our own design agreed with those obtained on the Motulevich—Shubin system [34].

5. Samples for Optical Investigations

The measurements of the optical constants and other characteristics were carried out with layers of tin and lead obtained by vacuum evaporation. Glass plates served as substrates. The layers for optical study were deposited on plates of different lengths, corresponding to particular angles of incidence of the light. Plates from 43 to 180 mm long were employed. The width of the plates was 30 mm. This enabled us to use working beams up to 24 mm in diameter without limitation. The thickness of the plates was 10 mm. The working surface of the plates was carefully polished (deviations from a plane were no greater than 0.5-0.6 μ). Special attention was paid to the surface cleanliness of the substrate, since, in the presence of the slightest traces of grease, salts, etc., the resultant lead or tin layers had spots or bluish tinges. Only samples without visible defects were used for the measurements.

The purity of the original metals was: tin, over 99.99%; lead, 99.999%. It should be noted that for original metals of this purity the quality of the evaporated samples was limited, not by chemical impurities, but by structural defects, crystallite boundaries, and so on.

Evaporation of both lead and tin was carried out from tantalum boats. Although the melting points of the two metals were fairly low (327° for Pb, 232°C for Sn), the evaporation temperatures were considerable (in order to secure an appreciable vapor tension [134] we required 720° for Pb and 1200° for Sn). The evaporation was carried out in a vacuum of $(4-7) \cdot 10^{-6}$ mm Hg. During the evaporation the vacuum under the belljar remained almost constant owing to the large volume of the jar and the high pumping speed (~500 liters/min). The substrates were placed at a distance of 25-30 cm from the evaporators. With this arrangement a reasonable uniformity of the coating was ensured. The two substrates were placed close to each other in order to make both samples the same. The substrates were at room temperature. Certain precautions were taken in the evaporation in order to avoid possible contamination of the samples. The tantalum boats were carefully cleaned, and were degassed by heating to temperatures higher than those employed in evaporating the working metals; the tin and lead were degassed by heating before evaporation; the upper layers of tin and lead were evaporated onto the screen; in preparing the samples only a limited proportion of the metal in the boat was evaporated.

The evaporating conditions, i.e., the temperature of the evaporator and the associated evaporation rate, the time of evaporation, the distance from the substrate, etc., were specially chosen. This was particularly important for tin. It is well known [134] that slow evaporation leads to the formation of very finely dispersed layers. However, with extremely rapid evaporation one obtains mat layers of tin; these scatter the light severely and have a poor reflection. In the optimum condition the evaporation time was 15-25 sec in order to obtain a layer ~1 μ thick.

All this enabled us to obtain fairly thick films with a very clean surface without any traces of mat structure or contamination. We shall see subsequently that the properties of the films employed coincided with those of the massive metal, i.e., our samples constituted polycrystalline layers with fairly large crystals. Clearly, evacuation in high vacuum is one of the best means of obtaining good surfaces for metals with low melting points.

6. Measurement of the Thickness, Density, Conductivity, Critical Temperature, and Hall Effect

At the same time as the samples intended for optical measurements, we prepared other samples for measuring the conductivity, density, and Hall effect. Thus, all these characteristics and optical constants were measured on completely identical layers of metal.

The samples for measuring the density, conductivity, and Hall effect were deposited on polished glass plates through special stencils.

1. For determining the density, we measured the weight and thickness of a layer of given area. The working layer was deposited onto the substrate in the form of a rectangle roughly 10 by 40 mm² in area. The area of the layer was measured to an accuracy considerably better than 1%. Weighing was carried out on a microbalance. For the thicknesses used, the weight of the tin samples was 1.5-3.5 mg, and that of the lead samples 3-7 mg. Weighing was carried out to an accuracy of 0.01 mg.

The sharp edge of the working layer passed roughly halfway through the substrate, forming a step. The height of this step gave the thickness of the layer. This was measured by an interference method. In order to avoid errors due to nonuniform phase differences (introduced by reflection from the glass and metal) when measuring the thickness, the whole plate was covered with an additional nontransparent layer of metal. (Or else the plate was covered with a layer of metal beforehand and the working layer was deposited on this. In this case, the substrate metal was the same as the main metal.) Figure 9 shows cross sections of samples prepared for thickness measurement.

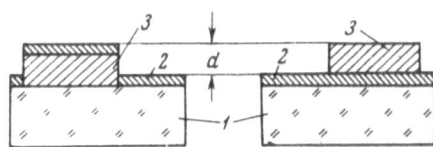

Fig. 9. Schematic section of a sample prepared for thickness measurement. d) Thickness of working layer; 1) glass substrate; 2) auxiliary layer of metal; 3) working layer.

The thickness of the layer was measured with an interference comparator. The arrangement of the measurements is indicated in Fig. 10. The interference comparator constituted a Michelson interferometer together with a monochromator. The interference was observed through the slit S_2. When working with a glow lamp the latter was placed near the slit S_1.

The plate with the deposited layer was placed in one of the arms of the interferometer. The plate formed a slight wedge with the virtual image of the standard interferometer mirror, so that the edge of this wedge was perpendicular to the boundary of the layer. In this way, a system of parallel lines (bands of equal thickness) was obtained. This system was

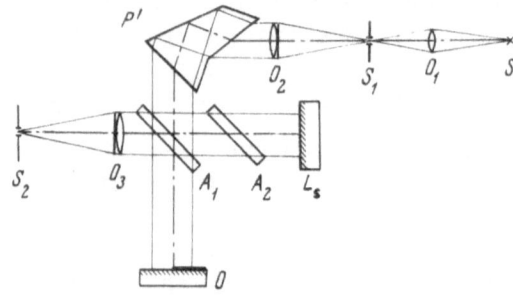

Fig. 10. Arrangement for thickness measurement. S)
source (helium lamp); O_1, O_2, O_3) objectives; P') constant-
deviation prism; A_1) plate with semitransparent coating;
A_2) compensating plate; L_S) standard mirror; O) sample.

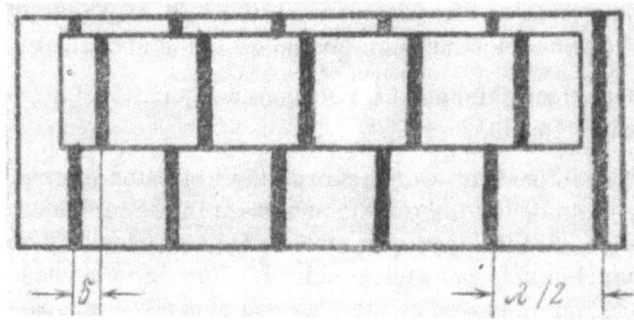

Fig. 11. System of bands of equal thickness used for determining the thickness of the samples (monochromatic
light); δ is the displacement between the two systems of
bands.

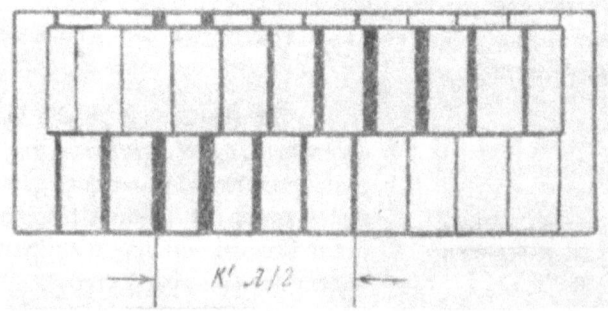

Fig. 12. System of bands of equal thickness obtained
for light of a wide spectral composition.

divided into two parts, displaced with respect to each other, by the boundary of the layer. The
value of the displacement characterized the thickness of the layer being measured. The form
of the interference picture thus obtained is illustrated in Fig. 11.

The distance between two neighboring dark lines corresponds to half a wavelength of the light employed. The thickness of the layers was greater than this quantity; hence, the displacement of the two systems of lines exceeded the distance between neighboring lines. In order to determine the value of this displacement, use might be made of measurements with different λ values. On using a helium lamp as source, six lines, from violet to red, could be used. In this case, for layers 0.5-1 μ thick, an accuracy of thickness measurements equal to 1 or 2% may easily be achieved. However, for our layers we were able to secure adequate accuracy, without requiring measurement at a single wavelength, by determining the whole number of bands from the interference picture obtained for light of a wide spectral composition. This greatly eased the measurements. The thickness of the layer is in this case given by $d = k'(\lambda/2) + \delta$, where k' is a whole number and δ is the fractional part of a half wavelength. The values of the displacement δ corresponding to fractions of a band were found with the help of the narrow spectral lines of the helium lamp. In this case, a displacement of $\frac{1}{10}$ of a band may be determined quite reliably. The whole number of bands k' may easily be reckoned by using a glow lamp as source. In this case, we used a wide slit so that the interference picture constituted a superposition of pictures corresponding to different λ. The line corresponding to the zero order of interference appeared white while those of the other orders were colored. For high orders of interference the whole picture became diffuse. The displacement of the zero-order lines gave the whole number of half waves incorporated in the thickness of our layer. Figure 12 shows the approximate form of the interference picture for the case k' = 4, $\delta = 0.3\lambda/2$.

The accuracy of determining the thickness was 3-4% for layers 0.6-0.9 μ thick; for thicker layers the accuracy was still higher. The error in measuring the density was determined by the error committed in measuring the thickness, being 2-4%. We investigated the dependence of the density on the thickness of the evaporated metal, preparing samples of various thicknesses for this purpose.

Within the limits of experimental accuracy there was no dependence of density on thickness for thicknesses of 0.5-1.2 μ.

2. The sample for measuring conductivity was a zigzag strip of metal ~0.5 mm wide and ~170 mm in total length. The resistance of such strips equalled 70-100 Ω for the metals under consideration at room temperature. Electrical contact with the deposited layer was established by means of Covar leads sealed into the glass.

For measuring the resistances we used bridge and potentiometric circuits. The use of the potentiometric circuit was particularly important at low temperatures, when the resistance of the sample was low. Sensitive galvanometers enabled us to make measurements with a low working current (20-50 μA or under). The low working currents were necessary, first, for measurements in the range of hydrogen temperatures, in view of the very strong dependence of the resistance of tin and lead on temperature in this region and, secondly, for measuring resistance in the neighborhood of the critical temperature.

The resistance of the samples was measured at room temperature and the temperatures of liquid nitrogen, hydrogen, and helium. For lead samples the resistance was measured as a function of temperature from the hydrogen temperature to the critical point. The accuracy in measuring the resistance was better than 1%. The samples had direct contact with the cooling liquid; hence, their temperature was determined by reference to the vapor pressure of the liquid in question. Temperatures below 4.2°K required for measuring the T_{cr} of tin were obtained by pumping out helium vapor and measured by reference to the vapor tension. Near the point at which the tin passed into the superconducting state, the accuracy of temperature measurement was about 0.01°. In order to obtain temperatures above 4.2°K when measuring the T_{cr} of lead and the temperature dependence of the resistance of lead, the samples were heated in helium vapor. The sample could be raised above 4.2°K by means of a heater. In this way, the whole

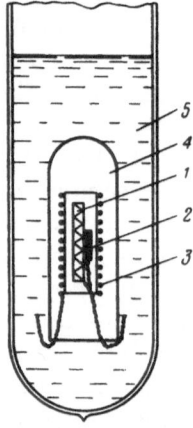

Fig. 13. Arrangement of a lead sample in helium for measuring the conductivity near the critical temperature. 1) Sample; 2) carbon thermometer; 3) heater; 4) vessel containing helium vapor; 5) liquid helium.

temperature range up to 20°K was covered. The manner in which the sample was kept in the helium is shown in Fig. 13.

For the thickness of the layer employed in the resistance measurements we used the value obtained when determining the density. This value gave the true thickness, since both substrates were situated at the same distance from the evaporator and very close to each other. In addition to this, we made control experiments in which we measured the thickness of the layer employed in the resistance measurements directly. There were no discrepancies.

From the measured values of resistance, thickness, length, and width of the layer, we found the specific conductivity. The accuracy of determining the specific conductivity was determined by the accuracy of measuring the thickness; it equalled 3–4%. We made some control experiments to study the effect of thickness on the specific conductivity. For layers 0.5–1.2 μ thick, no such effect was observed.

3. We determined the transition of the metal into the superconducting state by reference to the jump in resistance, employing the same samples as those used for determining the conductivity. In tin the transition occurred in the neighborhood of 3.8°K, while in lead it set in at 7.3°K. The temperature was measured with a carbon thermometer. The critical temperature was in both cases taken as the temperature close to the critical point (R_n) at which the resistance fell to half its value in the normal state. From the temperature/resistance curve in the neighborhood of the critical temperature T_{cr} we estimated the range of temperatures in which the transition into the superconducting state took place.

4. Measurement of the Hall effect is important in that it enables us to draw certain conclusions regarding the correspondence between the zone structure of the samples employed and the massive metal. The point is that the Hall effect is extremely sensitive to changes in the shape of the Fermi surface.

The sample for measuring the Hall effect was a strip of metal 60 mm long and 4 mm wide. The leads for measuring the Hall emf were arranged at distances of 30 mm from the current leads. Special measures were taken to ensure that the potential leads should as far as possible constitute equipotentials. The remaining slight parasitic voltage drop was compensated. The Hall effect was measured in a steady magnetic field of 8 kOe. The Hall emf was measured with a dc photoelectric amplifier with a sensitivity of 10^{-9} V/mm. We used a current density of between 1 and 50–70 A/mm² through the samples. The sign of the magnetic field was changed during the measurements. In view of the smallness of the Hall effect for lead, and particularly for tin, we had to use relatively high current densities and very sensitive amplifiers. In order to eliminate parasitic variable thermal emf we had to pay special attention to the thermostating of the whole apparatus. As control measurements for checking the apparatus we measured the Hall effect in copper samples. These samples constituted 0.1-mm thick foils. The resultant value of the Hall field agreed with the tabulated value [135] to an accuracy of 5%.

The error committed in determining the Hall field in lead was 10% and in tin, 30%. It should be noted that this large error was obtained after averaging the values obtained for different samples. In measuring the Hall emf for a particular sample, the error was much smaller.

The large spread in the values of the Hall field for different samples (particularly for tin) indicates that the Hall effect is very sensitive to small changes in the conditions of sample preparation. The other characteristics were not so sensitive in this respect.

CHAPTER III

RESULTS OF THE MEASUREMENTS

1. Results of Measuring the Optical Constants of Tin

Figures 14 and 15, and Table 1, give the optical constants of tin found from formulas (II.9), allowing for the dependence of the surface impedance on the angle of incidence of the light. The maximum value of the correction term was about 1% in the range $\lambda = 0.7$-$1\ \mu$, 0.3% in the range $\lambda = 2\ \mu$, and negligible in the range $\lambda > 2\ \mu$. The results given constitute the averages of many sets of measurements. Figures 14 and 15 give the total dependence of n and $\varkappa$ on λ, and also the short-wave region, shown separately. Measurements of the optical constants were carried out at three temperatures, 293, 78, and 4.2°K. The temperature dependence of n is considerable over the whole spectral range; the temperature dependence of $\varkappa$ is only large in the range $\lambda = 8$-$12\ \mu$. In the range $\lambda = 2$-$5\ \mu$, $\varkappa$ varies only slightly with temperature.

The error in determining n was 1-2% at room temperature and the temperature of liquid nitrogen, and 1-3% at helium temperature. The error in $\varkappa$ was 0.5-1 and 1-2%, respectively. The maximum accuracy was achieved in the central part of the spectral range, in the region ~3-10 μ. At the ends of the range studied, and in the neighborhood of the absorption band (~1-2 μ) the accuracy fell somewhat. For all λ, the error in $\varkappa$ was much smaller than that in n. The error was determined from the spread in the values obtained for different occasions of deposition. The error obtained in each individual series was still smaller.

Before this the optical constants of tin in the infrared had only been measured at room temperature [34, 83]. In these earlier papers no data were given regarding the other parameters of the layers studied. The results obtained in [34] only refer to the spectral range 1.3-6.3 μ. In this range the results agree very well with our own data. In [83], the author was apparently only interested in the qualitative dependence of $\varkappa^2 - n^2 + 1$ and $n\varkappa/\lambda$ on λ. The results were given in the form of a graph of $\log(\varkappa^2 - n^2 + 1)$ and $\log(n\varkappa/\lambda)$ as functions of $\log\lambda$, and n and $\varkappa$ can only be determined from these to a low accuracy. Hence, we can only report qualitative agreement between [83] and our own investigation as regards the dependence of n and $\varkappa$ on λ. In addition to this, the results of [83] relate to a single semitransparent layer.

At low temperatures only one quantity had earlier been measured for tin, the absorption (absorbing power) A. This quantity was measured in [20] at helium temperature, a black body at room temperature serving as source. The light was hardly monochromatized at all. The effective wavelength was 14 μ, and the absorbing power A = 1.24%. The measurements were made on a cast and electropolished tin sample. Agreement with our own results (A = 1.21° for the region 10-12 μ) was very good. Our values of the absorbing power (for normal incidence of the light) $A = \frac{4n}{(n+1)^2 + \varkappa^2}$ and the reflecting power of tin $R = 1 - A = \frac{(n-1)^2 + \varkappa^2}{(n+1)^2 + \varkappa^2}$ for three temperatures are shown in Fig. 16. The reflection coefficient of tin in the infrared part of the spectrum at helium temperatures reaches 99%. In Fig. 16 we notice the presence of a sharp

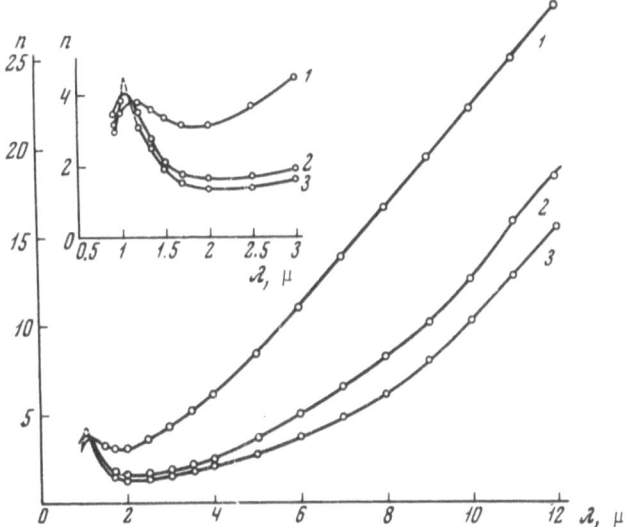

Fig. 14. Refractive index of tin as a function of wavelength
for various temperatures. T = 293, 78, and 4.2°K for
curves 1, 2, and 3, respectively.

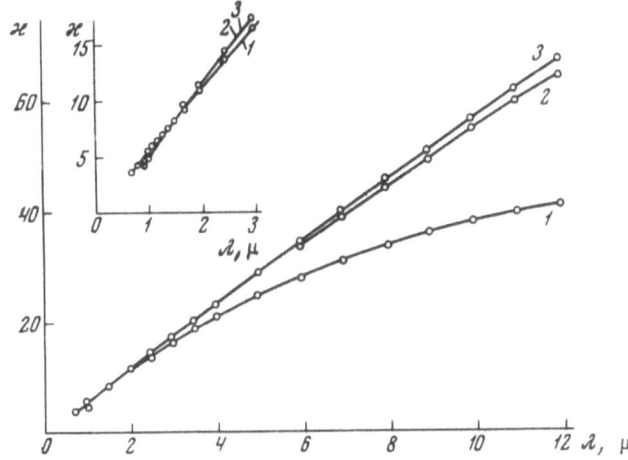

Fig. 15. Absorption index of tin as a function of wavelength
at various temperatures. Notation as in Fig. 14.

maximum for the absorbing power of tin in the region of 1-1.2 μ; this is associated with the internal photoeffect.

2. Results of Measuring the Optical

Constants of Lead

Figures 17 and 18 and Table 2 show the optical constants of lead, also obtained from formulas (II.9). The correction for the dependence of the surface impedance on the angle of incidence for lead was about 2% for $\lambda < 1\,\mu$, 0.5% for $\lambda \approx 1.5\,\mu$, and negligible for $\lambda \geq 2\,\mu$. The results were obtained by averaging many series of measurements. Figures 17 and 18 show the total dependence of n and $\varkappa$ on λ, while the short-wave part is shown separately. The measurements were made at 293, 78, and 4.2°K.

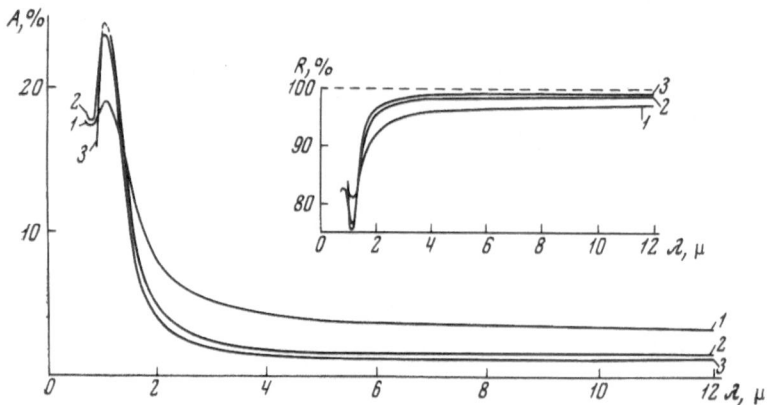

Fig. 16. Absorbing and reflecting powers of tin as functions of
wavelength at various temperatures. Notation as in Fig. 14.

Table 1. Optical Constants of Tin

λ, μ	293° K		78° K		4.2° K		λ, μ	293° K		78° K		4.2° K	
	n	$\varkappa$	n	$\varkappa$	n	$\varkappa$		n	$\varkappa$	n	$\varkappa$	n	$\varkappa$
0.73	2.18	6.29	2.24	6.19	—	—	3.5	5.27	20.5	2.13	21.1	1.95	21.1
0.8	2.40	6.62	2.27	6.42	—	—	4.0	6.19	23.2	2.46	24.2	2.13	24.2
0.93	3.15	7.28	3.43	7.17	2.95	7.62	5.0	8.49	28.5	3.75	29.7	2.75	30.0
0.99	3.44	7.34	3.92	6.94	3.70	7.15	6.0	11.0	33.1	4.97	35.5	3.73	35.8
1.2	3.76	7.63	3.53	6.45	3.05	5.98	7.0	13.8	37.1	6.51	41.4	4.89	41.6
1.35	3.57	8.04	2.76	6.99	2.55	6.64	8.0	16.6	40.6	8.17	47.0	6.05	47.4
1.5	3.31	8.67	2.09	7.98	1.99	7.80	9.0	19.3	43.8	10.0	51.7	7.90	53.3
1.7	3.13	9.88	1.75	9.29	1.51	9.35	10.0	22.0	46.4	12.4	55.8	10.1	58.7
2.0	3.10	11.8	1.65	11.4	1.38	11.4	11.0	24.8	49.0	15.7	59.8	12.6	63.4
2.5	3.63	14.8	1.69	14.6	1.39	14.6	12.0	27.8	51.6	18.2	63.8	15.3	67.0
3.0	4.41	17.8	1.88	18.0	1.58	18.0							

Table 2. Optical Constants of Lead

λ, μ	293° K		78° K		4.2° K		λ, μ	293° K		78° K		4.2° K	
	n	$\varkappa$	n	$\varkappa$	n	$\varkappa$		n	$\varkappa$	n	$\varkappa$	n	$\varkappa$
0.7	1.68	3.67	—	—	—	—	3.0	4.27	16.4	1.53	17.3	0.81	17.3
0.8	1.51	4.24	1.05	3.90	—	—	3.5	5.39	18.6	2.01	20.4	1.10	20.1
0.9	1.44	4.85	0.97	4.50	0.81	4.30	4.0	6.58	20.8	2.48	22.9	1.49	23.1
1.0	1.41	5.40	0.87	5.12	0.68	4.94	5.0	9.04	24.8	3.99	28.7	2.15	28.6
1.1	1.42	5.97	0.82	5.73	0.58	5.60	6.0	11.7	28.1	5.41	33.9	2.95	34.4
1.2	1.46	6.53	0.69	6.35	0.465	6.24	7.0	14.1	30.9	7.16	38.7	3.75	39.9
1.3	1.51	7.12	0.63	7.02	0.345	6.90	8.0	16.4	33.6	8.82	43.9	4.50	45.5
1.4	1.59	7.67	0.595	7.69	0.28	7.54	9.0	18.7	35.8	10.5	49.1	5.56	50.6
1.5	1.67	8.24	0.595	8.35	0.27	8.19	10.0	21.0	37.4	12.3	54.4	6.70	55.9
1.7	1.90	9.37	0.65	9.64	0.29	9.51	11.0	23.2	39.2	14.4	59.1	7.90	61.3
2.0	2.28	11.1	0.78	11.4	0.34	11.5	12.0	24.6	40.5	16.3	63.5	9.20	66.5
2.5	3.20	13.7	1.09	14.3	0.53	14.3							

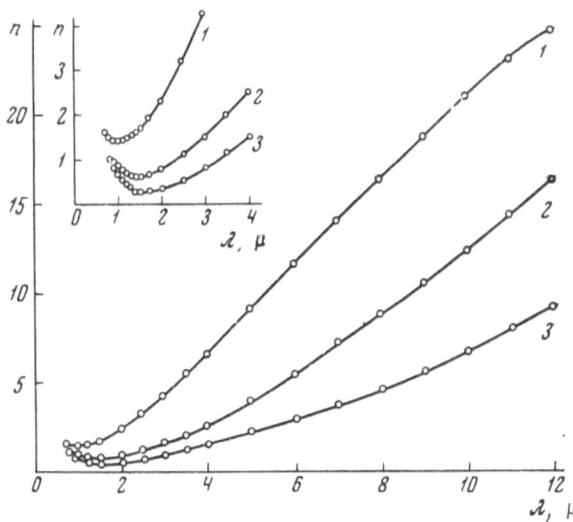

Fig. 17. Refractive index of lead as a function of wave-length at various temperatures. Notation as in Fig. 14.

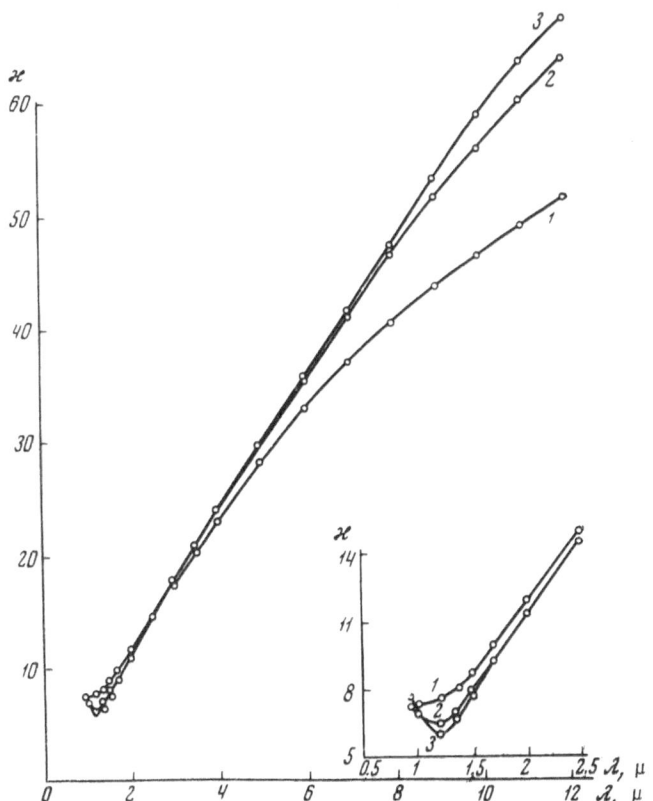

Fig. 18. Absorption index of lead as a function of wave-length at various temperatures. Notation as in Fig. 14.

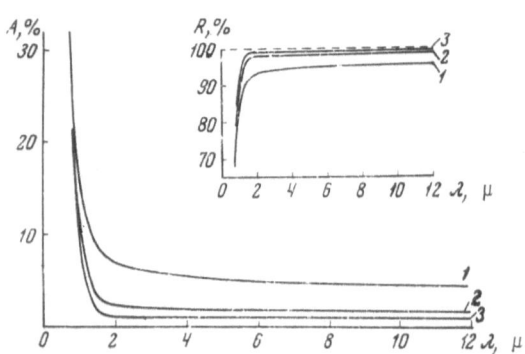

Fig. 19. Absorbing and reflecting powers of lead as functions of wavelength. Notation as in Fig. 14.

The temperature dependence of the optical constants of lead was reminiscent of that of tin.

The error in determining the optical constants of lead was 1-2% in n and 0.5-1% in $\varkappa$ at room and nitrogen temperatures, and 2-4% in n and 1-2% in $\varkappa$ at helium temperature. The maximum accuracy in determining n was achieved in the spectral range ~3-10 μ. At the ends of the wavelength range studied the error in n increased slightly. The maximum error in n at helium temperature reached 5% for $\lambda \approx 1.5\ \mu$. The accuracy in determining $\varkappa$ fell slightly at the very ends of the spectral range studied. In lead, as in tin, the main error arose from the averaging of the different series of measurements.

Earlier, the optical constants of lead had only been measured in the infrared at room temperature over the range 1.3-7 μ [34]. A comparison of our own results for this spectral range with those of [34] shows agreement for $\varkappa$ and a slight discrepancy for n. In [34] the values of n were slightly higher, indicating that our samples were rather better. There were no data relating to the electric and other properties of the samples in [34].

At low temperatures only the absorbing power of lead has previously been measured [20]. In [20] the absorbing power of cast and electropolished lead was measured at helium temperature. A black body at room temperature was used as source, with an effective radiation wavelength of 14 μ; this gave a value of A = 1.15%. Our own value of A = 0.83% in the range 10-12 μ suggests that the surfaces of our lead samples were of a better quality.

Figure 19 shows our values of the absorbing power of lead for normal incidence (A) together with the reflecting power R. The absorption coefficient of lead at helium temperature in the infrared exceeds 99%.

3. Effect of Oxidation and Annealing
on Optical Constants

It is well known [12] that the optical properties of metals may be considerably affected by various surface conditions: mechanical cold working, layers of chemical compounds, oxide layers. When studying metals deposited on a polished substrate in vacuo the most dangerous effect of this kind is the possible oxidation of the surface layer. In the present investigation we therefore measured the optical constants of both tin and lead immediately after deposition. The effect of oxidation was also specially studied. Investigation showed that the results of measurements made on the optical constants of tin and lead after remaining for a day in either vacuum or air were exactly the same as those made immediately after deposition. Slight changes in optical constants only set in after several days. It should be noted that after finishing the adjustment of the apparatus the sample mirrors remained in a helium atmosphere for the whole experiment at room temperature. In general, the oxidation of good tin and lead surfaces occurred extremely slowly. Mirrors kept for four months in air were no different in external form from freshly prepared ones. All this indicates that oxide films had no effects on the results of our experiments.

For tin mirrors we also checked the effect of annealing on the optical constants. The mirrors were heated in vacuum to 150 or 200°C for several hours. No change was observed in the optical constants of the tin samples after annealing.

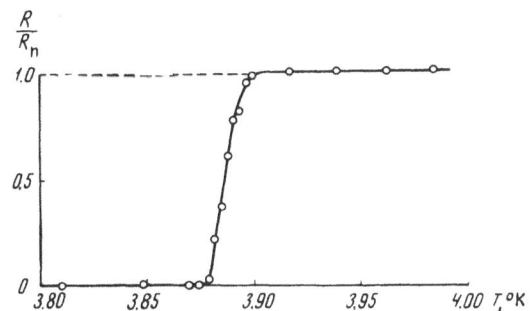

Fig. 20. Resistance of a tin sample as a function of temperature near T_{cr}.

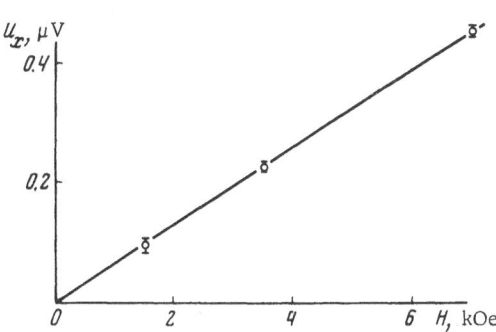

Fig. 21. Hall emf for a tin sample as a function of the magnetic field.

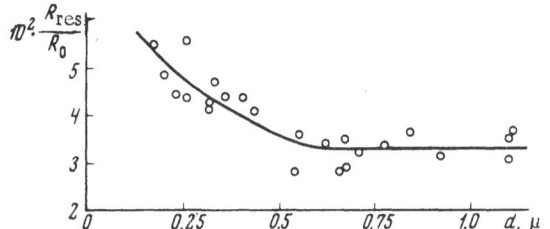

Fig. 22. Residual resistance of tin samples as a function of thickness.

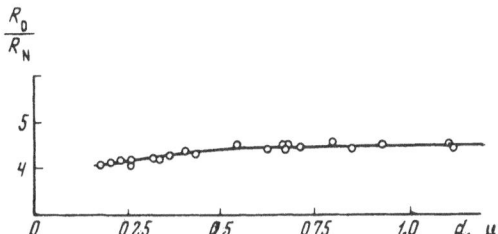

Fig. 23. Resistance ratio of tin samples at room temperature and nitrogen temperature as a function of thickness.

Table 3. Some Characteristics of Tin Layers

Characteristics	Layers used	Massive metal	Characteristics	Layers used	Massive metal
d, μ	0.5—1.1	∞	R_{res}/R_0	$3.5 \cdot 10^{-2}$	$1 \cdot 10^{-3}$
ρ, g/cm^3	7,2	7.28	T_{cr}, °K	3.88	3.73
σ_0, CGSE	$7.1 \cdot 10^{16}$	$7.8 \cdot 10^{16}$	R_x, $\mu\Omega \cdot$ cm/kOe	$+5.0 \cdot 10^{-4}$	$-4.1 \cdot 10^{-4}$
σ_N, CGSE	$3.1 \cdot 10^{17}$	$4.0 \cdot 10^{17}$	θ, °K	187	189
σ_H, CGSE	$1.6 \cdot 10^{18}$	$6.7 \cdot 10^{18}$			

4. Results of Measuring the Density, Conductivity, Critical Temperature, and Hall Effect of Tin

The results of our experimental determination of the electrical and other parameters of the tin layers employed, as well as the values of the same parameters for the massive metal taken from handbooks [135, 137-139], are given in Table 3.

The first line shows the thickness of the layers used, d. No dependence of the parameters studied on thickness was observed for the values of d indicated. This shows that no further increase in thickness is required.

Up to an accuracy of 1%, the density ρ of our layers coincided with that of the massive metal. Since the accuracy of measuring the density of the layers was rather under 1%, the value obtained may be regarded as coinciding with the density of the massive metal. The indicated density value was obtained by averaging the results of many experiments.

The values of conductivity σ_0, σ_N, and σ_H in the table respectively relate to the temperatures 293, 78, and 20.3°K.

The resistance of the tin layers fell 4.5 times on cooling to nitrogen temperature and 22.5 times on cooling to hydrogen temperature. The conductivity of our layers was: at room temperature, 91%, and at nitrogen temperature, 78% of the conductivity of the massive metal. The conductivity at liquid-hydrogen temperature was 22% of that of the massive metal at the same temperature. It should be noted that, according to handbooks [136–139], the values given for the conductivity of the massive metal differ by 20–30%. Table 3 gives the most reliable data. The value of the temperature-dependent part of the specific resistance of the layers under consideration is close to that of the massive metal at all temperatures. The slight difference is probably associated with the slightly lower Debye temperature of our samples as compared with the Debye temperature of massive tin. The residual resistance R_{res} in the table is given in the form of a ratio with respect to the resistance R_0 at 293°K, since this ratio was measured to a higher accuracy (~1%) than the absolute value of the conductivity. The residual resistance of tin was measured at liquid-helium temperature. The value remains constant between 4.2°K and the temperature corresponding to the transition into the superconducting state.

The critical temperature of the transition into the superconducting state T_{cr} was 3.88 ± 0.02°K for tin, and was independent of thickness, even for samples less than 0.2 μ thick. The critical temperature was close to the T_{cr} of the massive metal. The transition into the superconducting state was very sharp for the layers studied (ΔT = 0.01–0.03°; in the massive metal, ΔT = 0.01°), which indicated that the homogeneity of the samples employed was excellent. A typical curve representing the variation in sample resistance near T_{cr} is given in Fig. 20.

The value of the Hall constant of tin $R_X = E_X/jH$ gives the value of the Hall field E_X for a current density of j = 1 A/cm^2 and a magnetic field of H = 1 kOe. The plus sign indicates that the Hall effect of the evaporated tin samples has a hole character. The magnitude of the effect is extremely small, which corresponds to almost complete compensation of the effects respectively associated with the electrons and "holes," as in the massive metal. The samples studied had a slight excess of "hole" influences, whereas, in the massive metal (polycrystalline material) the electron influence was slightly greater. In tin single crystals the sign of the effect differs for different directions, although its magnitude remains just as small [135, 140–141]. In all cases the effect is almost vanishingly small. The magnitude of the Hall constant was independent of the current density over the whole range of current densities studied. The variation in the Hall field with varying magnetic field was also studied. Right up to 7 kOe the Hall field was proportional to the magnetic field (see Fig. 21, which gives the Hall emf U_X).

As indicated in the foregoing, none of the parameters found depended on thickness over the range studied. Layers thinner still, however, did show such a variation for a number of parameters. Figures 22 and 23 indicate the residual resistance of tin and the resistance ratio R_0/R_N as functions of thickness at 293 and 78°K. We see from the figures that at small thicknesses the residual resistance increases, while the ratio R_0/R_N falls. The variation ends at $d \approx 0.4$–0.5 μ. In order to eliminate any such influence of thickness, all our optical constants were measured with thicker layers.

The resultant characteristics of the tin layers used were very close to those of the massive metal. Since the thickness of the layers studied exceeded the depth of the skin layer and the electron free path by a factor of several tens, from the optical point of view the layers employed corresponded to massive specimens. Hence, all the optical results and the corresponding conclusions correspond to the massive metal. The slight deviation of the measured electrical

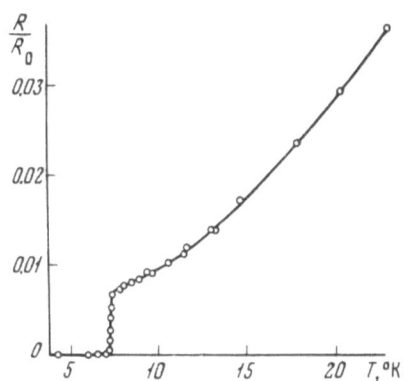

Fig. 24. Resistance of a lead sample as a function of temperature.

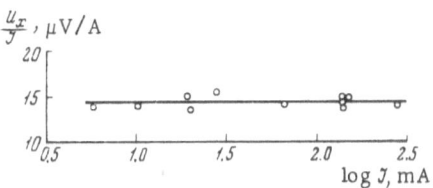

Fig. 25. Ratio of the Hall emf of a lead sample to the working current for various currents.

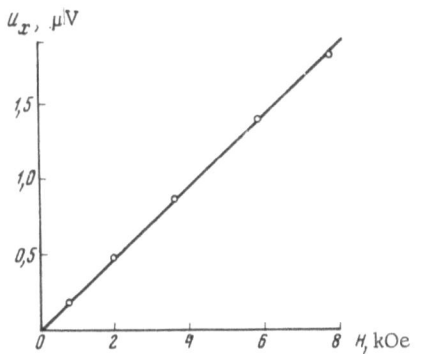

Fig. 26. Hall emf of a lead sample as a function of the magnetic field.

parameters of our layers from tabulated values simply indicates the necessity for a complex investigation of the properties of the material.

5. Results of Measuring the Density, Conductivity, Critical Temperature, and Hall Effect of Lead

The results obtained on determining the electrical and other characteristics of our lead layers, and also the corresponding values for the massive metal taken from handbooks [135-139] are given in Table 4 (the notation here agrees with that used in Table 3).

We studied samples from 0.6 to 1.6 μ thick. The parameters were independent of thickness for the layers examined. The majority of the measurements were therefore made on samples 0.6-1.2 μ thick. The density of our layers coincided with that of the massive metal. The final result was obtained by averaging the measurements for a very large number of samples.

Table 4 presents the values of σ_0, σ_N, σ_H, and also the resistance R_n of the samples at a temperature close to the critical value. The resistance of the lead layers studied fell by a factor of 4.15 on cooling to nitrogen temperature and by a factor of 34 on cooling to hydrogen temperature. The conductivity of our layers agreed with that of the massive metal at 293 and 78°K. At 20.3°K, the difference was about 7%.

Figure 24 shows the temperature dependence of the lead layers employed between 7.3 and 20°K. We see that at 7.3°K the residual resistance has not yet been reached. The value of R_n was 0.6% of R_0. An estimate of the residual resistance by extrapolation of the temperature/resistance curve to 0°K gives a value of $R_{res} \approx 5 \cdot 10^{-3} R_0$.

The value of the critical temperature obtained for our layers practically coincides with T_{cr} for the massive metal.

Table 4. Some Characteristics of Lead Layers

Characteristics	Layers used	Massive metal	Characteristics	Layers used	Massive metal
d, μ	0.6—1.6	∞	R_n/R_0	$6 \cdot 10^{-3}$	$1 \cdot 10^{-3}$
ρ, g/cm³	11,3	11,3	R_{res}/R_0	$\sim 5 \cdot 10^{-3}$	$\sim 10^{-3}$
σ_0 CGSE	$0.41 \cdot 10^{17}$	$0.41 \cdot 10^{17}$	T_{cr}, °K	7.3	7.2
σ_N, CGSE	$1.70 \cdot 10^{17}$	$1.70 \cdot 10^{17}$	R_x, $\mu\Omega \cdot$ cm/kOe	$+1.5 \cdot 10^{-3}$	$+1.0 \cdot 10^{-3}$
σ_H, CGSE	$1.4 \cdot 10^{18}$	$1.5 \cdot 10^{18}$	θ, °K	85	86

For the layers studied the Hall effect had a hole character, as in the case of the massive metal [135]. The value of the Hall constant for lead is also extremely small, indicating a considerable compensation between electron and "hole" effects. The value of R_x for lead, as indicated in Fig. 25, was independent of current density j right up to j = 70 A/mm². (Figure 25 gives the Hall emf U_x divided by the current I through the sample. The sample was 0.97 μ thick. The magnetic field H = 7 kOe.) The Hall field was proportional to the magnetic field H to a high accuracy up to H ~ 8 kOe (see Fig. 26, which gives the U_x/H relationship for the same sample; the working current I = 112 mA).

Table 4 also shows the Debye temperature, the determination of which will be indicated subsequently.

The characteristics thus obtained for the layers in question agreed with those obtained for the massive metal. Hence, as in the case of tin, all the optical data and the conclusions derived from these apply to massive samples.

CHAPTER IV

ANALYSIS OF EXPERIMENTAL RESULTS

1. Obtaining Microcharacteristics

The optical data obtained for tin and lead enable us to find their most important characteristics: the concentration of conduction electrons, the electron/phonon collision frequency, and the velocity of the electrons on the Fermi surface. We may also determine the Fermi energy, the density of levels near the Fermi surface, and the area of the Fermi surface for both these metals. These microcharacteristics are associated with other electronic properties of metals as well as their optical properties: the conductivity, electron specific heat, thermal conductivity, etc. Hence, these phenomena may also be used to determine them. It should be noted, however, that a great advantage of the optical method of determining the principal microcharacteristics is the fact that they may be obtained separately. On measuring the electron part of the specific heat, the conductivity, the surface absorption at radio frequencies, and other properties of metals, the separation of N and ν demands either additional experimental data or an assumption regarding the specific relationship between the quantities in question. In addition to this, the formulas giving the relationship between N and the parameters of the phenomena indicated usually depend much more on the particular model representations employed than do the optical formulas [12].

We see from Figs. 14-16 for tin, and 17-19 for lead, that the internal photoeffect has a considerable influence on the optical properties of these metals in the short-wave part of the spectrum. The quantities n and $\varkappa$ vary nonmonotonically with λ and an absorption band appears. An evaluation of the influence of the internal photoeffect will be given subsequently. Here we simply note that for $\lambda \gtrsim 2\text{-}2.5 \ \mu$ the effect of the internal photoeffect is negligibly small, so that we may consider that in this range the optical properties are determined by the conduction electrons. Hence, in order to obtain the microcharacteristics of tin and lead associated with the conduction electrons we used the spectral range between 3 and 12 μ. It will be shown later that in this spectral range both tin and lead show a weakly anomalous skin effect at all temperatures. Hence, in order to determine the microcharacteristics, we must use formulas (I.8).

The concentration of conduction electrons N and the electron collision frequency ν are explicitly separated in formulas (I.8), while the quantities β_1 and β_2 play the part of correction terms. This kind of description is justified by the fact that in the case of the normal skin effect, $\beta_1 \approx \beta_2 \approx 0$, and we obtain the usual normal skin-effect formulas.

Most frequently, $\varphi_1 < \varphi_2$ and $\beta_1 < \beta_2$, so that the correction is much more important in determining ν than in determining N. The quantity β_1 characterizes the difference between N and the electron concentration obtained from the formulas of the normal skin effect. Since, as indicated earlier, the quantity ε and the associated quantity N depend relatively weakly on the character of the skin effect, the correction β_1 is small in practically all cases. The quantity β_2 characterizes the surface losses, which are proportional to $\beta_2 \nu^{ef}$. The volume losses are proportional to ν defined by the second formula in (I.8). This formula loses its meaning when β_2 becomes of the order of unity, i.e., the skin effect becomes sharply anomalous.

The determination of the microcharacteristics N, ν, v and the quantities β_1 and β_2 for the two metals was carried out by reference to formulas (I.8) on the basis of the n and $\varkappa$ values obtained, using the method of successive approximations. Equations (I.8) may be solved exactly; however, for our particular case, this is more complicated. In the zero approximation we put $\beta_1 = \beta_2 = 0$ and determined the zero approximations for N and ν. These quantities were used for determining the first approximations to the corrections terms β_1 and β_2 with an assigned parameter v. Then formulas (I.8) were used to find the first approximations for N and ν and the second approximations for β_1 and β_2, etc. The third approximation practically coincided with the second. Being a parameter, the velocity v of the electrons on the Fermi surface was chosen in such a way as to make the right-hand side of the first expression in (I.8) independent of λ. For a more reliable determination of v, the method of least squares was employed.

By using the value found for N and the results of measuring the density ρ, we may obtain the number of conduction electrons associated with one atom:

$$\frac{N}{N_a} = \frac{NA}{N_0\rho}.$$

Here A is the atomic weight and N_0 is Avogadro's number.

Knowing the values of N and v, we may determine the density of levels near the Fermi surface [12]:

$$\left(\frac{dY}{dW}\right)_F = \frac{3}{2}\frac{N}{mv^2}. \tag{IV.1}$$

The electrical parameters of tin and lead measured at the same time as the optical parameters enable us to obtain a number of additional microcharacteristics of these metals. An important characteristic of the metals subsequently required for verifying theoretical conclusions regarding the temperature dependence of the electron collision frequency is the Debye temperature Θ. This characteristic was determined for our samples from the measured temperature dependence of the static conductivity. The characteristic temperature Θ_R thus obtained may differ from the Debye temperature Θ_D obtained by measuring the specific heat at low temperatures, although still being quite close to the latter [106]. The Gurzhi theory considered earlier, however, incorporates precisely this temperature Θ_R, which should thus be determined from the $\sigma(T)$ relationship. (In what follows this will simply be called the Debye temperature.) Calculation of this temperature was effected in the following way. The reduced resistance of the metal [106, 142] is

$$r = 1.056\frac{T}{\Theta}F\left(\frac{\Theta}{T}\right),$$

where $F(\Theta/T)$ is a function tabulated by Grüneisen. For the temperature range T_N to T_0 we obtain

$$\frac{r(T_N)}{r(T_0)} = \frac{T_N}{T_0}\frac{F\left(\dfrac{\Theta}{T_N}\right)}{F\left(\dfrac{\Theta}{T_0}\right)} = \frac{R_N - R_{res}}{R_0 - R_{res}}.$$

Here, R_0 and R_N are the real resistances for temperature T_0 and T_N, respectively, while R_{res} is the residual resistance. From this relation and the tables given in [106, 142] the temperature Θ may be found by the method of successive approximations, using the known T_0, T_N, R_0, R_N, and R_{res}. We determined the Debye temperature for the ranges 78 to 293 and 20.3 to 78°K.

As indicated in Section 3 of Chapter I, the collision frequency ν incorporates the electron—phonon collision frequency ν^{ef}, the electron—defect collision frequency ν^{ed}, and the interelectron collision frequency ν^{ee}, i.e., $\nu = \nu^{ef} + \nu^{ed} + \nu^{ee}$. The frequency ν found from optical data, as indicated by the solution of the quantum kinetic equation set out in Section 4 of Chapter I, in general differs from the static electron collision frequency ν^{st} which enters into the static conductivity and equals $\nu^{st} = \nu^{ef}_{cl} + \nu^{ed} + \nu^{ee}_{cl}$. The difference between these should be particularly substantial at low temperatures.

The set of optical data and the results obtained by measuring the static conductivity enable us to determine the classical electron-collision frequencies ν^{ef}_{cl} and ν^{ed}. The classical electron—phonon collision frequency ν^{ef}_{cl} for a given temperature and the electron—defect collision frequency ν^{ed} may be found from the relations

$$\left.\begin{aligned} \nu^{ef}_{cl} + \nu^{ed} &= \frac{e^2}{m}\frac{N}{\sigma(\Gamma)}, \\ \frac{\nu^{ed}}{\nu^{ef}_{cl} + \nu^{ed}} &= \frac{R_{res}}{R(\Gamma)}. \end{aligned}\right\} \tag{IV.2}$$

Here $\sigma(T)$ and $R(T)$ are the specific conductivity and resistance for a given temperature T. The electron concentration N is obtained from optical measurements.

In the expressions (IV.2) no account is taken of the classical interelectron collision frequency ν^{ee}_{cl}, since, as will be shown later, the effect of this on the static conductivity is negligibly small.

2. Microcharacteristics of Tin

The values of the quantities N, ν, β_1, and β_2 obtained for various wavelengths λ in the case of tin are presented in Table 5. The values of the microcharacteristics N, N/N_a, ν, ν^{ef}, the velocity of the electrons on the Fermi surface v, and the frequency ν^{ef}_{cl}, averaged over the range 3–12 μ, are given for three temperatures in Table 6.

The Debye temperature Θ found for tin equals 187°K between 78–293°K and 128°K between 20.3–78°K. The values of Θ obtained from the temperature dependence of the conductivity of massive tin are about 200°K for the range 78–293°K and 130–135°K for the range 20.3–78°K. The value of Θ was determined for a large number of samples. The error determined from the scatter in the values was ±3°K for both ranges.

The Debye temperature of our tin samples varied quite considerably with temperature. The same picture held for massive tin [106], for which there was a minimum value of Θ in the range 10–15°K, where the Debye temperature reached 127°K. Below 5°K the Debye temperature rose again to between 180 and 190°K.

A discussion of the results obtained for other microcharacteristics will be presented subsequently. Here we shall simply consider the accuracy of their determination. The errors in determining the microcharacteristics N and ν were obtained from the formulas

$$\begin{aligned} \Delta N &= \frac{\partial N}{\partial n}\Delta n + \frac{\partial N}{\partial \varkappa}\Delta \varkappa, \\ \Delta \nu &= \frac{\partial \nu}{\partial n}\Delta n + \frac{\partial \nu}{\partial \varkappa}\Delta \varkappa, \end{aligned}$$

Table 5. Values of N, ν, β_1, and β_2 for Various Temperatures and Wavelengths λ in the Case of Tin

λ, μ	293° K				78° K				4.2° K			
	$N \cdot 10^{-22}$, cm^{-3}	$\nu \cdot 10^{-14}$, sec^{-1}	$\beta_1 \cdot 10^2$	$\beta_2 \cdot 10^2$	$N \cdot 10^{-22}$, cm^{-3}	$\nu \cdot 10^{-14}$ sec^{-1}	$\beta_1 \cdot 10^2$	$\beta_2 \cdot 10^2$	$N \cdot 10^{-22}$, cm^{-3}	$\nu \cdot 10^{-14}$, sec^{-1}	$\beta_1 \cdot 10^2$	$\beta_2 \cdot 10^2$
2.5	4.69	3.81	0.40	3.34	3.97	1.63	0.18	7.82	3.91	1.33	0.12	8.0
3.0	4.73	3.19	0.48	3.97	4.16	1.19	0.19	10.7	4.12	0.99	0.13	10.7
3.5	4.70	2.84	0.58	4.37	4.20	1.05	0.20	12.8	4.17	0.88	0.16	11.9
4.0	4.66	2.59	0.68	4.77	4.22	0.83	0.22	14.8	4.19	0.71	0.16	14.4
5.0	4.75	2.36	0.93	5.17	4.14	0.83	0.34	14.6	4.13	0.58	0.19	17.2
6.0	4.76	2.26	1.2	5.26	4.16	0.76	0.44	15.8	4.11	0.54	0.26	18.0
7.0	4.79	2.24	1.6	5.11	4.22	0.73	0.58	16.4	4.12	0.53	0.33	18.6
8.0	4.79	2.24	1.9	4.91	4.24	0.71	0.72	16.8	4.13	0.50	0.40	19.5
9.0	4.78	2.23	2.2	4.75	4.15	0.71	0.91	16.6	4.20	0.52	0.55	18.8
10.0	4.77	2.26	2.5	4.48	4.06	0.76	1.2	15.4	4.23	0.55	0.73	17.8
11.0	4.83	2.30	2.9	4.24	4.11	0.84	1.6	13.8	4.21	0.60	0.96	16.6
12.0	5.00	2.36	3.2	3.98	4.10	0.87	2.0	12.3	4.11	0.64	1.2	15.1

Table 6. Microcharacteristics of Tin

Micro-characteristics	293° K	78° K	4.2° K
N, cm^{-3}	$4.8 \cdot 10^{22}$	$4.2 \cdot 10^{22}$	$4.15 \cdot 10^{22}$
N/N_a	1.30	1.1_4	1.1_2
ν, sec^{-1}	$2.26 \cdot 10^{14}$	$0.76 \cdot 10^{14}$	$0.56 \cdot 10^{14}$
ν^{ef}, sec^{-1}	$2.20 \cdot 10^{14}$	$0.70 \cdot 10^{14}$	$0.50 \cdot 10^{14}$
ν^{ef}_{cl}, sec^{-1}	$1.64 \cdot 10^{14}$	$0.28 \cdot 10^{14}$	—
, cm-sec^{-1}	$0.93 \cdot 10^8$	$1.0 \cdot 10^8$	$0.85 \cdot 10^8$
ν^{ed}, sec^{-1}	$6 \cdot 10^{12}$	$5 \cdot 10^{12}$	$5 \cdot 10^{12}$

Table 7. The Parameter $|\gamma|$ for Tin

λ, μ	293° K		78° K		4.2° K	
	$|\gamma|$	$|\gamma|^2$	$|\gamma|$	$|\gamma|^2$	$|\gamma|$	$|\gamma|^2$
2	0.012	$1.4 \cdot 10^{-4}$	0.014	$2.0 \cdot 10^{-4}$	0.012	$1.4 \cdot 10^{-4}$
3	0.019	$3.6 \cdot 10^{-4}$	0.022	$4.8 \cdot 10^{-4}$	0.019	$3.6 \cdot 10^{-4}$
5	0.029	$8.4 \cdot 10^{-4}$	0.036	$1.3 \cdot 10^{-3}$	0.032	$1.0 \cdot 10^{-3}$
10	0.038	$1.4 \cdot 10^{-3}$	0.066	$4.4 \cdot 10^{-3}$	0.060	$3.6 \cdot 10^{-3}$
12	0.038	$1.4 \cdot 10^{-3}$	0.072	$5.2 \cdot 10^{-3}$	0.067	$4.5 \cdot 10^{-3}$

where N and ν are understood to mean the quantities in expressions (I.8). By using the val
obtained for the errors in n and $\varkappa$, we may find the corresponding errors in N and ν; at roo
and nitrogen temperatures, in the middle of the spectral range studied, these reached 1-2%
the ends of the range they reached 3%, and at helium temperature, 2-3%. The maximum er
in N was about 4%, and in ν about 5%. The error in N averaged rather less than the error
this was because the former was mainly determined by the error in $\varkappa$ and the latter by tha

Expression (I.8) was obtained on the assumption that terms of the order $1/|\varepsilon'|$ could be neglected in comparison with unity. In the part of the spectrum considered for tin, $1/|\varepsilon'| \sim 3 \cdot 10^{-3} - 2 \cdot 10^{-4}$, and this neglect may be regarded as justified.

Table 7 gives the absolute values of the parameter γ for various wavelengths and three temperatures. We see from the table that for tin this parameter is in fact small over the whole spectral range studied and at all temperatures. In the very worst case the error due to the neglect of terms quadratic in γ is no greater than a few hundredths of a percent. Thus, the error in obtaining the microcharacteristics is determined by the accuracy of the experiments and not the approximations made in deriving formulas (I.8).

3. Microcharacteristics of Lead

For lead, the microcharacteristics N, ν, β_1, and β_2 corresponding to various wavelengths are given in Table 8. The microcharacteristics N, N/N_a, ν, ν^{ef}, the electron velocity v, and the frequency ν^{ef}_{cl}, are averaged over the range 2.5-12 μ for three temperatures in Table 9.

Since the temperature dependence of the resistance is the same for our samples as for massive lead, the values of Θ obtained from this relationship also coincide. On determining Θ we obtained values of $85 \pm 1.5°$K. The calculation was made for a large number of samples and the error determined from the spread in these values. We determined the temperature Θ for various temperature ranges, from room temperature to ~8°K. The Θ of lead depended very little on temperature. The value given for Θ is the average taken over the whole range from 8 to 293°K. It should be noted that, since $T \gg \Theta$ for lead at room temperature, the resistance should be proportional to T. However, there is a slight deviation from this law, in that a term $\sim T^2$ becomes appreciable [142]. This deviation may be associated partly with interelectron collisions, and partly with anharmonicity and other causes [106]. Hence, in order to determine Θ [142], we must separate out the linear term in the R(T) relationship. The correction for nonlinearity in lead is only important at $T \sim 300°$K and equals a few percent. The value of Θ for our samples was close to that of the Θ derived from specific-heat data [106].

Table 8. Values of N, ν, β_1, and β_2 for Various Temperatures and Wavelengths λ in the Case of Lead

λ, μ	293° K				78° K				4.2° K			
	$N \cdot 10^{-22}$, cm^{-3}	$\nu \cdot 10^{-14}$, sec^{-1}	$10^2 \cdot \beta_1$	$10^2 \cdot \beta_2$	$N \cdot 10^{-22}$, cm^{-3}	$\nu \cdot 10^{-14}$, sec^{-1}	$10^2 \cdot \beta_1$	$10^2 \cdot \beta_2$	$N \cdot 10^{-22}$, cm^{-3}	$\nu \cdot 10^{-14}$, sec^{-1}	$10^2 \cdot \beta_1$	$10^2 \cdot \beta_2$
1.5	3.85	5.12	0.18	2.2	3.56	1.64	0.06	8.0	3.38	0.73	0.02	11.2
1.7	3.87	4.52	0.20	2.6	3.67	1.35	0.07	9.7	3.54	0.58	0.02	14.0
2.0	3.90	3.91	0.24	2.9	3.68	1.15	0.08	11.2	3.70	0.46	0.02	17.3
2.5	3.95	3.61	0.34	3.2	3.72	1.02	0.10	12.5	3.66	0.46	0.03	17.1
3.0	4.09	3.40	0.46	3.3	3.80	0.98	0.15	13.0	3.73	0.49	0.05	16.4
3.5	4.06	3.30	0.59	3.4	3.91	0.94	0.19	13.8	3.71	0.50	0.07	16.3
4.0	4.08	3.22	0.72	3.4	3.79	0.90	0.24	14.1	3.77	0.51	0.09	15.9
5.0	4.10	3.09	0.99	3.4	3.89	0.92	0.40	13.9	3.72	0.47	0.13	16.9
6.0	4.13	3.10	1.3	3.2	3.86	0.89	0.56	14.2	3.75	0.45	0.17	17.8
7.0	4.07	3.06	1.6	3.0	3.80	0.90	0.74	13.8	3.73	0.42	0.21	18.8
8.0	4.03	2.98	1.8	3.0	3.82	0.86	0.91	14.4	3.72	0.38	0.24	20.4
9.0	4.02	2.98	2.1	2.8	3.84	0.81	1.1	15.1	3.66	0.37	0.30	20.4
10.0	4.04	3.08	2.4	2.6	3.89	0.77	1.3	15.8	3.65	0.36	0.36	20.7
11.0	4.08	3.13	2.6	2.4	3.90	0.78	1.5	16.0	3.66	0.36	0.42	21.1
12.0	3.88	3.04	2.8	2.4	3.86	0.75	1.7	16.3	3.64	0.35	0.49	21.3

The errors relating to the microcharacteristics N and ν of lead were determined in the same way as for tin. Over the greater part of the spectral range studied the error at room temperature and nitrogen temperature was 1-2% and at helium temperature 2-4%. At the ends of the range the error increased to between 2-3% and between 3-5%, respectively. For lead the error in N was rather smaller than the error in ν.

The quantity $1/|\varepsilon'|$ for lead in the part of the spectrum studied was $4 \cdot 10^{-3}$ to $2 \cdot 10^{-4}$. This is much smaller than unity and may justly be neglected.

The parameter $|\gamma|$ and the quantity $|\gamma|^2$ for lead are given in Table 10 for a number of wavelengths and three temperatures. We see from the table that for lead also the parameter γ is small over the whole spectral range under consideration and at all temperatures. The error due to the neglect of the following terms in γ is no greater than a few tenths of a percent in the very worst case. For lead as for tin, the accuracy of the determination of the microcharacteristics is determined by the experimental accuracy and not the approximations made in deriving formulas (I.8).

4. Character of the Skin Effect
for Tin and Lead

The character of the skin effect for a given metal is determined by the relation between the characteristics of the electrons (free path, velocity, etc.) and the characteristics of the electromagnetic field in the metal (depth of penetration, frequency, wavelength, etc.). This relation may be expressed in different ways; however, an essential condition determining the character of the skin effect is the ratio of surface absorption to volume absorption of light in the metal. The normal skin effect is operative when the surface losses are negligibly small compared with the volume losses; a sharply anomalous skin effect is operative in the opposite case.

Table 9. Microcharacteristics of Lead

Micro-characteristics	293° K	78° K	4.2° K
N, cm^{-3}	$4.0 \cdot 10^{22}$	$3.8 \cdot 10^{22}$	$3.7 \cdot 10^{22}$
N/N_a	1.2_1	1.1_3	1.1_2
ν, sec^{-1}	$3.06 \cdot 10^{14}$	$0.86 \cdot 10^{14}$	$0.42 \cdot 10^{14}$
ν^{el}, sec^{-1}	$3.05 \cdot 10^{14}$	$0.85 \cdot 10^{14}$	$0.40 \cdot 10^{14}$
ν''_{cl}, sec^{-1}	$2.50 \cdot 10^{14}$	$0.56 \cdot 10^{14}$	—
v, cm-sec^{-1}	$0.9 \cdot 10^8$	$1.0 \cdot 10^8$	$0.71 \cdot 10^8$
ν^{el}, sec^{-1}	$1 \cdot 10^{12}$	$1 \cdot 10^{12}$	$1 \cdot 10^{12}$

The question as to the conditions governing the particular character of the skin effect was considered in the accepted form in Section 1 of Chapter I [expression (I.7)]. In the theory of the weakly anomalous skin effect [19] these conditions are expressed in terms of the smallness of the corrections β_1 and β_2 in the normal skin effect. It should be noted that, since ε and N depend very little on the character of the skin effect, the related quantity β_1 should be small in all cases. The smallness of β_1 is thus rather a criterion for the applicability of the assumed representations to the metal in question. The quantity β_2 is the essential criterion for

Table 10. Parameter $|\gamma|$ and Value of $|\gamma|^2$ for Lead

λ, μ	293° K		78° K		4.2° K	
	$|\gamma|$	$|\gamma|^2$	$|\gamma|$	$|\gamma|^2$	$|\gamma|$	$|\gamma|^2$
2	0.012	$1.4 \cdot 10^{-4}$	0.014	$2.0 \cdot 10^{-4}$	0.010	$1.0 \cdot 10^{-4}$
3	0.014	$2.0 \cdot 10^{-4}$	0.021	$4.5 \cdot 10^{-4}$	0.015	$2.3 \cdot 10^{-4}$
5	0.023	$5.3 \cdot 10^{-4}$	0.035	$1.2 \cdot 10^{-3}$	0.025	$6.4 \cdot 10^{-4}$
10	0.025	$6.2 \cdot 10^{-4}$	0.064	$4.2 \cdot 10^{-3}$	0.049	$2.4 \cdot 10^{-3}$
12	0.024	$6.0 \cdot 10^{-4}$	0.074	$5.4 \cdot 10^{-3}$	0.058	$3.4 \cdot 10^{-3}$

Table 11. Relaxation Times, Free Paths of the Electrons, and Depths
of the Skin Layer in Tin

λ, μ	293° K			78° K			4.2° K		
	$\tau \cdot 10^{14}$, sec	$l \cdot 10^6$, cm	$\delta \cdot 10^6$, cm	$\tau \cdot 10^{14}$, sec	$l \cdot 10^6$, cm	$\delta \cdot 10^6$, cm	$\tau \cdot 10^{14}$, sec	$l \cdot 10^6$, cm	$\delta \cdot 10^6$, cm
1	0.04	0.04	2.2	0.03	0.03	2.2	0.04	0.03	2.2
1.5	0.09	0.08	2.8	0.15	0.15	3.0	0.15	0.13	3.1
2	0.19	0.18	2.7	0.38	0.38	2.8	0.45	0.38	2.8
3	0.31	0.29	2.7	0.84	0.84	2.7	1.0	0.85	2.7
5	0.42	0.39	2.8	1.2	1.2	2.7	1.7	1.4	2.6
8	0.45	0.42	3.1	1.4	1.4	2.7	2.0	1.7	2.7
10	0.44	0.41	3.4	1.3	1.3	2.9	1.8	1.5	2.7
12	0.42	0.39	3.7	1.2	1.2	3.0	1.6	1.4	2.9

the character of the skin effect. This quantity gives the proportion of surface losses with respect to volume losses. If, in fact, we use the expression for β_2 from (I.8), putting in the values
of angles φ_1 and φ_2, we easily obtain

$$\beta_2 = \frac{3}{8} \frac{v}{c} \frac{n^2 + \varkappa^2}{2n} \frac{1 + \frac{n}{\varkappa} \frac{v}{\omega}}{1 - v^2/\omega^2}.$$

Since, in the region of present interest, we usually have $n/\varkappa \lesssim 1$, $v/\omega \lesssim 1$ for metals, we find
that

$$\beta_2 \approx \frac{3}{8} \frac{v}{c} \frac{n^2 + \varkappa^2}{2n} \approx \frac{\frac{3}{4} \frac{v}{c}}{A},$$

where $A \approx 4n/(n^2 + \varkappa^2)$ is the absorbing power of the metal. The quantity $\frac{3}{4}$ (v/c), of course,
characterizes the surface losses [3, 12]. For all conditions $\beta_2 \geq 0$. For the sharply anomalous
skin effect $\beta_2 \to 1$. As noted earlier, in this case we must consider the next order in expressions (I.8).

It follows from Tables 5 and 8 that the skin effect is weakly anomalous at all temperatures
for both tin and lead in the part of the spectrum studied. The maximum value of β_2 occurs at
helium temperatures, where it reaches 15-20% for both metals. At room temperatures the value
of β_2 is 4-5% for tin and 2.5-3% for lead. In this case, the skin effect is almost normal. Thus,
for both metals we must use the formulas of the weakly anomalous skin effect when determining
the microcharacteristics. Smallness of the quantities β_1 and β_2 indicates great reliability in
determining the microcharacteristics from formulas (I.8).

In order to use the ordinary criteria for determining the character of the skin effect [con-
dition (I.7)], we must find the free path of the electrons l and the depth of the skin layer δ.

The free path of the electrons $l = v\tau = v/\nu$, where τ is the electron relaxation time. Using the values found for the collision frequencies, we obtain the value of l for tin and lead at
various temperatures and wavelengths (Tables 11 and 12). For the long-wave part of the spectrum the value of l is independent of λ. Table 13 gives the average values of l for this region;
Tables 11-13 contain the relaxation times τ.

The depth of the skin layer $\delta = c/\omega\varkappa$. The values of δ found for tin and lead at various λ
and three temperatures are given in Tables 11 and 12. The quantity δ depends rather weakly on
λ and temperature. The mean values of δ for the long-wave region are given in Table 13.

Table 12. Relaxation Times, Free Paths of the Electrons, and Depths
of the Skin Layer in Lead

λ, μ	293° K			78° K			4.2° K		
	$\tau \cdot 10^{14}$, sec	$l \cdot 10^6$, cm	$\delta \cdot 10^6$, cm	$\tau \cdot 10^{14}$, sec	$l \cdot 10^6$, cm	$\delta \cdot 10^6$, cm	$\tau \cdot 10^{14}$, sec	$l \cdot 10^6$, cm	$\delta \cdot 10^6$, cm
1	0.10	0.09	2.9	0.16	0.16	3.1	0.20	0.14	3.2
1.5	0.20	0.18	2.9	0.61	0.61	2.9	1.4	1.0	2.9
2	0.26	0.24	2.9	0.87	0.87	2.8	2.2	1.6	2.8
3	0.29	0.26	2.9	1.0	1.0	2.8	2.0	1.4	2.8
5	0.32	0.29	3.2	1.1	1.1	2.8	2.1	1.5	2.8
8	0.34	0.31	3.8	1.2	1.2	2.9	2.6	1.8	2.8
10	0.32	0.29	4.3	1.3	1.3	2.9	2.8	2.0	2.8
12	0.33	0.30	4.7	1.3	1.3	3.0	2.9	2.1	2.9

Table 13. Mean Values of l, δ, l/δ, and τ

Quantity	Tin			Lead		
	293° K	78° K	4.2° K	293° K	78° K	4.2° K
$l \cdot 10^6$, cm	0.40	1.3	1.5	0.29	1.2	1.8
$\delta \cdot 10^6$, cm	3.1	2.8	2.7	3.8	2.9	2.8
l/δ	0.13	0.46	0.56	0.08	0.40	0.64
$\tau \cdot 10^{14}$, sec	0.43	1.3	1.8	0.32	1.2	2.5

Table 14. Losses of Electromagnetic Radiation

Metal	293° K			78° K			4.2° K		
	A_f	A_Σ	A_d	A_f	A_Σ	A_d	A_f	A_Σ	A_d
Sn	0.93	0.05	0.02	0.79	0.15	0.06	0.75	0.17	0.08
Pb	0.97	0.03	0.00	0.84	0.15	0.01	0.79	0.19	0.02

We see from these tables that for both metals and all λ, $l < \delta$. However, whereas at room temperature $l/\delta \sim 0.1$, at low temperatures l is of the order of δ and the ratio l/δ exceeds $\frac{1}{2}$. Thus the first of conditions (I.7) is not satisfied, at any rate for nitrogen and helium temperatures. Since, in the part of the spectrum studied, $\varkappa$ is several times greater than n, this is the decisive factor. This may easily be verified by comparing the real and imaginary parts of the first term in Eq. (I.1). Hence, although the second condition in (I.7) is satisfied, it is not a criterion for the normal nature of the skin effect.

This consideration also shows that for both metals the skin effect is weakly anomalous at nitrogen and helium temperatures and almost normal at room temperature. In addition to this, even for $l/\delta \sim 0.1$, the anomaly correction (the coefficients β_1 and β_2) is about 4%. Hence, for the present case we should use the formulas of the weakly anomalous skin effect if the accuracy of the measurements exceeds the correction in question.

The resultant values of l and δ enable us to confirm the conclusion of Sections 4 and 5 in Chapter III to the effect that our samples are "optically massive." In fact, for the thicknesses d employed, the ratio d/$\delta \approx 20$–60. The attenuation of the light in the thickness of our samples equals e^{20}-e^{60}, or $\sim 10^{10}$-10^{25}. On the other hand, $l/d \sim 0.03$-0.005, and the existence of the second boundary of the layer in no way affects the electrons interacting with the light.

The set of optical and electrical measurements gives us the collision frequencies of electrons with impurities, defects, crystallite boundaries, etc.

$$\nu^{ed} = \frac{e^2}{m}\,\frac{N}{\sigma_0}\,\frac{R_{res}}{R_0}\,.$$

The values calculated for tin are given in Table 6, and for lead in Table 9. The temperature variation of ν^{ed} for tin is associated with the variation of N. We see that ν^{ed} is much smaller than the fundamental electron-collision frequencies for both metals, i.e., the collisions of electrons with impurities, defects, etc., have practically no influence on the optical properties of our samples.

Since we used chemically pure samples for the measurements, the frequency ν^{ed} is mainly associated with layer defects and crystallite boundaries. Hence, the results obtained enable us to estimate the dimensions of these crystallites

$$L \sim v/\nu^{ed}.$$

These dimensions are equal to $L \sim 0.15$-$0.2\ \mu$ for tin and $L \sim 0.6$-$0.8\ \mu$ for lead. Thus, for both metals the samples are of a polycrystalline nature, being made up of coarse crystals.

In Table 14 we compare the different losses of the electromagnetic field in our metals: A_f are associated with the electron/phonon interaction, A_Σ are surface losses, A_d are associated with the presence of impurities and defects. The surface losses, as indicated earlier, equal $\beta_2 A$, while $A_d : A_f = \nu^{ed} : \nu^{ef}$. The values of the losses are given in proportions of the total light absorption A. As values of A_f and A_Σ we take the averages over the range $\lambda = 4$-$12\ \mu$ for tin and $\lambda = 3$-$12\ \mu$ for lead. For lead at room temperature we have $A_d \approx 4 \cdot 10^{-3}$.

5. Concentration of Conduction Electrons

One of the principal quantities characterizing the electronic properties of metals is the concentration of conduction electrons N. The optical method of determining the electron concentration is one of the most reliable [12], while obtaining N from optical data is a most important problem in metal optics.

The physical meaning of the quantity N was given detailed consideration in [12]. It was shown in [12, 143] that the formula relating the value of N to the dielectric constant ε is in fact independent of model representations and is obtained in both the classical and quantum cases.

For tin the conduction-electron concentrations found at various λ are given in Table 5 and Fig. 27, and the mean values of N and N/N_a for various temperatures are given in Table 6. These results for lead are presented in Table 8 and Fig. 28, and in Table 9, respectively.

First of all we should note that for both metals N remains constant over a wide spectral range. As already noted, the relation between N and the optical constants depends very little on the character of the skin effect. It is precisely this which makes the determination of N from optical data so reliable. Hence, in general, we may determine N from n and $\varkappa$, not only by means of formulas (I.8), but also by means of the formulas of the normal or anomalous skin effects. For our case, the resultant N values differ by a few percent. This may clearly be seen from the values of the correction coefficients β_1, which range from a few tenths of a percent to a few percent in the case of both metals over the part of the spectrum studied.

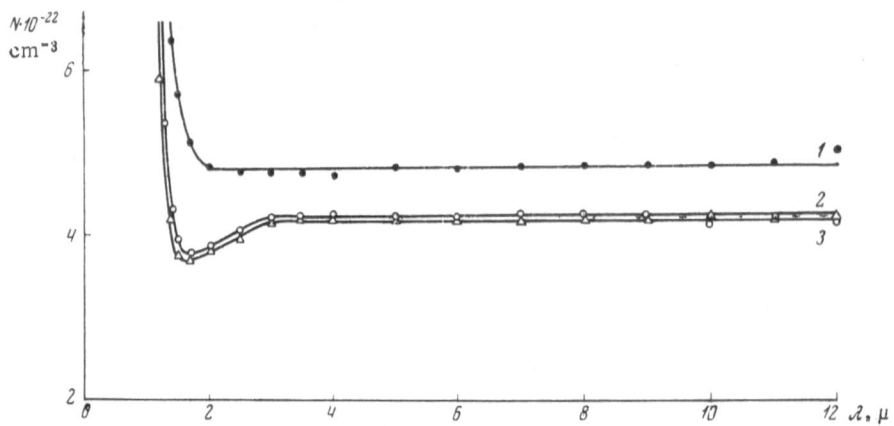

Fig. 27. Conduction-electron concentration of tin at various temperatures. Notation as in Fig. 14.

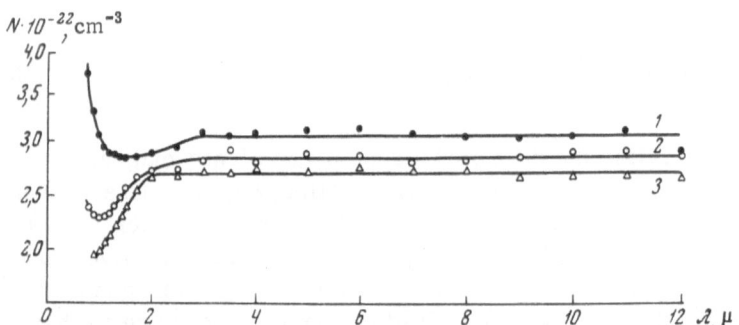

Fig. 28. Conduction-electron concentration of lead at various temperatures. Notation as in Fig. 14.

We see from Figs. 27 and 28 that in the short-wave part of the spectrum, $\lambda < 2\text{-}3\ \mu$, the value of N found from formula (1.8) starts depending on λ. In this part of the spectrum the internal photoeffect exerts a strong influence on the optical properties of tin and lead. Experimental data enable us to establish the boundaries of this region. This enables us to use only the data relating to the long-wave region in order to obtain the characteristics associated with the conduction electrons.

Our values of conduction-electron concentration may be compared with the values obtained by Chambers [105, 106] when measuring the surface impedance at radio frequencies. Chambers' values of $N/N_a = 1.12 \pm 0.14$ for tin and 1.24 ± 0.14 for lead agree closely with ours.

The concentration ratio N/N_a was determined in [144, 145] from measurements of resistance in thin tin and lead wires. The resultant values for tin of $N/N_a = 1.2\text{-}1.25$ agreed closely with our own data. For lead the value was $N/N_a = 0.64$, much lower than our own result.

These results show that both for tin and lead there is quite a wide divergence of frequencies in the infrared spectrum in which the concentration of conduction electrons may be determined. This conclusion is not trivial; it indicates that for these metals, despite the complexity of their Fermi surfaces, the ideas developed for "good" metals are still applicable.

The conduction-electron concentrations found for tin and lead vary with temperature as in Figs. 27 and 28 and Tables 6 and 9. For lead the variation in N with temperature is very slight and falls within the limits of measuring accuracy. However, the sign of the increment ΔN is preserved for all experiments and all wavelengths. This suggests that for lead there is a slight fall in N with decreasing temperature. For tin the concentration N changes more sharply. On reducing the temperature from 293°K to nitrogen temperature, the N of tin changes by 12.5%, which is much greater than experimental error. There is hardly any change in concentration on passing from nitrogen to helium temperature.

The reduction in concentration with falling temperature is only found in metals for which overlapping of the energy bands takes place. Both in tin and lead there is an even number of valence electrons. In the absence of band-overlapping the electrons would fill the last band completely and these elements would be dielectrics. In other words, the Fermi surface of these metals lies in several Brillouin zones. In the monovalent metals Cu, Au, and Ag, for which the Fermi surface lies in the first zone and there is no overlapping of bands, there are no changes in conduction-electron concentration with falling temperature [59-60].

At the present time there is no theoretical explanation for the phenomenon observed. We consider that the following may be a possible explanation. The change in the concentration of conduction electrons in tin and lead is associated with a change in the zone structure as temperature diminishes. Tin, in particular, has a deformed diamond lattice. Elements of the same subgroup, germanium, silicon, and α-tin, having a diamond lattice, are semiconductors. The metallic properties of β-tin and lead are determined by the overlapping of the energy bands. This overlapping evidently changes with changing temperature. The change in the width of the energy gap and concentration of free carriers with changing temperature is well known in the case of semiconductors [146-148]. For semiconductors having a lattice of the diamond type, germanium and silicon, this change is quite considerable. Experiments [149-152] show that for germanium and silicon the change in the energy gap reaches 10-12% on reducing the temperature from the room value to that of liquid helium. It should be noted [152] that the main change in the energy parameters of the zone structure takes place in the range ~300 to 100°K. The character of the changes in these parameters in elements with the diamond lattice and in the majority of other semiconductors is such that the energy gap widens, the overlapping of the bands diminishes, and the concentration of free carriers falls with falling temperature [146, 149, 153]. It is well known [99, 154] that this is one of the reasons for the fall in the conductivity of semiconductors at lower temperatures. Thus, in metals of the tin and lead type, a weak "semiconductor" type of temperature dependence of the resistance is superimposed on a stronger "metallic" type. Hence, it is natural to expect a change in the zone structure of tin and lead with changing temperature.

Our optical data in the short-wave region suggest precisely this kind of change in the parameters of the zone structure with temperature for tin (and evidently for lead). The long-wave boundary of the absorption band moves in the short-wave direction (small λ) with falling temperature, indicating a widening of the energy gap responsible for this band, i.e., a reduction in the overlapping of the bands as temperature falls.

Causes of the changes in zone-structure parameters in semiconductors with temperature include their volume expansion [146, 153] and a change in the lattice vibrations with temperature [155]. Calculations of the relation between the energy bands of elements with the diamond structure [156, 157] and the lattice constant show that for these elements a fall in lattice constant widens the energy gap and thus reduces the concentration of conduction electrons [146]. Evidently, the same mechanisms are responsible for the observed changes in the conduction-electron concentration of tin and lead.

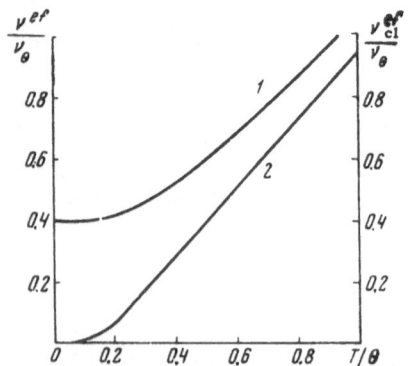

Fig. 29. Theoretical relation between
the frequency of electron—phonon collisions and the reduced temperature.
1) Relation between ν^{ef}/ν_Θ and T/Θ;
2) relation between ν^{ef}_{cl}/ν_Θ and T/Θ.

It should be noted that the optical study of metals
makes it fairly easy to observe changes in electron
concentration with temperature, since it enables one
to obtain the value of N for each temperature of interest. Other experiments either demand the use of low
temperatures (measurements of electron specific heat,
cyclotron resonance, van Alphen—de Haas effect, etc.)
or else require a set of data taken over a range of
temperatures in order to determine N (measurements
of surface impedance in the radio range), so that only
an average value of N is obtained.

6. Experimental Proof of the Gurzhi—Holstein Theory

The frequency of electron—phonon collisions obtained from optical data does not in general coincide
with the classical electron—phonon collision frequency.
As indicated in Section 4 of Chapter I, the electron—phonon collision frequency in the presence
of a high-energy light quantum $\nu^{ef} = \nu_\Theta(T/\Theta)\varphi(T)$, where $\varphi(T)$ may be found from (I.12) or
(I.13). At high temperatures $(T \gg \Theta)\varphi(T) \to 1$ with increasing T and $\nu^{ef} \to \nu^{ef}_{cl}$. For $T \to 0$,
$\varphi \to 2\Theta/5T$ and $\nu^{ef} \to \frac{2}{5}(\nu_\Theta)$; the classical frequency $\nu^{ef}_{cl} \sim T^5$. Thus, in contrast to the classical collision frequency, ν^{ef} remains finite right down to absolute zero, its value being quite con-
siderable.

This result and the form of the $\varphi(T)$ relationship were obtained by Gurzhi [100, 112]. The
limiting expression for ν^{ef} at low temperatures was obtained a little earlier by Holstein [158].
The temperature dependence of ν^{ef}/ν_Θ is shown in Fig. 29 (curve 1). The same figure shows
the classical relation for ν^{ef}_{cl}/ν_Θ (curve 2). Curves 1 and 2 are approximately linear at high
temperatures. The qualitative difference in the behavior of the curves only sets in for $T/\Theta \approx$
0.4. (We remember that the frequency ν^{ef} is related to the volume absorption of light in the
metal, which is determined by the electron—phonon interaction.)

The reasons for the disagreement between ν^{ef} and ν^{ef}_{cl} at low temperatures lie in the difference between the absorption mechanisms at different frequencies. The absorption of light by
the electrons of a metal is brought about by the generation or absorption of phonons, since only
on this condition may the laws of conservation of energy and momentum be satisfied. In the classical case (low frequencies $\hbar\omega \ll kT$, steady current), at low temperatures the electrons cannot
absorb "large" phonons, of which there are hardly any $(kT \ll k\Theta)$. The electrons do not have
enough energy for the emission of such phonons. Hence there is weak absorption, falling off
rapidly with diminishing temperature $(\sim T^5)$. For electromagnetic quanta of high energy, particularly when the conditions $\hbar\omega \gg k\Theta \gg kT$ are satisfied, there is a considerable deviation
from the classical case. In this case the electrons, having absorbed the quanta, pass out a long
way beyond the limits of the Fermi surface. As before, they cannot absorb "large" phonons but,
having greater energy, they may generate phonons over the whole spectrum. Absorption will
remain substantial right down to absolute zero temperature.

At high temperatures $(T \gg \Theta)$ the absorption of light in both cases will be determined both
by the generation and by the absorption of phonons from the whole spectrum; this will lead to a
high electron—phonon collision frequency and $\nu^{ef} \approx \nu^{ef}_{cl}$.

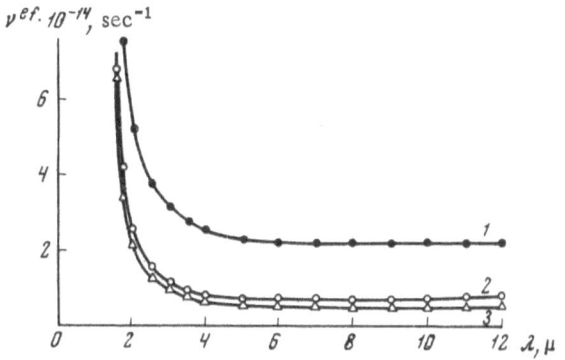

$\nu^{ef} \cdot 10^{-14}$, sec^{-1}

Fig. 30. Frequency of electron—phonon collisions as a function of wavelength in tin at various temperatures. Notation as in Fig. 14.

In order to verify the conclusions of the Gurzhi—Holstein theory, as indicated in Fig. 29, we must measure the frequency ν^{ef} at low temperatures. Since even for T ~ Θ, ν^{ef}, and ν^{ef}_{cl} differ only slightly in magnitude and temperature dependence, it is hard to give preference to any particular theory. In addition to this, in order to verify the theory in question we require that the electron—phonon interaction should be the decisive factor in absorption right down to the lowest temperatures. The anomalous skin effect may completely mask the phenomenon, since this also leads to finite absorption at low temperatures. In view of this, we cannot attempt the verification with metals of the first group, in which the skin effect is of a very anomalous nature, particularly at low temperatures, and the absorption is determined mainly by surface losses.

Interaction of electrons with impurities, structural defects, etc., also leads to finite absorption at low temperatures. Hence, we must make a special check of the quality of the samples used (for example, by measuring R_{res}/R_0).

The presence of interband transitions in the range studied may completely hide the absorption associated with the interaction between electrons and phonons.

Thus, in order to verify the theory, we must measure both optical constants n and $\varkappa$, since only then may we find the frequency of electron collisions, determine the character of the skin effect, and separate the surface and volume absorption. In addition to this, the measurements of n and $\varkappa$ should be carried out over a fairly wide spectral range in order to separate the influence of the conduction electrons from the possible influence of any other electron groups. Finally, in order to determine a number of the required parameters (ν^{ed}, Θ), we must study certain supplementary properties of the samples.

All these conditions were satisfied in the present investigation. In particular, we find that (as indicated in Section 4 of the present chapter), tin and lead are extremely suitable subjects for checking the theory, since their absorption even at helium temperatures is due almost entirely to electron—phonon interaction.

The Gurzhi—Holstein theory has not hitherto been proved experimentally. Attempts were made in [21, 62, 159] to separate the contributions corresponding to the anomalous skin effect and electron—phonon interaction, respectively, from the absorbing powers of Ag and Cu at helium temperatures, found experimentally. Since only one quantity (the absorbing power) was measured in these cases, so that neither the character of the skin effect nor the relation between surface and volume losses could be determined, the authors had to use other considerations and external data in order to secure the desired separation. In addition to this, as mentioned earlier, these metals were by no means suitable for attempting such a verification. The author of [159] finally came to the erroneous conclusion that the experimental data failed to agree with the Holstein results.

In order to verify the conclusions of the Gurzhi—Holstein theory we must separate the electron—phonon collision frequency ν^{ef}. (The absorbing power A contains not only ν^{ef} but other additional parameters; hence, strictly speaking, we should verify the dependence of ν^{ef} on T.) The frequency ν^{ef} may be determined from the frequency ν found from optical data and

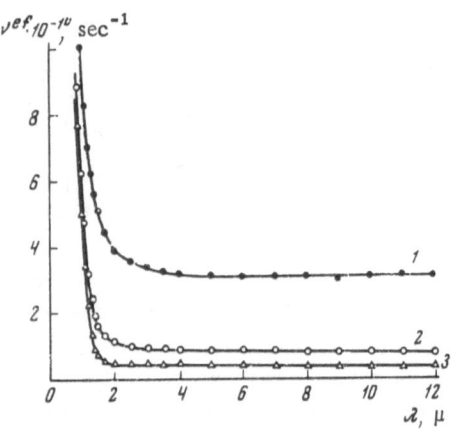

Fig. 31. Electron—phonon collision frequency as a function of wavelength for lead at various temperatures. Notation as in Fig. 14.

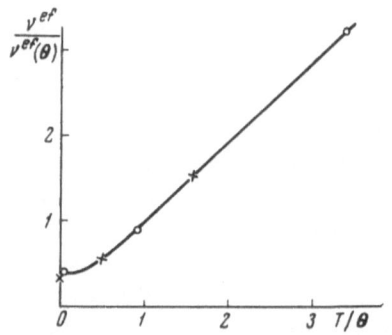

Fig. 32. Experimental relation between the electron—phonon collision frequency and the reduced temperature. Circles, lead; crosses, tin; solid curve, theoretical.

Table 15. Ratio of Electron—Phonon Collision Frequencies ν^{ef} (4.2°K) : ν^{ef} (78°K) : ν^{ef} (293°K)

Metal	Theory	Experiment
Sn	0.25 : 0.38 : 1	0.26 : 0.42 : 1
Pb	0.12 : 0.29 : 1	0.13 : 0.28 : 1

the frequency ν^{ed} (if the frequency of interelectron collisions may be neglected): $\nu^{ef} = \nu - \nu^{ed}$. The values of ν^{ef} thus obtained are shown as a function of λ in Fig. 30 for tin and Fig. 31 for lead. In the long-wave region $\lambda > 3\text{-}4\ \mu$ for both tin and lead this frequency is almost independent of the wavelength. The mean values of ν^{ef} for this range are given in Tables 6 and 9. In the short-wave region the internal photoeffect has a considerable influence on the electron collision frequency. The resultant electron—phonon collision frequencies enable us to verify the Gurzhi—Holstein theory regarding the way in which ν^{ef} varies with temperature. First of all, there is a qualitative agreement. The electron—phonon collision frequency remains quite large even at helium temperature, equal to $5 \cdot 10^{13}$ sec^{-1} for tin and $4 \cdot 10^{13}$ sec^{-1} for lead. For comparison, we mention that the classical electron—phonon collision frequency ν^{ef}_{cl} at this temperature would be $5 \cdot 10^{7}$ and 10^{10} sec^{-1}, respectively. The values obtained differ from these by three to six orders. We note that in good metals, for example gold [104], the electron—phonon collision frequency is no greater than $4 \cdot 10^{13}$ sec^{-1} even at room temperature, i.e., less than or equal to the limiting frequencies of collisions in the metals here employed.

The effect observed cannot be explained by surface losses. These losses are taken into account by the formulas of the weakly anomalous skin effect used (I.8). Since the surface losses are no greater than 20% for either metal, any possible error in the method used for taking these factors into account (associated with the fact that the formulas relate to a spherical Fermi surface and that they only contain the first order of the correction terms) entirely fails to alter the results. The observed effect cannot be explained either by the presence of impurities, defects, etc. The losses associated with these were accounted for in the residual resistance and constituted 8% for tin and 2% for lead. Hence, the slight inaccuracy in the calculation associated with the use of the Matthiessen rule has no effect on the results obtained.

For a quantitative verification of the theory, it is convenient to use the ratio of the ν^{ef} values at different temperatures. In this case all the undetermined coefficients [110], which are determined theoretically to a relatively low accuracy, drop out. The values of the ratios obtained theoretically and experimentally for temperatures of 293, 78, and 4.2°K are given in Table 15.

In the present investigation the parameter Θ was measured for the samples in question by an independent method. Hence, by finding the value of ν_Θ from optical data we may compare the experimental values of ν^{ef}/ν_Θ with the theoretical. We note that experimentally it is more convenient to determine not the value of ν_Θ but that of $\nu^{ef}(\Theta)$, since the relationship $\nu^{ef}(T)$ in the region of $T \gtrsim 0.5\Theta$ is linear to a fair accuracy. The theoretical value of $\nu^{ef}(\Theta) \approx 1.06\nu_\Theta$. The comparison is made in Fig. 32. The continuous curve gives the theoretical dependence of $\nu^{ef}/\nu^{ef}_{(\Theta)}$ on T/Θ. The points and crosses respectively represent the experimental values of $\nu^{ef}/\nu^{ef}(\Theta)$ for tin and lead.

We see from Table 15 and Fig. 32 that the experimental temperature dependence of the quantity ν^{ef} agrees closely with the theoretical for both metals. The limiting value (average for tin and lead)

$$\frac{\nu^{ef}(0^\circ \text{K})}{\nu^{ef}(\Theta)} = \frac{\nu^{ef}(4.2^\circ \text{K})}{\nu^{ef}(\Theta)} = 0.39,$$

which agrees with the theoretical value to within 3%.

Thus we have confirmed the conclusions of the quantum kinetic equation regarding the temperature dependence of the electron—phonon collision frequency in the presence of a high-energy light quantum. The set of optical and electrical data thus obtained enables us to verify one further conclusion of the theory. The theory [100] predicts that the parameter ν_Θ determined from optical measurements should agree with the parameter ν_Θ determined from the static conductivity (provided that such quantities as the constant of the electron—phonon interaction and others are identical in both the static and optical cases). However, for both metals the value of ν_Θ obtained from optical data is 25-30% higher than the value of ν_Θ obtained from the static conductivity. This means that for both metals there is a slight additional absorption of light not predicted by the theory. This absorption may be associated with interactions between the electrons and high-frequency lattice vibrations. In the theory under consideration, the law of dispersion is represented by a linear relationship between the frequency and momentum of the phonon. This only holds for low-frequency acoustic phonons. The existence of high-frequency lattice vibrations and the interaction of electrons with these is ignored in the theory. Meanwhile it is well known that interaction between electrons and high-frequency phonons may make a contribution to the absorption of light by semiconductors [160-162] if the energy of the quantum exceeds the energy of these phonons. In principle, such a contribution may occur in our present case. This contribution should vary relatively little with temperature [160].

7. Velocity of Electrons on the Fermi Surface and Other Microcharacteristics

The velocity of the electrons on the Fermi surface v is another fundamental quantity in the electron theory of metals. A knowledge of v is important in a number of fields of metal research. Optical investigations in principle enable us to determine v at the same time as N and ν. However, v only comes into the correction terms β_1 and β_2. Hence, it is only possible to determine v when the surface losses make up a considerable proportion of the volume losses. Thus, for any reliable sort of determination we require great accuracy in the measurement of the optical constants. The accuracy of the optical measurements made enabled us to determine v to an error of 15-20%. Values of v determined entirely from optical data for all temperatures are given in Table 6 for tin and Table 9 for lead. The exception is the velocity of electrons in lead at helium temperature. In this case, in view of the smallness of the coefficients β_1, any reliable determination of v from optical data was very difficult. Hence, in order to find v at this temperature we used the results of [163] relating to the electron specific heat of lead, and our own value of the conduction-electron concentration. In view of the relatively low accuracy of the determination of v, the values obtained at different temperatures coincide. Hence, the mean electron velocity on the Fermi surface is

$$v = 0.9 \cdot 10^8 \pm 0.15 \cdot 10^8 \text{ cm/sec for tin,}$$
$$v = 0.9 \cdot 10^8 \pm 0.2 \cdot 10^8 \text{ cm/sec for lead.}$$

For completely free electrons the mean values of v obtained from the expression v = $(3/\pi)^{1/3} \cdot N^{1/3} (\pi \hbar / m)$ will be

$$v = 1.26 \cdot 10^8 \text{ cm/sec for tin,}$$
$$v = 1.21 \cdot 10^8 \text{ cm/sec for lead.}$$

The experimentally determined values are rather lower. Our value of v for tin coincides with the value determined from the electron specific heat [163]. In the latter paper the author obtained $\gamma_e = 1.02 \cdot 10^3$ erg/cm$^3 \cdot$ deg^2, where γ_e is the coefficient of T in the electron part of the specific heat $C_e = \gamma_e T$. Using the expression $\gamma_e = \pi^2 k^2 N / mv^2$ and our own value of N for low temperatures, we find v = $0.92 \cdot 10^8$ cm/sec, agreeing closely with our direct value for v.

An analogous comparison carried out for room and nitrogen temperatures in the case of lead shows that our values of v were 20-25% higher than those obtained from the electron specific heat. This degree of agreement is quite satisfactory.

The resultant data enable us to determine a number of additional characteristics relating to the Fermi surface. For both metals the Fermi energy $W_F = mv^2/2 = 2.3 \pm 0.7$ eV. The extent of the total Fermi surface is

$$S_F = \frac{3}{2} \frac{h^3}{v} \frac{N}{m} = \begin{cases} (23 \pm 4) \cdot 10^{-38} \text{ g}^2 \cdot \text{cm}^2/\text{sec}^2 \text{ for tin,} \\ (22 \pm 4) \cdot 10^{-38} \text{ g}^2 \cdot \text{cm}^2/\text{sec}^2 \text{ for lead.} \end{cases}$$

The density of the levels near the Fermi surface found from formula (IV.1) is

$$\left(\frac{dY}{dW}\right)_F = \begin{cases} (1.4 \pm 0.4) \cdot 10^{22} \text{ cm}^{-3}/\text{eV for tin,} \\ (1.4 \pm 0.5) \cdot 10^{22} \text{ cm}^{-3}/\text{eV for lead,} \end{cases}$$

or the density of the levels associated with the atom

$$\frac{1}{N_a} \left(\frac{dY}{dW}\right)_F = \begin{cases} 0.4 \pm 0.1 \text{ eV}^{-1} \text{ for tin,} \\ 0.4 \pm 0.15 \text{ eV}^{-1} \text{ for lead.} \end{cases}$$

The values of W_F, S_F, and $(dY/dW)_F$ are the averages for the three temperatures. The values obtained for these quantities are quite close to the corresponding values for free electrons found on the basis of our conduction-electron concentration.

It is interesting to compare the resultant characteristics of tin and lead. The optical properties and their temperature dependence in the region in which the internal photoeffect is absent, the conduction—electron concentration, the phonon collision frequency, the velocity of the electrons on the Fermi surface, and other characteristics determined by the conduction electrons are very close for both metals. The fact that their crystal lattices are very different apparently only affects the N:T relationship.

It is also interesting to compare the characteristics of these metals with those of metals belonging to the first group, Au, Ag, Cu [59, 60, 82, 102, 104, 123]. A comparison of this kind reveals the following. 1) The concentration of conduction electrons associated with one atom is nearly the same for both groups of metals and is close to unity; 2) the collision frequencies of the first-group metals are about an order smaller than the corresponding frequencies for tin and lead; 3) the character of the skin effect in metals of the first group is anomalous and in tin and lead almost normal. We see from this comparison that the optical properties of the metals of the different groups differ much more than those of metals in the same group.

8. Effect of Interelectron Collisions on the Optical Constants of Tin and Lead

In the range $\lambda \lessgtr 2\,\mu$, the absorption of light by tin and lead is mainly determined by the internal photoeffect. In addition to the internal photoeffect, interelectron collisions may have a certain influence on the optical properties of the metal in this part of the spectrum. The frequency of interelectron collisions in the infrared part of the spectrum is determined from the expression [111]

$$\nu^{ee} = \nu_{cl}^{ee}\left[1 + \left(\frac{\hbar\omega}{2\pi kT}\right)^2\right].$$

Thus interelectron collisions are particularly important in the near-infrared region, where $\hbar\omega > 2\pi kT$. Naturally, this formula relates to the conduction electrons and is true for $\omega < \omega_0$ (before the onset of interband transitions). Hence optical investigations in the near-infrared enable us in principle to obtain the frequency of the interelectron collisions. It is extremely difficult to obtain the values of ν_{cl}^{ee} from the static conductivity owing to the smallness of ν_{cl}^{ee} in comparison with the other electron collision frequencies [107]. In order to separate ν_{cl}^{ee} out from the static conductivity, we must either have high temperatures, for which $\nu_{cl}^{ee} \sim T^2$ will be comparable with $\nu_{cl}^{ee} \sim T$ (but then other mechanisms may appear leading to a square-law temperature dependence of resistance, such as anharmonicity of the thermal vibrations, or a growth in defect concentration), or else very low temperatures for which $\nu_{cl}^{ee} \sim T^5$ (but for this very pure samples will be required). In the optical region the term $(\hbar\omega/2\pi kT)^2$ may exceed unity by two to five orders. Hence, the influence of interelectron collisions on the optical properties of metals should be much stronger than on the static conductivity.

Since for the majority of metals in the near-infrared there is in fact a term proportional to ω^2 in the electron collision frequency, the mistake is often made of ascribing this entirely to interelectron collisions. This question, however, demands more careful verification. This we may do by measuring the temperature dependence of the electron collision frequency. It follows from the expression for ν^{ee} that in a region in which $(\hbar\omega/2\pi kT)^2 \gg 1$ the frequency of the interelectron collisions is independent of temperature.

We may only verify this in the near-infrared, where it may be expected that the value of ν^{ee} will be large compared with other collision frequencies. Using the estimate of ν_{cl}^{ee} given in [107], we find for lead $\nu^{ee} \approx 2 \cdot 10^{13}$ sec^{-1} at $\lambda = 1\,\mu$ and $\nu^{ee} \approx 2 \cdot 10^{11}$ sec^{-1} at $\lambda = 10\,\mu$ (room temperature). Roughly the same values are obtained for tin. Since in tin and lead even at 4.2°K the electron—phonon collision frequency ν^{ef} is no smaller than $4 \cdot 10^{13}$ sec^{-1}, it is clear that interelectron collisions can only be appreciable for a λ shorter than 3-4 μ. In lead, however, the boundary of the internal photoeffect ω_0 lies at the start of the infrared region, which leads to a sharp rise in collision frequency at $\lambda \sim 1\,\mu$. Thus there remains an extremely narrow part of the spectrum in which interlectron collisions may be observed. In tin the absorption band is in general situated at $\lambda = 1$-1.2 μ, so that there is hardly any region in which the influence of interelectron collisions might be appreciable.

As indicated in Figs. 30 and 31, the experimental collision frequency for tin and lead in the short-wave region also has a term proportional to ω^2, i.e., it has the form predicted theoretically in the presence of interelectron collisions

$$\nu = \nu(T) + \frac{B}{T^2}\left(\frac{\hbar}{2\pi k}\right)^2 \omega^2. \qquad (IV.3)$$

Here $\nu(T)$ is the ω-independent part of the electron collisions and B is some coefficient. In the present investigation the collision frequency was measured for the first time over such a wide temperature range. It was found that the term quadratic in ω exhibited by tin and lead at room and nitrogen temperatures was not associated with interelectron collisions. If the second term in expression (IV.3) were determined by interelectron collisions, then $B = \nu_{cl}^{ee}$. In this way, $B \sim T^2$, and the quantity B/T^2 would not depend on T. Table 16 shows the values of the coefficient B for tin and lead obtained at various temperatures and also the theoretical value of ν_{cl}^{ee}. We see from Table 16 * that only for lead at helium temperature is the experimental value of B close to the theoretical. At higher temperatures for lead and at all temperatures for tin the experimental values of B are much higher than the theoretical values of ν_{cl}^{ee}. The ratio B/T^2, which ought to be independent of temperature, falls by a factor of 3 for tin and 9 for lead as the temperature falls from room value to that of liquid helium. This result indicates that at room and nitrogen temperatures the quadratic dependence of the electron collision frequency in the short-wave part of the spectrum is not associated with interelectron collisions.

Measurements at helium temperature give the upper limit to the frequency of interelectron collisions. Clearly the true interelectron-collision frequency is only given by the B of lead at 4.2°K (for tin even at helium temperature the value of B is probably too large owing to the effect of the neighboring band). If we consider that the value of B found for lead at 4.2°K is associated with interelectron collisions, then the corresponding value at room temperature should be $\nu_{cl}^{ee} = 6 \cdot 10^{11}$ sec^{-1}. The value found for B is 13 times larger.

Thus, interelectron collisions can only have an effect on the optical constants of the metals in question in the near-infrared part of the spectrum for fairly low temperatures. This result justifies neglecting ν^{ee} in the range $\lambda = 3$-$12\ \mu$, since it is there very small.

9. Separation of the Effects of

Interband Transitions

A measurement of the optical constants in the short-wave part of the spectrum, where the internal photoeffect has a dominant influence on the optical properties, enables us to obtain certain characteristics of band structure. In the present investigation we made an attempt to separate the effect on the optical properties of tin, of the electrons taking part in an interband transition.

In the short-wave region, $\varepsilon' = \varepsilon_0 + \varepsilon_e$. Here, ε_e and ε_0 are the complex dielectric constants associated respectively with the conduction electrons and the electrons participating in the interband transition. The dielectric constant ε_e may be found from the microcharacteristics determined in the long-wave region. For determining ε_e we used the formulas of the normal skin effect (I.6), which is justified in the small-λ region. Subtracting from the experimental value of ε' the value of ε_e obtained in this manner (real and imaginary parts separately), we may find ε_0 for various λ. Figures 33 and 34 give the values of Re ε_0 and Im ε_0 so obtained for nitrogen temperature. Analogous curves are obtained for other temperatures. The dependence of the two quantities on λ is reminiscent of the dispersion formula:

---

*In calculating ν_{cl}^{ee} the average values are taken for v.

Table 16. Separation of Interelectron Collisions

T, °K	Tin				Lead			
	B, sec^{-1} exptl.	ν_{cl}^{ee}, sec^{-1} theory	B/T^2, sec^{-1} · deg^{-2}, exptl.	ν_{cl}^{ee}/T^2, sec^{-1} · deg^{-2}, theory	B, sec^{-1} exptl.	ν_{cl}^{ee}, sec^{-1} theory	B/T^2, sec^{-1} · deg^{-2}, exptl.	ν_{cl}^{ee}/T^2, sec^{-1} · deg^{-2}, theory
293	$2 \cdot 10^{13}$	$4.9 \cdot 10^{11}$	$3 \cdot 10^8$	$0.6 \cdot 10^7$	$8 \cdot 10^{12}$	$5.6 \cdot 10^{11}$	$9 \cdot 10^7$	$0.7 \cdot 10^7$
78	$6 \cdot 10^{11}$	$3.8 \cdot 10^{10}$	$1 \cdot 10^8$	$0.6 \cdot 10^7$	$2 \cdot 10^{11}$	$4.0 \cdot 10^{10}$	$3 \cdot 10^7$	$0.7 \cdot 10^7$
4.2	$2 \cdot 10^9$	$1.1 \cdot 10^8$	$1 \cdot 10^8$	$0.6 \cdot 10^7$	$2 \cdot 10^8$	$1.2 \cdot 10^8$	$1 \cdot 10^7$	$0.7 \cdot 10^7$

Table 17. Some Characteristics of the Interband Transmission
of Tin

T, °K	$N_b \cdot 10^{-22}$, cm^{-3}	$\omega_0 \cdot 10^{15}$, sec^{-1}	$\gamma_b \cdot 10^{-15}$, sec^{-1}	$E_g^{(1)}$, eV	$E_g^{(2)}$, eV
293	5.2	1.8	2.2	1.2	1.2
78	3.2	1.9	1.0	1.25	1.2
4.2	1.9	2.0	0.6	1.3	1.3

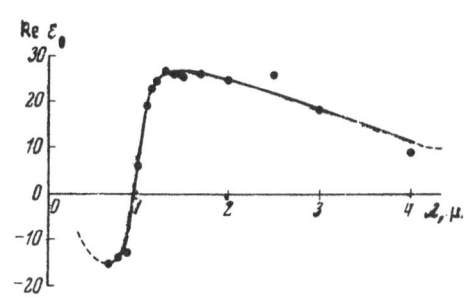

Fig. 33. Dependence of Re ε_0 on wavelength for Tin (T = 78°K).

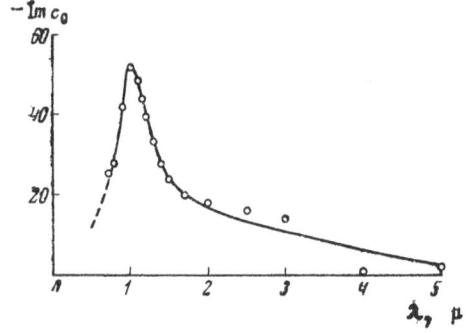

Fig. 34. Dependence of −Im ε_0 on wavelength for tin (T = 78°K).

Fig. 35. Dependence of Re $[1/(\varepsilon_0 - 1)]$ on ω^2 for tin (T = 78°K).

$$\varepsilon_0 = \operatorname{Re}\varepsilon_0 + i\operatorname{Im}\varepsilon_0 = 1 + \frac{4\pi e^2 N_b}{m(\omega_0^2 - \omega^2 + i\omega\gamma_b)}, \quad \text{(IV.4)}$$

where N_b, ω_0, and γ_b are parameters respectively characterizing the concentration of the electrons taking part in the interband transition, the frequency corresponding to the gap width, and the attenuation. Using expression (IV.4) and the resultant values of ε_0, we may find the characteristics of the electrons taking part in the interband transition. In determining the quantities N_b, γ_b, and ω_0, it is convenient to use the quantity $1/(\varepsilon_0 - 1) = (m/4\pi e^2 N_b)(\omega_0^2 - \omega^2 + i\omega\gamma_b)$. The quantities Re $[1/(\varepsilon_0 - 1)]$ and Im $[1/(\varepsilon_0 - 1)]$ should in fact depend linearly on ω^2 and ω, respectively, which enables us to determine the parameters N_b, ω_0, and γ_b quite easily.

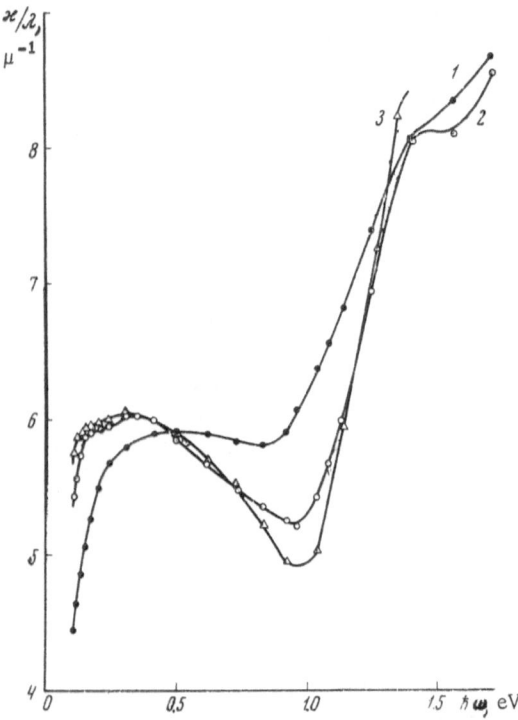

Fig. 36. Dependence of $\varkappa/\lambda$ on $\hbar\omega$ for tin. Notation as in Fig. 14.

Figure 35 shows the dependence of Re$[1/(\varepsilon_0 - 1)]$ on ω^2 for T = 78°K. We see that over the whole of the short-wave range the quantity Re$[1/(\varepsilon_0 - 1)]$ is linear to a high accuracy. Table 17 shows the values of N_b, γ_b, and ω_0 obtained for tin and also the value of the energy gap $E_g^{(1)} = \hbar\omega_0$. The wavelength corresponding to the frequency ω_0 is $\sim 1\ \mu$ for all temperatures.

It should be noted that according to formula (IV.4) the influence of the interband transition should extend right up to very long λ. We see, however, from experiment that this influence practically finishes at $\lambda \sim 3\text{-}4\ \mu$. Hence, the dispersion formula (IV.4) only roughly gives the dependence of ε_0 on λ in the short-wave region.

In order to determine the value of the energy gap we may also use the dependence of $\varkappa/\lambda$ on ω, as in the case of semiconductors [122]. Figure 36 shows this relationship for tin at various temperatures. The frequency corresponding to the interband transition was determined from the maximum slope of $\varkappa/\lambda$. For tin this method is quite valid, since, in the neighborhood of the transition the influence of the conduction electrons is small compared with that of the electrons taking part in the transition. Hence, the contribution of the conduction electrons may to a first approximation be neglected. The resultant values of the width of the forbidden band $E_g^{(2)}$ are given in the last column of Table 17. These agree very closely with the values found by the first method.

The use of low temperatures enables us to determine the parameters of the band structure more accurately and reliably. We see from Fig. 35 that at low temperatures it is easy to separate the influence of various groups of electrons on the optical properties.

For lead the frequency corresponding to the interband transition lies in a short-wave region. Hence, in order to obtain the corresponding characteristics we should have to make measurements in the visible part of the spectrum.

CONCLUSION

In this investigation we have carried out a complex study of the optical and other properties of tin and lead, developed a method of measuring the optical properties of metals at low temperatures, and also measured the optical constants of tin and lead in the spectral range 0.9-12 μ at temperatures of 293, 78, and 4.2°K.

For the same samples we have measured: 1) the density; 2) the conductivity and its temperature dependence; 3) the residual resistance; 4) the Hall effect; and, 5) the temperature corresponding to the transition into the superconducting state. We have also determined the Debye temperature.

The resultant data have enabled us to show that for lead and tin there is a weakly anomalous skin effect in the infrared part of the spectrum for all temperatures right down to that of liquid helium, and also to determine the most important microcharacteristics: the conduction-electron concentration, the velocity of the electrons on the Fermi surface, the electron—phonon collision frequency, the electron—impurity collision frequency, and a number of additional characteristics associated with the conduction electrons and the Fermi surface.

We have found that for tin and lead the conduction-electron concentration varies with temperature. The effect greatly exceeds experimental error in the case of tin. The change of concentration is much greater as the temperature changes from room value to that of liquid nitrogen than between nitrogen and helium temperatures. The effect is apparently associated with a change in the overlapping of the zones with changing temperature.

It has been established experimentally for the first time that even at helium temperatures the electron—phonon collision frequency in the presence of a high-energy quantum remains quite high ($5 \cdot 10^{13}$ sec^{-1} for tin and $4 \cdot 10^{13}$ sec^{-1} for lead).

We have compared the experimental data with the quantum theory of the phenomenon. The comparison shows that the experimental relationship between the electron—phonon collision frequency and the temperature agrees closely with the theoretical predictions.

Our complex investigation has enabled us to compare the temperature dependence of the electron—phonon collision frequency obtained from optical data with that given by the static conductivity. The comparison shows that in tin and lead there is a slight additional absorption of infrared radiation not predicted by theory. This absorption is probably associated with the interaction between the electrons and high-frequency lattice vibrations.

Our study of the optical properties of tin and lead in the short-wave part of the spectrum has also shown that the frequency of interelectron collisions in these metals can only play an appreciable part at the very lowest temperatures.

The characteristics of the electrons participating in the interband transition have been determined for the case of tin.

In conclusion, the author wishes to thank G. P. Motulevich most sincerely for direction of the work and systematic assistance; he is also very grateful to I. L. Fabelinskii and A. A. Shubin for useful discussions.

LITERATURE CITED

1. G. E. H. Reuter and E. H. Sondheimer, Proc. Roy. Soc., A195:336 (1948).
2. R. B. Dingle, Physica, 18:985 (1952).
3. T. Holstein, Phys. Rev., 88:1427 (1952).
4. R. B. Dingle, Physica, 19:331 (1953).
5. R. B. Dingle, Physica, 19:348 (1953).
6. R. B. Dingle, Physica, 19:729 (1953).
7. R. B. Dingle, Appl. Sci. Res., B3:69 (1953).
8. A. N. Gordon and E. H. Sondheimer, Appl. Sci. Res., B3:297 (1953).
9. V. L. Ginzburg, Dokl. Akad. Nauk SSSR, 97:999 (1954).
10. E. H. Sondheimer, Proc. Roy. Soc., A224:260 (1954).
11. A. B. Pippard, Proc. Roy. Soc., A224:273 (1954).
12. V. L. Ginzburg and G. P. Motulevich, Usp. Fiz. Nauk, 55:469 (1955).
13. M. I. Kaganov and V. V. Slezov, Zh. Éksper. i Teor. Fiz., 32:1496 (1957).
14. V. P. Silin, Zh. Eksper. i Teor. Fiz., 32:1282 (1957).
15. V. P. Silin, Zh. Eksper. i Teor. Fiz., 35:1001 (1958).
16. R. Engelman, Proc. Phys. Soc., 72:277 (1958).
17. J. G. Collins, Appl. Sci. Res., B7:1 (1958).
18. D. C. Mattis and J. Bardeen, Phys. Rev., 111:412 (1958).
19. G. P. Motulevich, Zh. Eksper. i Teor. Fiz., 46:287 (1964).
20. K. Ramanathan, Proc. Phys. Soc., A65:532 (1952).
21. M. Biondi, Phys. Rev., 102:964 (1956).
22. K. Försterling and V. Freedericksz, Ann. Phys., 40:201 (1913).
23. O. Bryan, J. Opt. Soc. Am., 26:127 (1936).
24. T. P. Kravets, Izv. Akad. Nauk SSSR, Ser. Fiz., 12:504 (1948).
25. J. Simon, J. Opt. Soc. Am., 41:336 (1951).
26. D. G. Avery, Proc. Phys. Soc., B65:425 (1952).
27. J. Bor, Proc. Phys. Soc., B65:753 (1952).
28. J. F. Archard, P. L. Clegg, and A. M. Taylor, Proc. Phys. Soc., B65:758 (1952).
29. G. K. T. Conn and G. K. Eaton, J. Opt. Soc. Am., 44:484 (1954).
30. G. K. T. Conn and G. K. Eaton, J. Opt. Soc. Am., 44:546 (1954).
31. J. R. Beattie, Phil. Mag., 46:235 (1955).
32. M. M. Noskov and B. A. Charikov, Opt. i Spektroskopiya, 1:1007 (1956).
33. N. N. Pribytkova, Opt. i Spektroskopiya, 2:623 (1957).
34. G. P. Motulevich and A. A. Shubin, Opt. i Spektroskopiya, 2:633 (1957).
35. N. E. Alekseevskii and E. V. Potapov, Zh. Eksper. i Teor. Fiz., 33:283 (1957).
36. N. G. Bakhshiev, Opt. i Spektroskopiya, 2:818 (1957).
37. I. N. Shklyarevskii and V. K. Miloslavskii, Opt. i Spektroskopiya, 3:361 (1957).
38. I. N. Shklyarevskii, Opt. i Spektroskopiya, 3:638 (1957).
39. M. P. Givens, Solid State Physics, 6:313 (1958).

40.   S. Roberts, Phys. Rev., 114:104 (1959).
41.   A. I. Golovashkin, G. P. Motulevich, and A. A. Shubin, Pribory i Tekhn. Eksper., No. 5, p. 74 (1960).
42.   V. G. Padalka and I. N. Shklyarevskii, Opt. i Spektroskopiya, 9:119 (1960).
43.   A. P. Prishivalko, Opt. i Spektroskopiya, 9:493 (1960).
44.   A. V. Sokolov, Optical Properties of Metals, Fizmatgiz, Moscow (1960).
45.   A. I. Golovashkin and G. P. Motulevich, Zh. Eksper. i Teor. Fiz., 44:398 (1963).
46.   L. G. Schulz, Advan. Phys., 6:102 (1957).
47.   R. Kretzmann, Ann. Phys., 37:303 (1940).
48.   J. R. Collins and R. O. Bock, Rev. Sci. Instr., 14:135 (1943).
49.   J. N. Hodgson, Proc. Phys. Soc., B67:269 (1954).
50.   L. A. Afanas'eva, M. M. Noskov, and V. I. Cherepanov, Fiz. Metal. i Metalloved., 1:566 (1955).
51.   L. D. Kislovskii, Opt. i Spektroskopiya, 5:66 (1958).
52.   A. P. Prishivalko, G. M. Gusak, and I. L. Onichek, Opt. i Spektroskopiya, 11:555 (1961).
53.   A. P. Prishivalko, Reflection of Light from Absorbing Media, Izd. Akad. Nauk BSSR, Minsk (1963).
54.   L. D. Landau and E. M. Livshits, Electrodynamics of Continuous Media, Gostekhteoretizdat, Moscow (1957).
55.   H. R. Philipp and E. A. Taft, Phys. Rev., 113:1002 (1959).
56.   M. P. Rimmer and D. L. Dexter, J. Appl. Phys., 31:775 (1960).
57.   H. Ehrenreich, H. R. Philipp, and B. Segall, Phys. Rev., 132:1918 (1963).
58.   F. C. Jahoda, Phys. Rev., 107:1261 (1957).
59.   V. G. Padalka and I. N. Shklyarevskii, Opt. i Spektroskopiya, 11:527 (1961).
60.   V. G. Padalka and I. N. Shklyarevskii, Opt. i Spektroskopiya, 12:291 (1962).
61.   H. E. Bennett, M. Silver, and E. J. Ashley, J. Opt. Soc. Am., 53:1089 (1963).
62.   M. Biondi, Phys. Rev., 96:534 (1954).
63.   I. N. Shklyarevskii, A. A. Avdeenko, and V. G. Padalka, Opt. i Spektroskopiya, 6:528 (1959).
64.   E. V. Potapov, Zh. Eksper. i Teor. Fiz., 47:464 (1964).
65.   W. A. Harrison, Phys. Rev., 116:555 (1959).
66.   W. A. Harrison, Phys. Rev., 118:1182 (1960).
67.   W. A. Harrison, Phys. Rev., 118:1190 (1960).
68.   A. V. Gold and M. G. Priestley, Phil. Mag., 5:1089 (1960).
69.   M. S. Khaiken and R. T. Mina, Zh. Eksper. i Teor. Fiz., 42:35 (1962).
70.   M. S. Khaiken, Zh. Eksper. i Teor. Fiz., 43:59 (1962).
71.   J. A. Rayne and B. S. Chandrasekhar, Phys. Rev., 125:1952 (1962).
72.   W. A. Harrison, Phys. Rev., 126:497 (1962).
73.   D. Zaiman, Usp. Fiz. Nauk, 78:291, 679 (1962); 79:319 (1963); 80:505, 665 (1963).
74.   N. A. Bezuglyi, A. A. Galkin, and A. I. Pushkin, Zh. Eksper. i Teor. Fiz., 44:71 (1963).
75.   R. T. Mina and M. S. Khaiken, Zh. Eksper. i Teor. Fiz., 45:1304 (1963).
76.   M. S. Khaiken, Author's Abstract of Dissertation, Institute of Physical Problems of the USSR Academy of Sciences, Moscow (1963).
77.   V. P. Haberezhnykh and V. P. Tolstoluzhskii, Zh. Eksper. i Teor. Fiz., 46:18 (1964).
78.   E. H. Sondheimer, Advan. Phys., 1:1 (1952).
79.   J. A. Rayne, Phys. Rev., 121:456 (1961).
80.   M. A. Biondi and J. A. Rayne, Phys. Rev., 115:1522 (1959).
81.   J. R. Beattie and G. K. T. Conn, Phil. Mag., 46:222 (1955).
82.   J. R. Beattie and G. K. T. Conn, Phil. Mag., 46:989 (1955).
83.   J. N. Hodgson, Proc. Phys. Soc., B68:593 (1955).
84.   L. G. Schulz, J. Opt. Soc. Am., 47:64 (1957).

85. G. P. Skornyakov and M. M. Kirillova, Opt. i Spektroskopiya, 6:248 (1959).
86. M. N. Markov and I. S. Lidstrem, Opt. i Spektroskopiya, 7:349 (1959).
87. J. N. Hodgson, Phil. Mag., 4:183 (1959).
88. A. I. Golovashkin, G. P. Motulevich, and A. A. Shubin, Zh. Eksper. i Teor. Fiz., 38:51 (1960).
89. I. N. Sklyarevskii and R. G. Yarovaya, Opt. i Spektroskopiya, 11:661 (1961).
90. G. P. Motulevich and A. A. Shubin, Zh. Eksper. i Teor. Fiz., 44:48 (1963).
91. M. M. Kirillova and B. A. Charikov, Fiz. Metal. i Metalloved., 15:315 (1963).
92. A. I. Golovashkin and G. P. Motulevich, Zh. Eksper. i Teor. Fiz., 46:460 (1964).
93. I. N. Shklyarevskii and R. G. Yarovaya, Opt. i Spektroskopiya, 16:85 (1964).
94. A. Yu. Eichis and G. P. Skornyakov, Opt. i Spektroskopiya, 16:159 (1964).
95. V. P. Kostyuk and I. N. Shklyarevskii, Opt. i Spektroskopiya, 16:304 (1964).
96. M. M. Kirillova and B. A. Charikov, Opt. i Spektroskopiya, 17:254 (1964).
97. M. M. Kirillova and B. A. Charikov, Fiz. Metal. i Metalloved., 16:205 (1963).
98. G. P. Motulevich, Zh. Eksper. i Teor. Fiz., 37:1770 (1959).
99. A. Wilson, Quantum Theory of Metals [Russian translation], Gostekhizdat, Moscow (1941).
100. R. N. Gurzhi, Zh. Eksper. i Teor. Fiz., 33:660 (1957).
101. M. Ya. Azbel' and E. A. Kaner, Zh. Eksper. i Teor. Fiz., 32:896 (1957).
102. J. R. Beattie, Physica, 23:898 (1957).
103. A. I. Golovashkin and G. P. Motulevich, Zh. Eksper. i Teor. Fiz., 47:64 (1964).
104. G. P. Motulevich and A. A. Shubin, Zh. Eksper. i Teor. Fiz., 47:840 (1964).
105. R. G. Chambers, Proc. Roy. Soc., A215:481 (1952).
106. A. I. Shal'nikov (editor), Low-Temperature Physics (Handbook) [Russian translation], IL, Moscow (1953).
107. V. L. Ginzburg and V. P. Silin, Zh. Eksper. i Teor. Fiz., 29:64 (1955).
108. N. N. Bogolyubov and K. P. Gurov, Zh. Eksper. i Teor. Fiz., 17:614 (1947).
109. Yu. P. Klimontovich and V. P. Silin, Zh. Eksper. i Teor. Fiz., 23:151 (1952).
110. R. N. Gurzhi, Zh. Eksper. i Teor. Fiz., 33:451 (1957).
111. R. N. Gurzhi, Zh. Eksper. i Teor. Fiz., 35:965 (1958).
112. R. N. Gurzhi, Author's Abstract of Dissertation, Khar'kov State Univ. (1958).
113. M. Born, Optics [Russian translation], ONTI, Khar'kov-Kiev (1937).
114. M. Ya. Azbel' and M. I. Kaganov, Dokl. Akad. Nauk SSSR, 95:41 (1954).
115. L. P. Pitaevskii, Zh. Eksper. i Teor. Fiz., 34:942 (1958).
116. V. P. Silin, Zh. Eksper. i Teor. Fiz., 34:707 (1958).
117. M. Ya. Azbel', Zh. Eksper. i Teor. Fiz., 34:766 (1958).
118. P. A. Wolff, Phys. Rev., 116:544 (1959).
119. M. Suffczynski, Phys. Stat. Sol., 4:3 (1964).
120. S. Roberts, Phys. Rev., 100:1667 (1955).
121. M. Suffczynski, Phys. Rev., 117:663 (1960).
122. T. S. Moss, Optical Properties of Semiconductors [Russian translation], IL, Moscow (1961). [Academic Press, New York.]
123. S. Roberts, Phys. Rev., 118:1609 (1960).
124. G. S. Landsberg, Optics, Gostekhteoretizdat, Moscow (1952).
125. A. Bolotin, A. N. Voloshinskii, M. M. Kirillova, M. M. Noskov, A. V. Sokolov, and B. A. Charikov, Fiz. Metal. i Metalloved., 13:823 (1962).
126. A. M. Bonch-Bruevich, Use of Radio Tubes in Experimental Physics, Gostekhizdat, Moscow (1955).
127. A. Elliott, E. J. Ambrose, and R. Temple, J. Opt. Soc. Am., 38:212 (1948).
128. Infrared Spectroscopy, article in: Physical Encyclopedic Dictionary, Gos. Nauchn. Izd. "Sovetskaya Entsiklopediya," Moscow (1962).

129. H. M. Randall, R. G. Fowler, N. Fuson, and J. R. Dangl, Infrared Determination of Organic Structures, D. Van Nostrand Co., New York (1949).
130. A. N. Aleksandrov and V. A. Nikitin, Usp. Fiz. Nauk, 56:3 (1955).
131. A. S. Toporets, Monochromators, Gostekhizdat, Moscow (1955).
132. G. K. T. Conn and G. K. Eaton, J. Opt. Soc. Am., 44:477 (1954).
133. M. Czerny, Z. Phys., 26:182 (1924).
134. L. Holland, Vacuum Deposition of Thin Films, Chapman and Hall, London (1956).
135. D. Zeeman, Electrons and Phonons [Russian translation], IL, Moscow (1962).
136. D. Kaye and T. Laby, Tables of Physical and Chemical Constants [Russian translation], Fizmatgiz, Moscow (1962).
137. Charles D. Hodgman et al. (editors), Handbook of Chemistry and Physics, 67th edition, Chemical Rubber, Cleveland, Ohio.
138. C. J. Smithells, Metals Reference Book, Butterworths Scientific Publications, London (1955).
139. J. G. Dorfman and S. E. Frisch (editors), Handbook of Physical Constants [Russian translation], ONTI, Moscow (1937).
140. V. N. Kachinskii, Dokl. Akad. Nauk SSSR, 135:818 (1960).
141. V. N. Kachinskii, Zh. Eksper. i Teor. Fiz., 43:1158 (1962).
142. E. Grüneisen, Ann. Phys., 16:530 (1933).
143. S. V. Vonsovskii, Izv. Akad. Nauk SSSR, Ser. Fiz., 12:337 (1948).
144. B. N. Aleksandrov, Zh. Eksper. i Teor. Fiz., 43:399 (1962).
145. B. N. Aleksandrov and M. I. Kaganov, Zh. Eksper. i Teor. Fiz., 41:1333 (1961).
146. J. Bardeen and W. Shockley, Phys. Rev., 80:72 (1950).
147. L. R. Godefroy and P. Aigrain, Proceedings of the International Conference on Physics of Semiconductors, Exeter (1962).
148. L. V. Keldysh and Yu. V. Kopaev, Fiz. Tverd. Tela, 6:2971 (1964).
149. G. L. Pearson and J. Bardeen, Phys. Rev., 80:72 (1950).
150. M. Becker and H. Y. Fan, Phys. Rev., 76:1531 (1949).
151. M. Becker and H. Y. Fan, Phys. Rev., 78:301 (1950).
152. G. G. Macfarlane and V. Roberts, Phys. Rev., 97:1714 (1955).
153. W. Shockley and J. Bardeen, Phys. Rev., 77:407 (1950).
154. D. A. Wright, Semiconductors [Russian translation], IL, Moscow (1957). [Wiley, New York.]
155. H. Y. Fan, Phys. Rev., 78:808 (1950).
156. G. E. Kimball, J. Chem. Phys., 3:560 (1935).
157. J. F. Mullaney, Phys. Rev., 66:326 (1944).
158. T. Holstein, Phys. Rev., 96:535 (1954).
159. J. A. Rayne, Phys. Rev. Letters, 3:512 (1959).
160. J. Bardeen, Phys. Rev., 79:216 (1950).
161. H. Y. Fan, W. Spitzer, and R. J. Collins, Phys. Rev., 101:566 (1956).
162. H. J. G. Meyer, Phys. Rev., 112:298 (1958).
163. J. Eisenstein, Rev. Mod. Phys., 26:277 (1954).

STUDY OF THE SPECTRUM
OF THERMAL AND STIMULATED
MOLECULAR SCATTERING OF LIGHT IN LIQUIDS*

V. S. STARUNOV

---

*Abbreviated text of a dissertation in pursuit of the degree of Candidate of Physicomathematical Sciences. Defended January 3, 1966, at the P. N. Lebedev Physical Institute of the Academy of Sciences of the USSR. Scientific director: I. L. Fabelinskii.

INTRODUCTION

The experimental and theoretical study of the spectrum characterizing the molecular scattering of light in crystals, amorphous solids, and liquids has yielded extensive information [1-7] regarding the kinetics of thermal fluctuations and the acoustic and other properties of these media at frequencies so high as to be inaccessible to other methods of investigation.

The difference between the spectra characterizing the exciting and scattered light respectively lies in the fact that the scattered light is modulated as a result of the time variation of the fluctuations in the medium.

In this study we shall be concerned with two main groups of fluctuations: those of density and those of anisotropy.

Fluctuations of density may in turn be regarded as consisting of fluctuations of pressure (adiabatic or isentropic fluctuations of density) and fluctuations of entropy (isobaric fluctuations of density).

The fluctuations of anisotropy may be regarded as consisting of fluctuations of deformation (strain) and fluctuations in the orientation of anisotropic molecules.

At the present time there is a fairly comprehensive phenomenological theory of light scattering at density fluctuations [1, 4, 6-9]. In particular, this phenomenological theory gives a quantitative expression for the intensity distribution in the spectrum of light scattered at density fluctuations.

The modulation of the scattered light resulting from changes in the pressure fluctuations with time (or, in other words, the modulation of the scattered light wave resulting from a change in the refractive index of the medium due to the propagation of an elastic wave) leads to the appearance of frequency-displaced components (the Mandelshtam-Brillouin components) in the spectrum of the scattered light.

The modulation of the scattered light by time-varying isobaric fluctuations of density leads to the appearance of an undisplaced (central) component in the spectrum of the scattered light.

The theoretical formulas describing the fine-structure spectrum enable us to find the velocity of hypersound of a frequency of $\sim 10^{10}$ cps by reference to the mutual disposition of the Mandelshtam-Brillouin components, and the absorption coefficient of the hypersound from the widths of these components.

The velocity of hypersound has been determined earlier [6, 10-12]. It was precisely these measurements which revealed the dispersion of the velocity of sound in a number of liquids characterized by large volume and small shear coefficients of viscosity.

In developing these earlier investigations, we took a new methodical step in that, instead of the ordinary sources of exciting light, we made use of a gas laser. The advantages of this new light source lay mainly in its strictly directional properties, its high intensity, and the extremely small line width of the stimulated emission.

In our investigations, we used the λ = 6328 Å line of the stimulated emission of an Ne—He gas laser. This line was so narrow that its use enabled us to increase the accuracy achieved in determining the velocity of hypersound by an order of magnitude, so that the residual error was no more than 0.5% of the measured quantity.

A still more important result of the use of the Ne—He gas laser for exciting the fine structure of the Rayleigh line lay in the fact that we were able to determine the true width of the Mandelshtam—Brillouin components and thus find the absorption coefficient of hypersound for a number of liquids [13, 14] (see Chapter V).

We were the first to carry out such measurements [13]; with ordinary gas-discharge light sources measurements of this kind had earlier proved impossible [6, 15] because the width of the exciting line was tens of times greater than the true width due to the attenuation of the hypersound.

The measurement of the dispersion of the velocity of sound and the determination of the absorption coefficient of hypersound is certainly of quite independent interest. We feel that the results obtained are particularly important because they enable us to verify the relaxation theory [16, 17] directly. Such a direct comparison of the relaxation theory with experimental data was impossible before we had secured our simultaneous results on the absorption and velocity dispersion of sound.

This comparison of theory with experiment showed that the simplified version of the relaxation theory with a single relaxation time gave a good description of the propagation of sound in such liquids as benzene, carbon disulfide, and carbon tetrachloride. There were liquids, however, such as chloroform, for which this simple form was inapplicable.

A study of the interaction between powerful light fluxes emitted by ruby lasers and matter led to the development of a new field of physics: nonlinear optics. One of the most important nonlinear optical effects is the stimulated Mandelshtam—Brillouin scattering (see, for example, [1, 18]) recently observed by American physicists in crystals [19] and liquids [20, 21]. This scattering was also observed in our own experiments with glasses and supercooled liquids [22] (see Chapter VI).

We also observed [23] a new phenomenon: the stimulated scattering of light on the wing of the Rayleigh line; this will be discussed later. A study of this phenomenon may give a great deal of valuable information regarding the character of thermal motion in various media and the nonlinear optical and acoustic characteristics of the medium.

Stimulated Mandelshtam—Brillouin scattering results in the simultaneous creation of one or several lasers at the displaced frequency together with powerful hypersound generators. In our experiments the power of the hypersound generated was measured as several kilowatts.

Our stimulated Mandelshtam—Brillouin scattering in silicate glasses and supercooled liquids offers a new method of studying these media and settles the long dispute as to whether fine structure of the scattered light exists in silicate glasses [1]. It should be emphasized that many attempts to find fine structure in silicate glasses in ordinary thermal scattering have ended in failure. The discovery of stimulated Mandelshtam—Brillouin scattering in silicate glasses clearly shows that the reason for the earlier failure lies not in the nature of the phenomenon but in the experimental technique.

As mentioned earlier, there is at the present time a phenomenological theory fairly fully describing the spectrum of the light scattered at density fluctuations. As regards the theory of light scattering due to anisotropy fluctuations (the "wing" of the Rayleigh line), things are not so favorable. Existing phenomenological theories of the wing [8, 9] fail to give a full description of the light scattered at anisotropy fluctuations. We were therefore compelled to set up a phenomenological theory of the wing based on simple models of molecular motion [24, 25]. By using the Fokker—Planck equation in considering the rotational motion of molecules over long periods of time (rotational diffusion) and solving the stochastic equation with due allowance for the rotational vibrations of the molecules between two reorientations, we were able to obtain formulas for the intensity distribution in the part of the wing immediately adjacent to the undisplaced scattering line (diffusion wing) and for the rest of the wing (see Chapter II). In order to verify the theory we made some special experiments on the distribution of intensity over the wing of the Rayleigh line for several low-viscosity liquids with a small shear viscosity.

Our theoretical formulas and experimental data enables us to find a number of important characteristics of the thermal rotatory motion of the molecules in liquids. In particular, we were able for the first time to follow the relaxation of the internal-friction (viscosity) coefficient and find its relaxation time for a number of liquids (see Chapter IV). These processes involving the relaxation of internal friction are associated with the kinetics of molecular motion for time intervals of the order of the period of the elastic vibrations of the molecule in the field of its immediate neighbors.

By studying the spectral intensity distribution in the wing of the Rayleigh line we also found the Debye relaxation time and determined the limiting frequencies of the elastic vibrations of the molecules in certain liquids. Although the results obtained are important, one is nevertheless conscious of the need to make fuller use of the information contained in the depolarized-scattering spectrum (Rayleigh line wing) regarding molecular motion. However, any attempt to construct a complete theory of the wing runs into the same difficulties as the kinetic theory of liquids, which is as yet far from perfect. Our first attempt to construct a semiphenomenological theory of the wing should not be regarded in any way as complete or final, but rather as an initial step. Even this first step, however, has enabled us to give a physical interpretation to various parts of the wing of the Rayleigh line and bring some clarity into the problem as a whole.

Our new phenomenon, the stimulated scattering of light on the wing of the Rayleigh line, arises as a result of the nonlinear interaction of the exciting and scattered radiation with the anisotropic molecules of the medium. As a result of this interaction we obtain waves of anisotropy of different frequencies. An anisotropy wave with a frequency of $1/\tau$ (τ is the relaxation time of the anisotropy) will have the largest amplitude. This explains the presence of a maximum in the emission of the stimulated wing (on exceeding a certain threshold value of the intensity of the exciting light). It may well be expected that a detailed study of this phenomenon will give new information regarding the character of the thermal motion of the molecules in liquids and the special features characterizing the interaction of powerful light fluxes with matter.

STATE OF THEORETICAL AND EXPERIMENTAL KNOWLEDGE REGARDING THE SPECTRAL COMPOSITION OF THE MOLECULAR SCATTERING OF LIGHT

1. Scattering of Light at Fluctuations of Anisotropy in Liquids (Thermal Scattering)

After the spectrum of the depolarized scattering of light in liquids had been observed in [26, 27], a large number of investigations were devoted to the further study of this phenomenon in various substances. The nature of the phenomenon has nevertheless remained uncertain up to the present time. This is partly confirmed by the fact that even recently there have been some papers [28, 29] reporting the "discovery" of maxima in the wing of the Rayleigh line (ordinary thermal scattering) in low-viscosity liquids. Actually, these maxima have no relation to the wing of the Rayleigh line but are a result of the incorrect analysis of experimental data. Apparently the authors of these papers based their treatment on the views of the first investigators (see, for example, [27]), who considered that the Rayleigh-line wing was a Raman spectrum arising as a result of the retarded rotation of anisotropic molecules. Here the liquid was considered as a compressed gas and the wing of the Rayleigh line as the remains of a Raman rotational spectrum.

The inadequacy of this point of view has been repeatedly indicated [6] and it can hardly be taken seriously nowadays. After Leontovich [8] had set up a quantitative phenomenological theory of the Rayleigh-line wing, ideas on the nature of this wing became more specific and definite. According to the Leontovich theory, the Rayleigh-line wing is due to the scattering of light at fluctuations of anisotropy, while the spectral composition of the scattering is determined by the time law governing the resorption of the anisotropy fluctuations. In this theory the anisotropy fluctuations are associated with fluctuations of the strain tensor.

As indicated by Leontovich, this theory is applicable a long way from the Mandelshtam—Brillouin doublet; for the wing it gives an intensity—frequency distribution of the dispersion type.

Experiment shows [6, 24, 25, 39] that the whole wing of the Rayleigh line is not described by a single dispersion expression. Three sections may be (approximately) distinguished in the wing. First (for low-viscosity liquids) there is a rapid fall in intensity (from 0 to 3-15 cm^{-1}) which may be described by a dispersion distribution with a specific half-width, then comes a slower fall in intensity (between 3 to 15 and 50 to 100 cm^{-1}), and finally in the remoter parts of the wing the intensity again starts falling rapidly.

Within the framework of the Leontovich calculation, Zhivlyuk [31] considered the effect of inertia by writing the kinetic equation in the form *

$$a\ddot{S}_{\alpha\beta} + \zeta\dot{S}_{\alpha\beta} + gS_{\alpha\beta} = \zeta\ddot{u}'_{\alpha\beta}. \tag{I.1}$$

Here $S_{\alpha\beta}$ is the anisotropy tensor characterizing the deviation of the medium from the isotropic state; $u'_{\alpha\beta} = u_{\alpha\beta} - \frac{1}{3}\delta_{\alpha\beta}u_{\alpha\beta}$; $u_{\alpha\beta}$ is the strain tensor, ζ is the internal friction, g is the elastic constant, and a is a parameter characterizing the inertia.

By using (I.1) we may obtain an expression for the spectral distribution of intensity in the wing, taking the form

$$\mathscr{I} = \text{const} \frac{kT}{g} \frac{\tau}{\left(1 - \frac{a\omega^2}{g}\right)^2 + \omega^2\tau^2}, \tag{I.2}$$

where $\tau = \zeta/g$. Expression (I.2) transforms into the Leontovich formula [8] at $a = 0$.

Thus, if we consider that the middle part of the wing (from 10 to between 70 and 100 cm^{-1}) is described by the Leontovich theory, then the allowance for inertia in the kinetic equation at least qualitatively explains the rapid fall in intensity in the remote parts of the wing. It should be mentioned that in (I.1) the friction (viscosity) is considered as independent of frequency. This assumption can only be justified at low frequencies.

As noted by Fabelinskii [6, 30], a relaxation of shear viscosity may appear at the frequencies corresponding to the remote part of the wing (~10^{13} cps). The possibility of this is indicated by experimental data relating to the temperature dependence of the spectral composition of the wing obtained by Fabelinskii [6, 30]. If in fact we consider that the molecules in the liquid execute a rotational Brownian movement obeying the diffusion equation, then, in this case, the intensity distribution of the light scattered at fluctuations of anisotropy will be of a dispersion nature [32]

$$\mathscr{I} = \frac{4}{45\pi} \frac{\tau_a}{1 + \omega^2\tau_a^2}, \tag{I.3}$$

where τ_a is the anisotropy relaxation time [8, 32] (from Sivukhin [40]):

$$\tau_a = \frac{1}{3}\tau_D = \frac{4}{3}\frac{\pi r^3\eta}{k\Gamma}. \tag{I.4}$$

Here, τ_D is the so-called Debye relaxation time, r is the radius of the molecule, and η is the viscosity of the medium.

It follows from (I.3) that the half-width of the Rayleigh line wing is associated with the anisotropy relaxation time by the relation

$$\Delta v = \frac{1}{\pi c\tau_a} \text{ (cm}^{-1}\text{)}. \tag{I.5}$$

Experiment shows [6, 24, 33-35] that relation (I.5) is excellently satisfied for a large number of low-viscosity liquids if we measure Δv by reference to the nearest part of the wing; immediately

---

*According to Leontovich and Zhivlyuk, the ζ and g in (I.1) are replaced by a parameter $\tau = \zeta/g$. The equation may be obtained in the form of [31] by dividing (I.1) by ζ. We found the above form better.

adjacent to the undisplaced line. This kind of broadening (I.5) should also be observed* in depolarized Raman lines [36, 37]; there also there is good agreement between theory and experiment [38]. The width of the near part of the wing, moreover, varies in inverse proportion to the viscosity, as indicated by (I.4) and (I.5).

It also follows from the Leontovich theory that the relaxation time should be proportional to the viscosity. The remote part of the wing behaves differently [6, 30]. As viscosity varies this part narrows much more slowly than the viscosity increases. This indeed gave rise to the suggestion that there might be a relaxation of viscosity at frequencies corresponding to the remote part of the Rayleigh-line wing.

Rytov, in a general correlation theory [9], took the relaxation of viscosity (and the elastic modulus) into account, writing the complex shear modulus $\tilde{\mu}$ in the form

$$\tilde{\mu} = \mu(\omega^2) + i\omega\eta(\omega^2) = \mu_\infty \frac{i\omega}{i\omega + 1/\tau_\eta}, \tag{I.6}$$

and thus $\eta = \eta_0/(1 + \omega^2\tau_\eta^2)$, $\tau_\eta = \eta_0/\mu_\infty$. Considering that the deviation of the dielectric constant from its mean value is associated with the strain tensor by means of the function

$$\varepsilon_{\alpha\beta} = X(\omega) u'_{\alpha\beta}, \tag{I.7}$$

Rytov obtained the following expression for the intensity of the shear wing:

$$\mathcal{I} = \text{const} \frac{kT}{4\pi} \left| \frac{X(\omega)}{\tilde{\mu}(\omega)} \right|^2 \eta(\omega^2). \tag{I.8}$$

In (I.8), the unknown function $X(\omega)$ still remains. It is not hard to see that, within the framework of the Leontovich theory (allowing for the inertial term in the kinetic equation), and making the natural assumption that $\varepsilon_{\alpha\beta} \sim S_{\alpha\beta}$, we obtain the following expression for the function $X(\omega)$:

$$X(\omega) = \frac{\zeta/g}{1 - \frac{a}{g}\omega^2 + i\omega\zeta/g} \tag{I.9}$$

and for the intensity in the wing of the Rayleigh line we obtain

$$\mathcal{I} = \text{const} \frac{kT}{g} \frac{\zeta}{\eta_0} \frac{\zeta/g}{(1 - \frac{a}{g}\omega^2)^2 + \omega^2(\zeta/g)^2}. \tag{I.10}$$

Here we have not considered the fact that ζ and g in (I.1) may also depend on the frequency. Actually, ζ, characterizing the friction in the resorption of the induced anisotropy, and g, characterizing the elasticity of the medium for sudden (sharp) deviations from the isotropic state in the medium, may be considered as functions of frequency with no less justification than the shear viscosity η and the shear modulus μ. It is natural that ζ and η and g and μ should differ considerably in view of the fact that ζ and g are characteristics of the action of the medium on the molecules during their oriented motion, while η and μ are characteristics of their translational motion (shear strains).

The theories of Leontovich and Rytov only allow for that part of the anisotropy fluctuation which is associated with the strain tensor (the so-called shear wing). Hence, the theory does not consider [39] the rotational Brownian motion, such as rotatory jumps of the molecules from one position (potential well) into another position of equilibrium orientation. Leontovich showed

---

*There are a number of other causes of broadening in Raman lines; however, temperature experiments enable us to distinguish the broadening due to the rotational Brownian motion.

[8] that, if such jumps of the molecules were considered as a kind of rotational diffusion, then the relaxation time of the anisotropy should be expressed by Sivukhin's formula (I.4) [40]. Valiev and Éskin [32] confirmed this by direct calculation, using the equation of rotational diffusion. This calculation is, strictly speaking, applicable to a macroscopic sphere of radius r, since, in obtaining (I.4), the Stokes formula for the coefficient of friction of a sphere moving in a viscous medium

$$\zeta = 8\pi r^3 \eta \qquad (I.11)$$

was used. Hence, the use of (I.11) and hence (I.4) for molecules may only be justified by its practical success.

Furthermore, the boundaries of applicability of expression (I.3) are not at all clear, since the diffusion equation used in deriving this expression does not contain the limiting transition to gases. A more general solution of the problem, including the limiting transition to gases, and the boundaries of applicability of (I.3) [with the half-width (I.5)] may be obtained by using the Fokker—Planck equation [41, 42] (see Chapter II).

Let us consider one further qualitative theory of the remote region of the Rayleigh-line wing. Gross and Vuks [43, 43a] assumed that the remote part of the wing (further than 10-20 cm^{-1}) owed its existence to the spreading of low-frequency lines observed in the corresponding molecular crystals in the same spectral region as the wing. Here it is considered that "quasi-crystalline" groupings are preserved in a liquid, and that so are the vibrations (of the oriented-wave type) characterizing the crystal lattice of the corresponding crystal.

We consider that the effect of rotational molecular oscillations on the Rayleigh-line wing in a liquid can hardly now be disputed, but it must be remembered that these oscillations are completely disordered, and in the majority of low-viscosity liquids there are not only no quasi-crystalline groupings but no short-range orientation order [5].

The foregoing brief review of existing ideas on the origin of the Rayleigh-line wing shows that none of the existing theories describes the phenomenon completely.

We have accordingly attempted to interpret the spectral composition of the Rayleigh-line wing quantitatively and qualitatively, basing our treatment on simple model considerations, in which we have used Leontovich's [8] assumption of the relaxation origin of the wing, Fabelinskii's idea regarding the influence of viscosity relaxation on the spectral intensity distribution in the wing [6], and the Gross and Vuks theory regarding the possible relation between the Rayleigh-line wing and the low-frequency spectrum of the corresponding molecular crystals [43]. We have started from the idea that the rotational jumps of molecules from one position of orientational equilibrium to another makes the main contribution to the part of the wing nearest to the undisplaced line. This part of the wing may be described by expressions (I.3) and (I.4).

The aperiodic rotational oscillations of the molecules in the intervals between the orientational jumps are principally responsible for the appearance of the remote part of the wing, since these movements are more rapid than the orientational jumps. The oscillations of the molecules are associated with the appearance of elasticity when they deviate from the equilibrium orientation, and hence with the shear strains. It is in this way that we establish a relation between the shear wing, described by the phenomenological theory, and the wing obtained on considering the scattering of light by molecules executing rotational oscillations.

It may be expected that this theory (see Chapter II) and our experimental results (see Chapter IV) will serve to develop physically and mathematically more comprehensive ideas regarding the nature of the Rayleigh-line wing.

We cannot here give detailed consideration to the large number of experiments devoted to the study of the Rayleigh-line wing hitherto published; we must mention, however, that the most important of these [6, 30, 33-35, 44] are devoted to the quantitative study of the near part of the wing and the qualitative analysis of the remoter regions. To a considerable extent these papers have been concerned with verifying the applicability of relation (I.4) to a large number of low-viscosity liquids over a wide range of temperatures.* At the present time the problem is one of understanding the origin of the wing as a whole over the whole range of frequencies accessible to observation.

2. Fine Structure of the Molecular-Scattering Lines of Liquids

The fine structure of the lines representing the molecular scattering of light predicted by Mandelshtam [45] and Brillouin [46], and observed in liquids by Gross [47] enables us to study the acoustic properties of substances at frequencies of $\sim 10^{10}$ cps. This range of hypersonic frequencies is hardly accessible to ultrasonic technology at the present time.† Furthermore, in this range of frequencies the hydrodynamic theory of the propagation of sonic vibrations becomes inapplicable, while the use of the relaxation theory lacks direct experimental proof. Hence, a study of the laws governing the propagation of high-frequency sonic vibrations in condensed media is important both in the scientific and practical respects. Interest in this field of investigation has heightened recently because the new phenomenon of stimulated Mandelshtam—Brillouin scattering has been discovered [19-22], and this is much more promising for the study of the acoustic properties of matter and for practical use.

The scattering of light at adiabatic fluctuations of density may be considered as the diffraction of light at thermal elastic Debye waves propagating in a condensed medium. This kind of consideration leads to the conclusion that the light scattered at adiabatic fluctuations of density will be modulated by the frequency f of the thermal Debye wave. The wave vectors of the exciting light $\mathbf{k}_0$, the scattered light $\mathbf{k}_p$, and the sound wave $\mathbf{q}$ should satisfy the well-known Bragg—Wulff relation

$$\mathbf{k}_0 - \mathbf{k}_p = \pm \mathbf{q}. \tag{I.12}$$

The displaced components (Mandelshtam—Brillouin components) are shifted relative to the undisplaced line by a distance $\Delta\nu$, given by

$$\Delta\nu = \pm f = \pm 2n\nu \frac{V}{c} \sin\frac{\vartheta}{2}. \tag{I.13}$$

Here, n is the refractive index of the medium, ν is the frequency of the exciting light, V is the velocity of sound, c is the velocity of light, and ϑ is the scattering angle.

Thus, the spectroscopic determination of $\Delta\nu$ offers the possibility of finding the frequency of the hypersound f and by means of formula (I.13) its phase velocity of propagation V. On propagation in a liquid, particularly at high frequencies, an elastic wave shows considerable attenuation. This attenuation, firstly, leads to a broadening of the Mandelshtam—Brillouin components, and secondly, slightly changes the position of their maximum. Solution of the problem

---

*Relation (I.4) can only claim agreement with experiment in order of magnitude. In the majority of cases for low-viscosity liquids the calculated and measured values of τ_a differ by not more than 30%.

†Only in [48, 49] are there any results on the measurement of the acoustic properties of liquids by means of artificially generated hypersound.

regarding the spectral composition of the scattering at adiabatic density fluctuations (using the hydrodynamic equation for the propagation of sound in a dissipative medium) leads to the following* expression for the intensity of the scattered light [1]:

$$\mathcal{I} = \frac{\frac{\Omega_0^2}{\pi}\delta}{(\Omega_0^2 - \omega^2)^2 + \delta^2\omega^2} \, . \tag{I.14}$$

Here, $\Omega = 2\pi f$; ω is the angular frequency, reckoned from the frequency of the exciting light; $\delta = 2\alpha V$; α is the amplitude of the hypersound attenuation coefficient.

By using (I.14), it is quite easy to find the position of the maxima of the fine-structure components:

$$\omega_{\max} = \pm\,\Omega_0\left(1 - \frac{\delta^2}{2\Omega_0^2}\right)^{1/2} \tag{I.15}$$

and the half-width of each component

$$\delta\omega \approx \delta\left(1 + \frac{1}{8}\frac{\delta^2}{\Omega_0^2} + \cdots\right) \quad (\delta < \Omega_0), \tag{I.16'}$$

or for $\delta \ll \Omega_0$,

$$\delta\nu \approx \frac{\alpha V}{\pi c}\ (\text{cm}^{-1}). \tag{I.16''}$$

It follows from expression (I.14) that each Mandelshtam—Brillouin component is slightly asymmetric, in such a way that the inner part of the contour is wider than the outer (away from the undisplaced line). The distance from the maximum of the component to the point at which the intensity has been halved will be: from the outer side of the contour,

$$\delta\omega_1 \approx \frac{\delta}{2}\left(1 - \frac{1}{4}\frac{\delta}{\Omega_0} + \frac{1}{8}\frac{\delta^2}{\Omega_0^2} - \cdots\right), \tag{I.17'}$$

from the inner side of the contour,

$$\delta\omega_2 \approx \frac{\delta}{2}\left(1 + \frac{1}{4}\frac{\delta}{\Omega_0} + \frac{1}{8}\frac{\delta^2}{\Omega_0^2} + \cdots\right), \tag{I.17''}$$

in which $\delta\omega = \delta\omega_1 + \delta\omega_2$ $(\delta < \Omega_0)$.

With the present accuracy (~10%) of measuring the width of the fine-structure components for the liquids studied in the present investigation, this asymmetry and the correction of $\sim \frac{1}{8}(\delta^2/\Omega_0^2)$ in expression (I.16') may be neglected and we may use formula (I.16''). However, the influence of absorption on the position of the maximum [formula (I.15)] cannot be neglected in certain cases if we remember that the accuracy of measuring the position of the maximum, using narrow-lined lasers as source of the exciting light, is about 0.5%.

Thus, by measuring the distance between the fine-structure components and their half-width, we may determine the velocity, absorption, and frequency of the hypersound.

The hydrodynamic theory [50] for the absorption coefficient of sound propagating in a condensed medium gives the following expression †

---

*The expression given for the spectral intensity distribution in the fine-structure components is normalized to unity.

† The absorption due to thermal conductivity is negligibly low [6] and thus the corresponding term is omitted from (I.18).

$$\alpha = \alpha_\eta + \alpha_{\eta'} = \frac{\omega^2}{2V_0^3\rho}\left(\frac{4}{3}\eta + \eta'\right). \tag{I.18}$$

Here V_0 is the velocity of sound (subsequently, V_0 will mean the velocity of sound at a low frequency), ρ is the density of the medium, and η and η' are the shear and volume viscosities of the sample.

If the data relating to the absorption of hypersound represented by formula (I.18) are extrapolated to the frequency Ω_0 of the hypersonic waves responsible for the appearance of the fine structure, then (I.16) gives a value for the width of the Mandelshtam—Brillouin components greater than the distance between them, so that the fine structure should not be observed at all. Experiment, however, reveals a very clear fine structure.

This contradiction was resolved by the relaxation theory of Mandelshtam and Leontovich [16], which gave the following expression for the absorption coefficient of sound due to volume viscosity:

$$\alpha_{\eta'} = \frac{\omega^2\tau\,(V_\infty^2 - V_0^2)}{2V_0^3(1 + \omega^2\tau^2)}, \tag{I.19'}$$

where τ is the relaxation time of the volume viscosity coefficient, and V_∞ is the velocity of hypersound at an infinite frequency. On comparing $\alpha_{\eta'}$ from (I.18) and (I.19') at low frequencies $\omega\tau \ll 1$, where they ought to coincide, we find an expression for $\alpha_{\eta'}$ (valid at any frequency) differing from (I.19'):

$$\alpha_{\eta'} = \frac{\omega^2}{2\rho V_0^3}\frac{\eta_0}{1 + \omega^2\tau^2}. \tag{I.19''}$$

We see from (I.19'') and (I.18) that even at high frequencies we may use formula (I.18) if we replace the static value of the volume viscosity η_0' by η', relaxing with a time τ

$$\eta' = \frac{\eta_0'}{1 + \omega^2\tau^2}.$$

The relaxation theory assumes the existence of a dispersion of the velocity of sound in the liquid ($V_\infty \neq V_0$).

If we know V_0, V_∞, τ, and ω, then the velocity of hypersound at a given frequency ω may be found from the expression

$$\frac{V_\omega^2 - V_0^2}{V_\infty^2 - V_0^2} = \frac{\omega^2\tau^2}{1 + \omega^2\tau^2}. \tag{I.20}$$

In practice, the opposite occurs: knowing V_ω (from measuring the distance between the components), we find τ for the liquid in question from the expression*

$$\tau = \frac{Z + (Z^2 - 4\omega^2)^{1/2}}{2\omega^2}, \tag{I.21}$$

where

$$Z = \frac{\omega^2\eta_0'}{\rho\,(V_\omega^2 - V_0^2)},$$

or, more simply (but less accurately),

---

*This expression for τ was obtained from (I.19'), (I.19''), and (I.20).

$$\tau \approx \frac{\eta'}{2V_0^2 \rho} \frac{V}{\Delta V},$$ (I.22)

where ΔV is the dispersion of the velocity of sound.

Thus, τ may be determined from two independent measurements: the absorption and the velocity dispersion of sound. By measuring the width of the fine-structure components and determining the amplitude of the hypersound attenuation factor from (I.16), we may use expression (I.18) to find $\alpha_{\eta'}$ and then (I.19") to find τ. On the other hand, τ may be determined from (I.21) by measuring $\Delta\nu$ and determining V_ω from (I.13), or if necessary from (I.15). By comparing the values of τ obtained by the two independent measurements, we may verify the validity of the version of relaxation theory with one relaxation time, for the particular liquid in question.

The dispersion of the velocity of sound predicted by the relaxation theory was observed in a number of liquids (benzene, carbon disulfide, carbon tetrachloride) by Fabelinskii [6], who measured the distance between the fine-structure components (the dispersion was 9-12%). Soon after, an analogous result was obtained for several other liquids [10-12]. Recently, progress in ultrasonic techniques has made it possible to measure the velocity and absorption of sound in some liquids at a frequency of 10^8-10^9 cps [48, 49, 51]. These measurements also confirm the existence of dispersion of the velocity of sound in certain liquids.

Thus, one of the main points of the relaxation theory has been shown to agree closely with experiment. However, for a final check on the quantitative relationships of the relaxation theory it was necessary to measure the absorption of hypersonic waves at the frequencies for which velocity dispersion occurred, since the relaxation theory gives a relation between the velocity dispersion and the absorption of hypersound. Knowing the velocity dispersion of hypersound, and using the relaxation-theory formulas, Fabelinskii estimated the expected broadening for a number of liquids and obtained a value of the order of $\delta\nu \sim 10^{-2}$ cm^{-1} [6]. Such small quantities could not be measured earlier owing to the great width of the exciting spectral line. The appearance of gas lasers enabled us to carry out such measurements [13, 14] and show [14] that for a number of liquids the relations of the relaxation theory with one relaxation time were satisfied to a fair accuracy.

Expression (I.14) gives the intensity distribution of the Mandelshtam—Brillouin components and does not contain the component not displaced in frequency, which is due to the scattering of light at isobaric density fluctuations. The existence of this component in the scattered light was first indicated by Mandelshtam [45]. The intrinsic width of this line is determined by the thermal diffusivity of the medium; according to Fabelinskii [6] it should be two orders smaller than that of the Mandelshtam—Brillouin component. Up to the present time the width of the central component of light scattering in pure liquids has not been measured. Such measurements may be very important, since a knowledge of the spectral distribution of intensity in the central component and in the Mandelshtam—Brillouin components enables us to find the space—time correlation functions for the liquids [52].

It should be noted that the measurement of the width of the Mandelshtam—Brillouin components and the velocity of hypersound in liquids with a large shear viscosity as functions of temperature may prove very important in the theory of very viscous liquids, particularly for the Isakovich—Chaban theory presently under discussion [53].

3. Mandelshtam—Brillouin Stimulated Scattering

The intensity of the molecular scattering of light at fluctuations of density and anisotropy increases linearly with increasing linear dimension of the scattering volume. In contrast to this ordinary thermal scattering, the intensity of the stimulated Mandelshtam—Brillouin scattering increases very nonlinearly with increasing linear dimensions of the scattering volume in the direction of the scattered light. This phenomenon sets in after the intensity of the exciting light has exceeded a certain threshold value, which differs for different samples.

The phenomenon of stimulated Mandelshtam—Brillouin scattering was first observed in quartz and sapphire crystals by Chiao, Townes, and Stoicheff [19]. Soon after, the phenomenon was also observed in liquids [20, 21].

The threshold values of the intensity of the exciting light necessary to observe stimulated Mandelshtam—Brillouin scattering are, for example, 30 MW/cm^2 for carbon disulfide and 1200 MW/cm^2 for benzene [20]. At the present time such intensities may be achieved by focusing the radiation of a giant-pulse ruby laser.

The development of stimulated scattering may be explained by the fact that, in the electric field of a high-intensity light wave an important part is played by the effect of the exciting and scattered waves on the character of the motion in the medium; elastic waves of considerable amplitude arise in this medium at the difference frequency of the exciting and scattered light.

In order to describe the stimulated Mandelshtam—Brillouin scattering we must carry out a simultaneous solution of the Maxwell and hydrodynamic equations, allowing for the nonlinearity of the medium resulting from the high intensity of the exciting light. The nonlinearity in the medium is associated with the fact that, under the influence of the electric field, electrostriction develops, causing a relative change of volume

$$\frac{\Delta V}{V} = \frac{1}{8\pi} \beta_s \left(\rho \, \frac{\partial \varepsilon}{\partial \rho} \right) \mathbf{E}^2. \tag{I.23}$$

Here, β_S is the adiabatic compressibility of the medium.

The solution of the problem of stimulated Mandelshtam—Brillouin scattering was given by Akhmanov and Chin Dong Ah [54] (see also [1, 18, 55]).

On studying stimulated scattering in liquids, we may find several Stokes or anti-Stokes components [20-22]. This is evidently associated with a successive-scattering mechanism. When in fact light scattered at 0 or 180° is being studied, the light scattered backwards, after suffering a frequency shift of $\pm\Omega_0$, returns to the ruby, where it is amplified, returns once more to the scattering medium, and produces further scattering, now with a frequency shift of $\pm 2\Omega_0$, etc. As a result of this mechanism of successive scattering, several tens of displaced components may be obtained. The question as to how far the second, third, and so on, Mandelshtam—Brillouin components are associated with a nonlinearity of higher order than quadratic at the moment remains open. In order to follow this up, it is of particular interest to study the time sequence in which the components of stimulated Mandelshtam—Brillouin scattering appear.

CHAPTER II

THEORY OF THE SPECTRAL COMPOSITION
OF THE DEPOLARIZED SCATTERING
OF LIGHT IN LIQUIDS

1. Relation Between the Spectral Composition
of the Scattered Light and the Rotational
Thermal Motion of the Molecules in the Liquid

In dealing with this problem we shall start from the assumption that the "wing" of the Rayleigh line arises as a result of the scattering of light at fluctuations of anisotropy, i.e., scattering at random deviations from the isotropic distribution of the axes of the anisotropic molecules. The time factor in the resorption of the anisotropy fluctuations determines the spectral composition of the anisotropic scattering. The mathematical relation between the spectral composition of the scattered light and the time characteristics of the attenuation of the fluctuation is established by introducing a time-correlation function. Let us introduce the tensor S_{ik} characterizing the deviation of the distribution of molecular axes in the medium from the isotropic state; we define this tensor so that its mean value over the volume is zero (medium isotropic on average). The tensor S_{ik} should be a function of the angles determining the orientation of the axes of the molecules relative to the stationary (laboratory) system of coordinates x, y, z (indices i, k, etc.).

In the molecules we set up a molecularly stationary system of coordinates ξ, η, ζ, the axes of which coincide with the principal axes of the polarizability ellipsoid. In subsequent calculations we shall only be concerned with molecules of the symmetrical gyrostat type. In this case, it is convenient to take the following functions of angles as the tensor S_{ik} [56]:

$$S_{ik} = \sigma_i \sigma_k - \frac{1}{3} \delta_{ik}.$$

(II.1)

Here, σ_i and σ_k are the direction cosines determining the position of the axis of symmetry of the molecule relative to the stationary system of coordinates. Let us transform to a spherical coordinate system ϑ and φ. Then the scattering intensity of the light polarized perpendicularly to the plane of scattering ($\mathcal{I}_z$) and in the plane of scattering ($\mathcal{I}_x$), will be determined by the time correlation functions [4, 8, 25, 56] of

$$S_z = \cos^2 \vartheta - \frac{1}{3}, \quad S_x = \sin \vartheta \cos \vartheta \cos \varphi,$$

(II.2)

i.e.,*

$$\mathcal{I}_z = \frac{1}{\pi} \int\limits_0^\infty \langle S_z(t_0) S_z(t') \rangle \, e^{-i\omega t} \, dt,$$

$$\mathcal{I}_x = \frac{1}{\pi} \int\limits_0^\infty \langle S_x(t_0) S_x(t') \rangle \, e^{-i\omega t} \, dt. \tag{II.3}$$

The process is considered stationary, i.e., the time correlation functions

$$f_z(t) = \langle S_z(t_0) S_z(t') \rangle \quad \text{and} \quad f_x(t) = \langle S_x(t_0) S_x(t') \rangle$$

only depend on $t = t' - t_0$.

In order to find the time correlation function $f(t) = \langle B(t_0)A(t_0 + t) \rangle$ (stationary process) for $\langle A \rangle = \langle B \rangle = 0$ we may follow Vladimirskii [57] by calculating the mean $\langle a_\alpha \rangle = \int A(X)\rho(X, t, \alpha)dX$ for the motion of the system with initial conditions corresponding to the equilibrium of this system with an additional energy $\alpha B(X)$ at $t = 0$. Then [57],

$$f(t) = \langle A(t_0) B(t_0 + t) \rangle = -kT \left[\frac{\partial}{\partial \alpha} a_\alpha(t) \right]_{\alpha=0}. \tag{II.4}$$

Here X are canonical variables, $\rho(X, t)$ is the phase density, α is a parameter by means of which the additional energy is introduced, so that the Hamiltonian for $t = 0$ will have the form

$$H_1(X) = H(X) + \alpha B(X), \tag{II.5}$$

where $H(X)$ is the Hamiltonian of the system in equilibrium and outside the field of external forces.

We shall use this approach in the next section when calculating the time-correlation functions on the basis of the Fokker—Planck equation; it is convenient to do this when a time equation for the probability density exists. If we have an equation directly describing the time law for the resorption of the anisotropy fluctuation, we may use simpler methods for determining the time correlation function (or more precisely, the required spectral density of the correlation function). These methods are well known [58, 59] and we shall not set them out here.

In order to find the time-correlation functions we must have an equation describing the time law of resorption of the anisotropy fluctuation, or an equation describing the rotational motion of the molecules. Unfortunately, we have no equation describing the rotational motion of the molecules simultaneously for short and fairly long time intervals. Starting from modern concepts of the kinetic properties of simple liquids [2, 3, 60], we shall consider that the rotational motion of the molecules consists of comparatively rare jumps of molecules from one position of equilibrium to another. In the intervals between the jumps the molecules execute elastic rotational oscillations due to random collisions with neighboring molecules. If on average the transformation of a molecule from one state to another occurs in a time τ_l, then over

---

*Here we omit the factor

$$\sim (\alpha_1 - \alpha_2)^2 \, \omega_0^4 / R^2 c^4,$$

which is not essential for subsequent arguments; ω_0 is the frequency of the exciting light, c is the velocity of light, α_1 and $\alpha_2 = \alpha_3$ are the principal polarizabilities of the molecule, R is the distance from the scattering entity to the point of observation.

periods of $t \gg \tau_l$ we may justifiably think of this as smooth motion and describe it as rotational (or rotatory) diffusion. The estimates of Valiev and Agishev [61], in fact, show that from the theoretical point of view this smoothing is valid. These estimates give angles of the order of several degrees for the jumps.

The so-called Debye relaxation times calculated from the equation of rotational diffusion and measured experimentally [62] agree closely, further confirming the validity of the equation of rotational diffusion. It is quite clear that this equation cannot be valid for time intervals $t < \tau_l$, so that at frequencies of $\omega > 1/\tau_l$ we must not use the diffusion equations to describe the spectral composition of anisotropic scattering.

For time intervals $t < \tau_l$ the motion bears the character of elastic oscillation, a frictional force acting on the molecules. The coefficient of friction for the elastic oscillations of the molecules must be considered as relaxing, i.e., depending on frequency in the spectral description. During periods of time smaller than the period of the elastic oscillations, the molecules are unable to exchange moments of momentum with their neighbors, and hence there is no frictional force; for longer intervals the ordinary frictional force acts on the molecules. This means that we may consider the coefficient of friction as depending on time (in the time equation) or on frequency in the spectral description.

On the basis of the foregoing argument, we shall solve the problem as to the influence of the rotational motion of the molecules on the spectral composition of anisotropic scattering for two spectral ranges: for the range $\omega > 1/\tau_l$ (using the equation describing the elastic oscillations of the molecules with relaxing friction) and for the range $\omega < 1/\tau_l$ (using the equation of rotational diffusion).

2. Effect of Rotational Diffusion of Molecules on the Spectral Composition of the Depolarized Scattering of Light [24, 25]

In this section we shall calculate the spectral distribution of the intensity of depolarized scattering resulting from the modulation of the scattered light by the rotational Brownian motion of the anisotropic molecules. In this calculation the probability densities required for obtaining the correlation functions $f_Z(t)$ and $f_X(t)$ (see Section 1) obey the well-known Fokker—Planck equation [41, 42]. As a result of the calculation we obtain general equations from which we may readily determine the intensity distribution in the spectrum of depolarized scattering, both in liquids with a fairly high viscosity (see later) and in the case of rarefied gases. The intermediate case of low viscosities cannot be solved in analytical form, but in principle a solution may be effected with the help of a computer, using a simplified equation.

Let us consider a linear molecule, the position of which is given by spherical coordinates ϑ and φ. The Fokker—Planck equation for the probability density* $w(t, \vartheta, V_\vartheta, V_\varphi)$ in the space of the angles and angular velocities of the molecules V_ϑ and V_φ has the form [42]

$$\frac{\partial w}{\partial t} + V_\vartheta \frac{\partial w}{\partial V_\vartheta} + \operatorname{ctg} \vartheta \left(V_\varphi \frac{\partial w}{\partial V_\vartheta} - V_\vartheta \frac{\partial w}{\partial V_\varphi} \right) = \beta \frac{\partial}{\partial V_\vartheta} \left(V_\vartheta w + \frac{kT}{I} \frac{\partial w}{\partial V_\vartheta} \right) + \beta \frac{\partial}{\partial V_\varphi} \left(V_\varphi w + \frac{kT}{I} \frac{\partial w}{\partial V_\varphi} \right). \quad \text{(II.6)}$$

According to the method of calculating the correlation functions set out in the previous paragraph, we put the initial condition in the form

$$w(0, \vartheta, V_\vartheta, V_\varphi) \approx \frac{1}{2} \left[1 - \frac{\alpha}{kT} S_z(\vartheta) \right] \exp \left[- \frac{I(V_\vartheta^2 + V_\varphi^2)}{2kT} \right]. \quad \text{(II.7)}$$

_____

*For simplicity we shall consider only the correlation function $f_Z(t) = \langle S_Z(t_0) S_Z(t') \rangle$; then the dependence on φ need not be taken into account.

In (II.6) and (II.7), $\beta = \zeta / I$ (ζ is the coefficient of internal friction, I is the moment of inertia of the molecule). In order to find $\langle S_z \rangle_\alpha$ we must carry out an averaging over the velocities V_ϑ and V_φ. Hence, it is more convenient to transform from the function w(t, ϑ, V_ϑ, V_φ) to the function ψ.

$$\psi(t, \vartheta, u_\vartheta, u_\varphi) = \exp\left[\frac{kT}{2I}(u_\vartheta^2 + u_\varphi^2)\iint w \exp\left(-iu_\vartheta V_\vartheta - iu_\varphi V_\varphi\right)dV_\vartheta\, dV_\varphi\right], \qquad (II.8)$$

and then put

$$u_\vartheta = u \cos\chi, \quad u_\varphi = u \sin\chi. \qquad (II.9)$$

Then, for averaging over the velocities we must take u = 0. Transforming (II.8) and (II.9) we find

$$\psi = \psi_0 + \psi_1(t, u)\cos^2\vartheta - i\psi_2(t, u)\sin\vartheta\cos\vartheta + \frac{1}{2}\psi_3(t, u)\sin^2\vartheta\sin^2\chi + \ldots. \qquad (II.10)$$

Hence, for finding ψ_1, ψ_2, and ψ_3 we obtain the following system of differential equations:

$$\left.\begin{array}{l}
\dfrac{\partial\psi_1}{\partial t} + 2\dfrac{\partial\psi_2}{\partial u} + \dfrac{\psi_2}{u} - \dfrac{2kT}{I}u\psi_2 = -\beta u \dfrac{\partial\psi_1}{\partial u}, \\[2mm]
\dfrac{\partial\psi_2}{\partial t} + 2\dfrac{\partial\psi_1}{\partial u} + \dfrac{\psi_3}{u} - \dfrac{2kT}{I}u\psi_1 = -\beta u \dfrac{\partial\psi_2}{\partial u}, \\[2mm]
\dfrac{\partial\psi_3}{\partial t} + 2\dfrac{\partial\psi_2}{\partial u} - 2\dfrac{\psi_2}{u} - \dfrac{2kT}{I}u\psi_2 = -\beta u \dfrac{\partial\psi_3}{\partial u}
\end{array}\right\} \qquad (II.11)$$

with initial conditions t = 0, $\psi_1 = -\alpha/2kT$, $\psi_2 = \psi_3 = 0$. The system of equations (II.11) may be solved for small u. After simple but cumbersome calculations we obtain an expression for the correlation function

$$f_z(t) = \frac{1}{15}\frac{I}{kT}\int_0^\infty \lambda R(\lambda, t)\exp\left(-\frac{I}{2kT}\lambda^2\right)d\lambda, \qquad (II.12)$$

where the function $R(\lambda, t)$ in turn is determined by the equation

$$\frac{d^3R}{dt^3} + 3\beta\frac{d^2R}{dt^2} + 2\beta^2\frac{dR}{dt} + 4\lambda^2 e^{-2\beta t}\frac{dR}{dt} + \frac{12\beta kT}{I}R = 0. \qquad (II.13)$$

For t = 0

$$R = \frac{4}{3}, \quad \frac{dR}{dt} = 0, \quad \frac{d^2R}{dt^2} = -4\lambda^2. \qquad (II.14)$$

For $\beta = 0$ (rarefied gas) we obtain the following expression for the spectral density*

$$I_z = \frac{4}{3}\mathcal{J}_x = \frac{1}{120}\frac{I}{kT}|\omega|\exp\left(-\frac{\mathcal{J}}{8kT}\omega^2\right) + \frac{1}{45}\delta(\omega), \qquad (II.15)$$

equivalent (apart from an unimportant factor) to the earlier [63] expression for the envelope of the Raman rotational band. Thus, the general solution of (II.12) [together with (II.13)] contains the limiting case of a rarefied gas. We note that the earlier calculations of the depolarized scattering of light in liquids [8, 39] did not allow this transition.

---

*Here we make no allowance for other causes of broadening; hence, in the rarefied gas the Rayleigh line is obtained in the form of a delta function.

For the case of a fairly high viscosity, and on condition that

$$\beta^2 \gg \frac{4kT}{I}, \quad \omega < 2\beta, \tag{II.16}$$

which are usually satisfied for ordinary liquids at reasonably low temperatures, we obtain the following for the intensity of the scattered light:

$$\mathcal{I}_z = \frac{4}{3}\mathcal{I}_x = \frac{I\beta/6kT}{\left(1 - \frac{I}{4kT}\omega^2\right)^2 + \omega^2\left(\frac{I\beta}{6kT}\right)^2}. \tag{II.17}$$

It is quite easy to see that in (II.17), $I\beta/6kT = \zeta/6kT = \frac{1}{3}\tau_D$, where τ_D is the so-called Debye relaxation time [62].

The spectral distribution of scattered light thus obtained differs from the dispersion distribution obtained earlier [32] (from the simple equation of rotational diffusion) by the presence of the inertial term in the denominator. The fact that here we have obtained criterion (II.16) for the applicability of (II.17) is important. In addition to this, it should be remembered that expression (II.17) is applicable in the case of a real liquid to the same degree as is the concept of the rotational Brownian motion in the case of a liquid molecule.

In order to obtain the intensity distribution in the case of a compressed gas or liquid at high temperature [when (II.16) is not satisfied], we must find an exact solution of Eq. (II.13) and then take the integrals (II.3) and (II.12). This solution should give a gradual transformation from distribution (II.15) to (II.17) with increasing viscosity. On passing from high to low viscosities, Eq. (II.17) shows that the wing under consideration first starts broadening [at least while (II.16) is satisfied] and then as viscosity falls further (compressed gas, high-temperature liquid) the central part of the wing starts contracting, and rotational maxima start forming on either side of the center.

At the limit, for a compressed gas, we obtain a narrow line in the center, with rotational maxima on either side of it. The character of the distribution (II.15) should in general be preserved at frequencies $\omega > 2\beta$ even for liquids.

In real liquids, at reasonably low temperatures, these frequencies lie beyond the range covered by the wing; for example, in the case of CS_2 at room temperature, $2\beta \sim 250\ cm^{-1}$.

It follows from expression (II.17) that, for small values of β there may be a maximum at a frequency of

$$\omega_{max} = \pm 2\sqrt{\frac{kT}{I}\left(1 - \frac{I\beta^2}{9kT}\right)} \tag{II.18}$$

in the depolarized-scattering intensity distribution. For liquids at reasonably low temperatures the maximum should not appear, since, usually,

$$\beta^2 \gg \frac{9kT}{I}.$$

For large β the influence of the inertial term in the denominator of (II.17) can only appear at frequencies $\omega \sim \beta$, which for ordinary low-viscosity liquids lie beyond the range of the wing. Hence, on satisfying conditions (II.16), we may use the dispersion formula (I.3) for the spectral distribution of intensity in the wing [neglecting the inertial term in (II.17)], with a relaxation time

$$\tau_a = \frac{\zeta}{6kT}. \tag{II.19}$$

If we take the Stokes formula (I.11) for the coefficient of internal friction, then (II.19) transforms into the ordinary expression for the anisotropy relaxation time (I.4). As already mentioned in Chapter I, we cannot regard it as strictly valid to use the hydrodynamic Stokes theory, which was developed and justified for the motion of a macroscopic sphere in a viscous liquid, in connection with the molecule; only practical confirmation will justify such a step.

3. Modulation of Scattered Light by the Rotational Oscillations of the Molecules in a Liquid

Let us consider the rotational motion of molecules over time intervals $t < \tau_l$, where τ_l are the time intervals between two jumps of orientation. A molecule situated in the fields of its surrounding neighbors executes elastic (but apparently aperiodic) oscillations as a result of random collisions. If the equilibrium position of the molecule is determined by spherical coordinates ϑ and φ, then for a small deviation $\Delta\vartheta$ of the axis of the molecule from the position of equilibrium we may write the equation of motion in the form*

$$I\frac{d^2\Delta\vartheta}{dt^2} + \zeta\frac{d\Delta\vartheta}{dt} + g\Delta\vartheta = M(t). \tag{II.20}$$

Here, ζ is the internal friction, g is the coefficient of elasticity, and M(t) is the random force. The oscillations of the molecule produce a change in $S_z(t)$ equal to

$$\Delta S_z = -2\sin\vartheta\cos\vartheta\Delta\vartheta. \tag{II.21}$$

The spectral densities of the correlation functions $\langle\Delta S_z(t_0)\Delta S_z(t_0 + t)\rangle$ and $\langle\Delta\vartheta(t_0)\Delta\vartheta(t_0 + t)\rangle$ are related by the equation

$$\langle(\Delta S_z)^2\rangle_\omega = 4\langle\sin^2\vartheta\cos^2\vartheta\rangle\langle(\Delta\vartheta)^2\rangle_\omega. \tag{II.22}$$

From the stochastic equation (II.20) we easily find $\langle(\Delta\vartheta)^2\rangle$ [58, 59], and, after averaging over the angles in (II.22), we obtain

$$\mathcal{I}_z = \frac{4}{3}\mathcal{I}_x = \frac{4}{45\pi}\frac{6kT}{g}\frac{\zeta/g}{\left(1 - \frac{I}{g}\omega^2\right)^2 + \omega^2(\zeta/g)^2}. \tag{II.23}$$

It was assumed in the foregoing that the internal friction and coefficient of elasticity in (II.20) were independent of time, and hence, in formula (II.23), these quantities are independent of frequency. As already mentioned in Section 1 of Chapter I, in the part of the wing corresponding to molecular motion over time intervals $\sim 10^{-13}$ sec (frequency $\sim 10^{13}$ cps), dispersion of the viscosity (and hence the elastic coefficient) is quite possible.

Here we shall simply start from the fact that during the rotatory motion of the molecules over infinitesimally small intervals of time, the effect of the frictional force on the molecule may be neglected. In this case, only the elastic force will act on the molecule. On considering the motion of the molecule for intervals of time t greater than the characteristic time τ_ζ, the molecule will be acted upon by both friction and the elastic force. Considering all this, the equation of motion will be written in the following form:

---

*We consider $f_z(t) = \langle S_z(t_0)S_z(t_0 + t)\rangle$. Then we need only have an equation describing the change in the angle between the axis of the molecule and a selected direction. For $\Delta\varphi$ we may set down a similar equation.

$$I \frac{d^2 \Delta \vartheta}{dt^2} + \zeta_0 \frac{d \Delta \vartheta}{dt} + g_0 \Delta \vartheta = M(t), \quad \text{for large t,}$$

$$I \frac{d^2 \Delta \vartheta}{dt^2} + g_\infty \Delta \vartheta = M(t), \quad \text{for small t.} \tag{II.24}$$

Then for the scattered-light intensity we obtain the same expression (II.23), but with frequency-dependent parameters

$$\zeta = \frac{\zeta_0}{1 + \omega^2 \tau_\zeta^2}, \quad g = g_0 + \tau_\zeta \zeta_0 \frac{\omega^2}{1 + \omega^2 \tau_\zeta^2}. \tag{II.25}$$

In (II.24) and (II.25), ζ_0 and g_0 are the friction and elastic constant at low frequencies, τ_ζ is the relaxation time of the internal friction, and g_∞ is the elastic constant at high frequencies,

$$g_\infty = g_0 + \frac{\zeta_0}{\tau_\zeta}. \tag{II.26}$$

On considering the motion of the molecule for short time intervals, the statistical action of the surrounding neighbors appears as a force tending to return the orientation to the equilibrium position. However, for large time intervals $t \sim \tau_l$, when the effect of the neighboring molecules tends to turn the original molecule into a different equilibrium orientation, this statistical force of the surrounding molecules changes sign, i.e., on average, the elastic force in Eq. (II.24) should change sign in a time τ_l. Then, after passing into the new equilibrium position a restoring force again acts on the molecule until the next reorientation is due. The foregoing argument means that for liquids we must regard g_0 as negative. Then, according to (II.25), g will only be greater than zero for high frequencies, starting from a particular critical value — the frequency at which $g(\omega_l = 0)$ must be identified with the frequency of reorientation of the molecules from one equilibrium orientation to another, i.e., $\omega_l \sim 1/\tau_l$. Naturally, formulas (II.23) and (II.25) are only applicable for $\omega > \omega_l$, when $g(\omega) > 0$.

Both of the mechanisms considered, the Brownian rotational motion and the elastic oscillations of the molecules, may be realized in a liquid. For short intervals of time, shorter than the time spent by the molecule in the potential well $t < \tau_l$, the motion of the molecule may be considered as elastic oscillations and described by Eq. (II.24). This means that at frequencies $\omega > 1/\tau_l$ the intensity distribution in the wing will have the form (II.23) with parameters (II.25).

Suppose that on average the molecule changes from one potential well to another in a time interval $t \sim \tau_l$. Then over periods of $t \gg \tau_l$ the whole set of such transitions may tentatively be regarded as a rotational diffusion, neglecting the slight oscillations of the molecules between transitions for such periods of time. This means that for $\omega \ll 1/\tau_l$ we may use expression (II.17). Naturally we must still satisfy (II.16).

The region of the transition $\omega_l \sim 1/\tau_l$ from the distribution (II.17) to the distribution (II.28) with parameters (II.25) should tend toward the undisplaced line as viscosity rises, since τ_l plays the part of the time spent by the molecules in the potential well. For a very large viscosity the diffusion wing should be drawn toward the center, and we shall only have the wing of elastic oscillations, (II.23) and (II.25) showing that for fairly long internal-friction relaxation times we may expect the appearance (on both sides of the central line) of a pair of depolarized maxima at a frequency of

$$\omega_{\max} \sim \left(\frac{g}{I} \right)^{1/2}. \tag{II.27}$$

Naturally we have only obtained one pair of maxima here, since we have considered a linear molecule with one moment of inertia. However, for large internal-friction relaxation times, corresponding to large values of static viscosity, the foregoing theory may be regarded as inapplicable, since it takes no account of the possible formation of associations or short-range orientational order.

The author acknowledges that his theory of light scattering at anisotropy fluctuations rests on coarse molecular models and is far fron contemporary theory. Despite its numerous failings, partly indicated in the foregoing, this theory nevertheless offers a physical interpretation for the whole Rayleigh-line wing in low-viscosity liquids, and thus clarifies the nature of the spectrum of the depolarized scattering of light. It is therefore of special interest to apply the theory to the interpretation of experimental results, including those already in existence, and also those obtained specially in an attempt to verify the theory.

The molecular-theory formulas, together with the experimental results, enable us to find some important characteristics of the medium at frequencies of $\sim 10^{13}$ cps. The results of an analysis of all the experimental data relating to the spectral composition of the Rayleigh-line wing will be presented in Chapter IV.

4. Stimulated Scattering of Light on the
Wing of the Rayleigh Line

In the thermal scattering of light, the Rayleigh-line wing is due to the modulation of the scattered light resulting from fluctuations in the orientations of the anisotropic molecules which are resorbed in the course of time. The weak field of the exciting light wave from an ordinary light source has such a slight effect on the anisotropy of the medium that we need not consider it. Things are different when the scattering is excited by the focused light of a giant pulse from an optical quantum generator (laser). In this case, the electric field of the light wave is so large that, together with the thermal-scattering field, it leads to the development of an orienting force

[3] $f \sim \frac{\alpha_1 - \alpha_2}{kT} E^2$ (α_1, $\alpha_2 = \alpha_3$ are the principal polarizabilities of the molecule, while **E** is the sum of the fields of the exciting and scattering light), the low-frequency component of which produces an anisotropy of the medium as a result of the orientations of the anisotropic molecules.

Let us consider what effect will be exerted on the light scattering by the orientations of the anisotropic molecules under the strong electric field of the light wave.

We shall indicate the orientation of the axes of the molecules (symmetric gyrostat) relative to a stationary coordinate system (indices i, k) by the angles ϑ_i, ϑ_k. Let us introduce an anisotropy tensor S_{ik} characterizing the deviation of the distribution of molecular axes from the isotropic distribution [3]

$$S_{ik} = \langle \cos \vartheta_i \cos \vartheta_k \rangle - \frac{1}{3} \delta_{ik}. \tag{II.28}$$

If the field is absent, then, since the medium is isotropic,

$$S_{ii} = 0. \tag{II.29}$$

The change in anisotropy arising under the influence of the force f_{ik} (the force remains for the time being indefinite) obeys the equation

$$b\ddot{S}_{ik} + \tau \dot{S}_{ik} + S_{ik} = f_{ik}. \tag{II.30}$$

Here b is an inertial parameter, τ is the effective relaxation time of the anisotropy arising under the influence of the field.

We note that the phenomenological equation (II.30) describes both the diffusion mechanism of the resorption of the developing anisotropy and the mechanism of elastic oscillations. The difference between these mechanisms appears simply in the specification of the parameters τ and b. For the moment we shall regard these parameters as effective values.

If f_{ik} is independent of time, then from (II.30) we obtain

$$S_{ik} = f_{ik}. \tag{II.31}$$

Let us put [3] in general form

$$f_{ik} = \lambda \left(E_i E_k - \frac{\sigma}{3} \delta_{ik} E^2 \right). \tag{II.32}$$

From (II.29) and (II.31) we easily find $\sigma = 1$ and, hence,

$$f_{ik} = \lambda \left(E_i E_k - \frac{1}{3} \delta_{ik} E^2 \right). \tag{II.33}$$

Let us suppose that a wave $\mathbf{E}$ polarized along the z axis propagates in the medium in the x direction, so that $\mathbf{E} = E_z = E$. Then we are interested in the component of the tensor S_{ik} characterizing the orientation of the molecules relative to the z axis:

$$\varepsilon_{11} = S = \langle \cos^2 \vartheta \rangle - \frac{1}{3}.$$

The energy of a single molecule in the field of the light wave is

$$\varepsilon = -\frac{1}{2} (\alpha_1 - \alpha_2) E^2 \cos^2 \vartheta - \frac{1}{2} \alpha_2 E^2. \tag{II.34}$$

Let us find the mean value $\langle \cos^2 \vartheta \rangle$:

$$\langle \cos^2 \vartheta \rangle = \frac{\int_0^\pi \cos^2 \vartheta \exp\left(-\frac{\varepsilon}{kT}\right) \sin \vartheta \, d\vartheta}{\int_0^\pi \exp\left(-\frac{\varepsilon}{kT}\right) \sin \vartheta \, d\vartheta}. \tag{II.35}$$

If we satisfy the condition *

$$\frac{\alpha_1 - \alpha_2}{kT} E^2 \ll 1, \tag{II.36}$$

then from (II.35) we obtain

$$\langle \cos^2 \vartheta \rangle - \frac{1}{3} \approx \frac{4}{45} \frac{\alpha_1 - \alpha_2}{kT} E^2. \tag{II.37}$$

Comparing (II.31) and (II.37) we find that

$$f_{11} = \frac{4}{45} \frac{\alpha_1 - \alpha_2}{kT} E^2. \tag{II.38}$$

Thus, finally, from (II.30) we obtain an equation of the following form [on satisfying condition (II.36)]

$$b\ddot{S} + \tau\dot{S} + S = \frac{4}{45} \frac{\alpha_1 - \alpha_2}{kT} E^2. \tag{II.39}$$

*This condition is satisfied up to fields of 10^7 V/cm, for example, for molecules of carbon disulfide, nitrobenzene, benzene, etc.

Maxwell's equations for the propagation of electromagnetic waves in a nonlinear medium give

$$\frac{\partial^2 D}{\partial t^2} + c^2 \, \text{rot rot } \mathbf{E} = 0, \tag{II.40}$$

where

$$\mathbf{D} = \mathbf{E} + 4\pi \, (\mathbf{P}^l + \mathbf{P}^{nl}). \tag{II.41}$$

Here, $\mathbf{P}^l$ is the linear part of the polarization, $\mathbf{P}^{nl}$ is the nonlinear increment to the polarization due to the orienting effect of the electric field of the light wave:

$$\mathbf{P}^l = \frac{\widetilde{\varepsilon} - 1}{4\pi}, \tag{II.42}$$

$$\mathbf{P}^{nl} = \frac{1}{4\pi}\left(\frac{\partial \varepsilon}{\partial S}\right) S. \tag{II.43}$$

The complex dielectric constant is given by the expression

$$\widetilde{\varepsilon} = \varepsilon' - i\varepsilon'', \quad \varepsilon' = n^2 - \varkappa^2, \quad \varepsilon'' = 2n\varkappa, \tag{II.44}$$

$$\varkappa = \frac{k_\omega c}{2\omega_0}, \tag{II.45}$$

where k_ω is the peak attenuation coefficient of the light. Thus, for the polarized light wave we obtain the following equation from (II.40) to (II.43):

$$\varepsilon' \frac{\partial^2 E}{\partial t^2} - i\varepsilon'' \frac{\partial^2 E}{\partial t^2} - c^2\left(\frac{\partial^2 E}{\partial x^2} + \frac{\partial^2 E}{\partial y^2}\right) = -A \frac{\partial^2}{\partial t^2}(SE). \tag{II.46}$$

Here we have introduced the notation

$$A = \left(\frac{\partial \mathscr{E}}{\partial S}\right).$$

The system of equations (II.39) and (II.46) is the desired system of nonlinear differential equations. For solving the problem of the stimulated scattering of light on the Rayleigh-line wing we must solve this system simultaneously. We may seek an approximate solution to the system of equations (II.39) and (II.46) (in analogy with the case of stimulated Mandelshtam—Brillouin scattering [1, 54]) by the method of the small Van der Pol parameter [64, 65]. Here we assume that E constitutes the superposition of an exciting field E_0 with a frequency ω_0 and scattered fields of the Stokes and anti-Stokes types, respectively characterized by E_{1l}, ω_{1l} and E_{2l}, ω_{2l}. Thus we suppose that the scattering is excited by monochromatic light, and the scattered light may be regarded as a spectrum of frequencies ω_{1l} and ω_{2l} with Fourier amplitudes E_{1l} and E_{2l} ($l = 1, 2, \ldots$). The amplitudes E_0, E_{1l}, and E_{2l} are considered as slowly varying in space. The solution of the system of equations (II.39) and (II.46) will be sought in the form

$$S = \sum_l S_l e^{i\,(\Omega_l t - \mathbf{q}_l \mathbf{r})} + \text{c.c.}$$

$$E = \sum_l \sum_{\sigma=0}^{\sigma=2} E_{\sigma l} e^{i\,(\omega_{\sigma l} t - \mathbf{k}\,\mathbf{r})} + \text{c.c.}$$

$$(\text{for} \quad \sigma = 0, \ l = 1). \tag{II.47}$$

Putting (II.47) into (II.39) and neglecting the terms $E_{1l'}E_{1l''}$, $E_{2l}E_{2l''}$, etc. compared with $E_0 E_{1l}$, $E_0 E_{2l}$ (we assume $E_0 \gg E_{1l}$, $E_0 \gg E_{2l}$), and satisfying the conditions

$$\omega_{1l} = \omega_0 - \Omega_l, \ \omega_{2l} = \omega_0 + \Omega_l,$$

$$k_{1l} = k_0 - q_l, \ k_{2l} = k_0 + q_l \qquad \text{(II.48)}$$

we obtain the following expression for S_l:

$$S_l \approx \frac{4}{45} \frac{(\alpha_1 - \alpha_2)}{kT} w_l (E_0 E_{1l}^* + E_0^* E_{2l}). \qquad \text{(II.49)}$$

Here,

$$w_l = w_{1l} - i w_{2l} = \frac{1 - b\Omega_l^2}{(1 - b\Omega_l^2)^2 + \Omega_l^2 \tau^2} - i \frac{\Omega_l \tau}{(1 - b\Omega_l^2)^2 + \Omega_l^2 \tau^2}. \qquad \text{(II.50)}$$

From (II.46), after satisfying (II.48), we obtain a system of so-called contracted equations

$$\frac{\partial E_0}{\partial x} + \delta_0 E_0 + i\gamma_0 (S_l E_{1l}^* + S_l^* E_{2l}) = 0,$$

$$\frac{\partial E_{1l}}{\partial x} + \delta_1 E_{1l} + i\gamma_1 S_l^* E_0 = 0,$$

$$\frac{\partial E_{2l}}{\partial x} + \delta_2 E_{2l} + i\gamma_2 S_l E_0 = 0, \qquad \text{(II.51)}$$

where

$$\delta_\sigma = \frac{\varepsilon'' \omega_\sigma^2}{2c^2 |k_\sigma| \cos (k_\sigma x)},$$

$$\gamma_\sigma = \frac{A\omega_\sigma^2}{2c^2 |k_\sigma| \cos (k_\sigma x)}$$

$$(\sigma = 0, 1, 2). \qquad \text{(II.52)}$$

We shall solve the problem for the case in which $E_0 \gg E_{1l}$, $E_0 \gg E_{2l}$, when we may put $E_0 \sim \text{const.}$ At the very beginning of the development of the process this approximation is fully justified, and instead of (II.51) we may solve the system of equations

$$\frac{\partial E_{1l}}{\partial x} + \delta_1 E_{1l} + i\gamma_1 S_l^* E_0 = 0,$$

$$\frac{\partial E_{2l}}{\partial x} + \delta_2 E_{2l} + i\gamma_2 S_l E_0 = 0. \qquad \text{(II.53)}$$

Let us confine attention to the approximation in which we may consider that the Stokes and anti-Stokes components of the scattered light do not interact. Then (II.53) decomposes into two independent equations. From (II.49) and the first of equations (II.53), it is quite easy to obtain the following expression for the Stokes wave:

$$|E_{1l}(x)|^2 = |E_1(0)|^2 \exp \left\{ \left[-\delta_1 + \frac{4}{45} \frac{\alpha_1 - \alpha_2}{kT} \gamma_1 w_{2l} |E_0|^2 \right] x \right\}. \qquad \text{(II.54)}$$

It follows from (II.54) that on satisfying the condition

$$|E_0|^2 \geqslant \frac{45}{4} \frac{\varepsilon''}{A} \frac{kT}{\alpha_1 - \alpha_2} \frac{(1 - b\Omega_l^2)^2 + \Omega_l^2 \tau^2}{\Omega_l \tau} \qquad \text{(II.55)}$$

the Stokes part of the scattered light will grow exponentially on propagating in a scattering medium consisting of anisotropic molecules. The anti-Stokes component of scattered light is a falling (attenuating) one in this approximation. Thus, on exceeding the threshold value (II.55) of the power of the exciting light we obtain stimulated scattering of the light of the Rayleigh-line wing in the Stokes region. It follows from (II.55) that, if for simplicity we neglect the inertia term (b = 0), the threshold value of $|E_0|^2$ has a minimum at the frequency $\Omega_l = 1/\tau$. This means that the stimulated scattering of light on the Rayleigh-line wing is most easily excited at a frequency corresponding to the half-width of the ordinary thermal Rayleigh-line wing.

It follows from the foregoing that, in contrast to the ordinary thermal scattering, in which the intensity falls off monotonically on both sides of the exciting line, in stimulated scattering, on exceeding the threshold value of the pumping wave, only the Stokes wing should be observed, a maximum appearing in the wing at a frequency of $\Omega = 1/\tau$. For the threshold we may approximately take the expression

$$\frac{|E_0|^2_{\text{thr}}}{8\pi} = \frac{45}{32\pi} \frac{nk_\omega \, (kT)}{|\, \mathbf{k}_1\,|\frac{\partial \varepsilon}{\partial S}\, (\alpha_1 - \alpha_2)} \frac{1 + \Omega^2\tau^2}{\Omega\tau}. \tag{II.56}$$

Generally speaking, (II.56) is obtained on the assumption that the law of resorption of the anisotropy developed is given by the equation

$$\dot{S} + \frac{1}{\tau}\, S = 0. \tag{II.57}$$

It is easy to show that, on satisfying (II.57), the Rayleigh-line wing in ordinary thermal scattering will have a dispersion distribution with a half-width $2\Omega = 2/\tau$. Actually, in thermal scattering, the wing falls off monotonically, but in accordance with a very complicated law. This means that (II.56) is only a coarse approximation to reality. Allowance for the inertial term in the kinetic equation leads to expression (II.55) for the threshold. However, as may clearly be seen from the foregoing discussion, this is insufficient to yield a complete description of the phenomenon of light scattering, since it is still necessary to allow for the dispersion of the parameters in the kinetic equation. We shall not make a full calculation here. Conditions (II.55) and (II.56) may be modified by writing, for example (see Sections 2 and 3 of this chapter):

$$\Omega\mathscr{I}\,(\Omega) = \frac{1}{\pi}\, \frac{\Omega\tau}{1 + \Omega^2\tau^2}, \tag{II.58}$$

where $\mathscr{I}\,(\Omega)$ is the intensity distribution in the ordinary, thermal depolarized scattering, normalized to unity.* Then, instead of (II.55) or (II.56) we obtain, finally

$$\frac{|E_0|^2_{\text{thr}}}{8\pi} = \frac{45}{32\pi^2}\, \frac{nk_\omega \,(kT)}{(\alpha_1 - \alpha_2)\,|\,\mathbf{k}_1\,|\frac{\partial \varepsilon}{\partial S}\, \Omega\mathscr{I}\,(\Omega)}. \tag{II.59}$$

Estimates show that the threshold value $|E_0|^2$ for stimulated Mandelshtam—Brillouin scattering is lower than for stimulated scattering of the light of the Rayleigh-line wing. This makes it difficult to carry out numerical estimates based on (II.59), since k_ω is largely determined by losses involving stimulated Mandelshtam—Brillouin scattering and the stimulated Raman scattering of light.

---

*The quantity $\mathscr{I}(\Omega)$ is expressed in seconds. We could, as in [23], define $\mathscr{I}\,(\Omega)$ so as to make it dimensionless. In this case, τ will come into the result, and this is not very convenient.

Finally, we note that for small scattering-angles there may be interaction between the Stokes, anti-Stokes, and laser radiation, both the Stokes and anti-Stokes parts of the wing being amplified with the same amplification factor. In contrast to the case considered in the foregoing, the maximum of the amplification factor for this three-photon interaction should be at $\Omega = 0$. We shall not consider this interaction here, since only the stimulated scattering of the Stokes wing of the Rayleigh line is observed experimentally.

CHAPTER III

METHODS OF STUDYING THE
SPECTRAL COMPOSITION OF SCATTERED LIGHT

1. Method of Studying the Spectral Composition
of the Wing of the Rayleigh Line
(Ordinary Thermal Scattering)

In studying the spectral intensity distribution in the wing of the Rayleigh line, we used a photoelectric method of recording the spectra. As spectral apparatus we used a DFS-12 diffraction spectrograph with two 600 lines/mm replica gratings working in the second order. The apparatus had a dispersion of 4.6 Å/mm. For slits of ~2-4 μ the half-width of the apparatus function, together with the intrinsic worth of the Hg 4358 Å line, was ~0.6 cm^{-1}. The main arrangement of the apparatus is shown in Fig. 1.

The illuminating part of the apparatus and the vessels for the scattering liquid were the same as described earlier [6, 30]. The construction of these was arranged so that only light scattered by the liquid should fall into the spectral system, not parasitic light from the walls of the vessel. This was a most important requirement in studying the near part of the Rayleigh-line wing, directly adjacent to the undisplaced line.

When studying the near part of the wing it is also important to eliminate the isotropic, polarized (z polarization) part of the scattered light. We therefore studied the near part of the wing (up to 20 or 50 cm^{-1} from the undisplaced line) in x polarization (Nicol N in Fig. 1), using source aperture limiters A, and the remoter part of the wing without aperture limiters or Nicol. This enabled us to carry out measurements in the wing to 130 or 150 cm^{-1}. As already noted earlier, for exciting the scattered light we used the blue mercury-line λ = 4358 Å emitted by low-pressure mercury lamps [66].

Two such lamps were used in our illuminating system. The light scattered by the liquid passed through the diaphragms D_1 and D_2, the interrupter O, the condenser L_1, and the Nicol N to the entrance slit S_1 of the spectrograph.

The part of the wing spectrum separated by the exit slit of the spectral apparatus fell on a photomultiplier. We used one of the EMI 6094B type. After passing through the preamplifier and the amplifier with the phase detector, the signal was recorded on an EPP-09 automatic recorder. The source of the reference voltage for the phase-detector amplifier was a photoresistor illuminated by light from a glow lamp. This light was interrupted at the same frequency (415 cps) as the main beam of scattered light, by the interrupter O.

In order to reduce the effect of a change in the intensity of the source during the recording of the wing (1-2 h), a circuit for automatically regulating the amplification factor of the preamplifier in accordance with the source intensity was employed. This regulation was effected by means of two FEU-1 photomultipliers placed over the mercury lamps and supplying

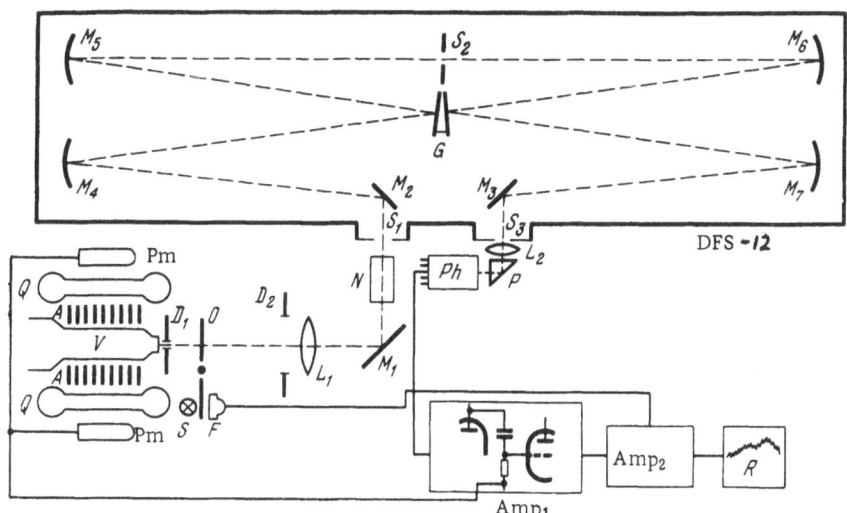

Fig. 1. Main arrangement of the apparatus for studying the wing of the Rayleigh-scattering line. V) Vessel containing sample liquid; Q) low-pressure mercury lamps [66]; A) source aperture limiters; D_1, D_2) scattered-light aperture limiters; O) interrupter; I_1) condenser; M_1, M_2, M_3) plane mirrors; N) Nicol; S_1, S_2, S_3) entrance, middle, and exit slits of the spectral system; M_4-M_7) parabolic mirrors of spectral system; G) two replicas (600 lines/mm); L_2) cylindrical lens; P) total internal-reflection prism; Ph) photomultiplier; Amp_1, Amp_2) preamplifier and amplifier with phase detector; R) self-recorder; Pm) photomultipliers of the system automatically regulating amplification factor of amplifier according to intensity of source; F) photoresistor giving reference signal for the amplifier with the phase detector; S) glow lamp (the light from this illuminates the photoresistor after modulation by the interrupter).

a bias voltage to the grid of the second tube of the preamplifier. This arrangement made it possible to compensate ~80% of the change in source intensity, which was quite enough for our purposes. We carefully verified the linearity of the electrical circuit over the whole working range. In addition to this, for benzene, with its wide, strong wing, we recorded the spectra both in the x component, using the Nicol N and the source aperture limiters A, and in unpolarized light without these. The contours coincided in both cases within the limits of experimental error.

Figure 2 shows the visible contour of the Rayleigh-line wing of benzene according to our own results and those of [39, 67, 68]. The slight discrepancies in the results are due not only to the difficulties of studying the wing quantitatively, especially by the photographic method, but also to the fact that the apparatus used in different cases had different apparatus functions, which naturally led to a slight redistribution of the intensity in the wing.

Considering that the apparatus function of the apparatus had a considerable effect on the wing contour, we studied the contour of the apparatus function of our apparatus (together with that of the exciting line λ = 4358 Å) in order to allow for this influence. This contour is fairly well described by a Voigt function; depending on the slit width, it may be closer either to a dispersion or to a Gaussian distribution. Considering the fact that the near part of the wing has a dispersion distribution [6,30], and knowing the characteristics of the apparatus contour describing the Voigt function and the half-width of the visible contour, we may use the results of [69]

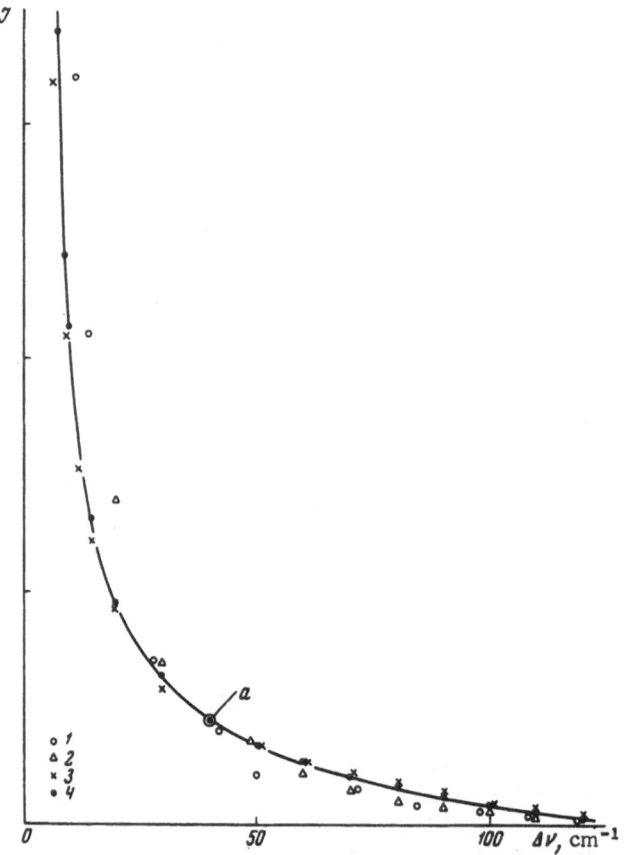

Fig. 2. Visible contour of the Rayleigh-line wing in benzene. Results of: 1) [67]; 2) [68]; 3) [29]; 4) the author.

to determine the true half-width of the Rayleigh-scattering line wing. We carried out a similar analysis for the near part of the wing for all our liquids.

It is hard to allow for the apparatus function over the whole wing because the wing has a very complex contour. We took no account of the effect of the apparatus function on the re-moter parts of the wing (further than 10–25 cm^{-1}). This cannot lead to any great errors, since, after passing the dispersion section (nearest to the undisplaced line) the intensity falls very slowly, this region lying at a distance of 10–20 times the half-width of the apparatus function from the wing maximum. A study of the spectral properties of the DFS–12 showed that the grat-ings had weak satellites (~1% of the intensity at the maximum of the exciting line) at frequencies of 12.5, 25, and 50 cm^{-1}. In studying the contour of the Rayleigh line-wing for low-viscosity li-quids, the intensity of the satellites was no greater than the noise in the recording part of the apparatus. For control purposes, after each recording of the Rayleigh-line wing contour, a re-cording of the exciting line with an intensity 1.5–2 times that of the Rayleigh-line wing was made. An example of the recording of the wing contour of acetic acid is given in Fig. 3.

We also studied the temperature dependence of the spectral intensity distribution in the remoter part of the wing. During these investigations, the vessel containing the scattering li-quid was kept in a Dewar, and either hot air or the vapor of intensively evaporated nitrogen was passed through the space between the vessel and the Dewar walls.

Our system, comprising a phase-detector amplifier and a specially selected photomulti-plier, ensured high stability of the whole recording part of the apparatus.

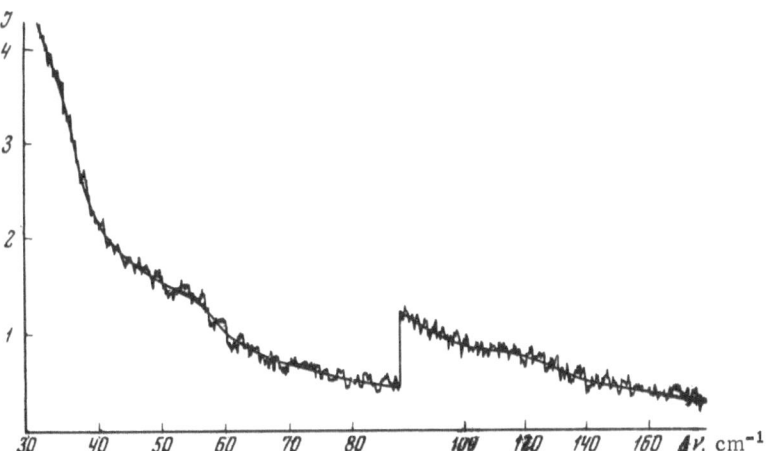

Fig. 3. Example of a recording of the contour of the Rayleigh-
line wing of acetic acid.

The results which will be presented in Chapter IV were obtained by averaging several recordings (not less than three) on the self-recorder; the spread of the points obtained by successive recordings in the middle of the Rayleigh-line wing (50-70 cm^{-1}) was no greater than 5%.

As already indicated, in studying the part of the wing close to the undisplaced line we had
to find the half-width of the contour of the scattered light. The experimentally observed contour F(x) is a convolution of the true contour φ(y) and the apparatus contour (together with the
contour of the exciting line) A(x'), so that

$$F\left(x\right) = \int\limits_{-\infty}^{\infty+} \varphi\left(y\right) A\left(x - y\right) dy. \tag{III.1}$$

For our apparatus, the apparatus contour was described quite satisfactorily by a Voigt
function (convolution of a dispersion and Gaussian distribution). Considering that the near part
of the wing has a dispersion distribution, we may use the graphs of [69] to find the half-width of
the true distribution by measuring the half-width and the width at 0.1 of the height of the apparatus contour, and also the half-width of the true contour. We note that the influence of the
Gaussian component in the apparatus function only slightly exceeded the accuracy of our measurements. This follows from the fact that the half-width of the true contour (determined in the
manner just described on the basis of the results of [69]) and the half-width found by simple
subtraction of the half-width of the apparatus function from the half-width of the visible contour
(which is permissible even if the true and apparatus contours have a dispersion distribution)
differ by no more than 20%, even in the case of such a liquid as acetic acid, with its narrow
Rayleigh-line wing.

In studying the remoter part of the Rayleigh-line wing we had to find two parameters from
the experimental intensity/frequency relation: the relaxation time of the internal friction τ_ζ
and the coefficient g_0. Knowing these parameters, we were able to compare the experimental
intensity/frequency relation: distribution calculated from formulas (II.23) and (II.25). The
means by which τ_ζ and g_0 were determined from the experimental results will be described in
Chapter IV. Here we should note, however, that only for the case of liquids consisting of molecules with one moment of inertia (carbon disulfide, chloroform), for which, generally speaking,
relation (II.23) was derived, is the high accuracy of the determination of τ_ζ and g_0 guaranteed

by good agreement between the contour obtained experimentally and that calculated on the basis of relations (II.23) and (II.25).

For CS_2 and chloroform, these contours differ even if τ_ζ and g_0 vary by 10-15% from the values given in Chapter IV. This determines the accuracy of the values of τ_ζ and g_0 found for these liquids. As regards the other liquids, clearly this accuracy cannot be guaranteed for them.

2. Experimental Apparatus for Studying the Fine-Structure Components of the Rayleigh Line (Thermal Scattering)

In studying the fine-structure components of the Rayleigh-scattering line for liquids we set ourselves the problem of simultaneously measuring the distance between the components, the spectral width of the Mandelshtam—Brillouin components, and the ratio of the intensity of the central component to the intensity of the Mandelshtam—Brillouin components. A problem of this complexity imposes heavy demands on the monochromatic properties of the source of exciting light and the resolving power of the optical apparatus. Figure 4 shows the arrangement of the apparatus for studying the fine structure of the Rayleigh line. The source of exciting light was the laser Z working with an Ne—He mixture ($\lambda = 6328$ Å), having a power of 1-5 mW. The vessel V holding the scattering liquid was placed inside the laser. As compared with the arrangement in which the vessel containing the scattering liquid was placed outside the resonator, the arrangement of Fig. 4 gives a great gain in the intensity of the exciting light. The laser tube containing the active material had one window set at the Brewster angle so that the exciting light was polarized in a plane perpendicular to the scattering plane. The scattered light was observed in a direction perpendicular to the direction of propagation of the exciting light. The vessel V for the sample liquid was made of quartz. The windows of the vessel in the path of the exciting light were set at the Brewster angle. The spherical mirror M_1 and the plane mirror M_2 of the resonator had 13- to 15-layer dielectric coatings. The spectrum of the scattered light was analyzed with a Fabry—Perot interferometer (F-P) with an 8-mm thickness of the separating ring (range of dispersion 0.625 cm^{-1}). The interferometer mirrors had multilayer dielectric coatings. A high-quality camera objective with a focal length of $f = 415$ mm projected an image of the spectrum into the isothermal chamber B maintaining a constant temperature during the exposure (accuracy 0.03°C). The resolving power of the optical apparatus was $\sim 4 \cdot 10^6$. The half-width of the apparatus function together with the half-width of the exciting line equalled ~ 0.02 cm^{-1}.

In measuring the width of the Mandelshtam—Brillouin components one must have a strictly limited aperture of the exciting and scattered light beams. Let us estimate what influence a finite aperture will have on the width of the components. We easily see from (I.13) that for an aperture angle of $\pm \delta \vartheta$ the components will be spread by an amount

$$\pm \delta(\Delta \nu) = \pm n\nu \frac{V}{c} \cos \frac{\vartheta}{2} \delta \vartheta, \tag{III.2}$$

or, using (I.13) again, we find that for a scattering angle of $\vartheta = 90°$

$$\pm \frac{\delta(\Delta \nu)}{\Delta \nu} = \pm \frac{1}{2} \delta \vartheta. \tag{III.3}$$

If we desire that the spread of the components due to the finite aperture should be smaller than ± 0.002 cm^{-1}, then (since $\Delta \nu \sim 0.2$ cm^{-1}) we must have $\pm \delta \vartheta \leq 0.02$ and the total aperture angle $2\delta \vartheta = 0.04$. In our apparatus the scattered light was limited by the diaphragms D_1 and D_2

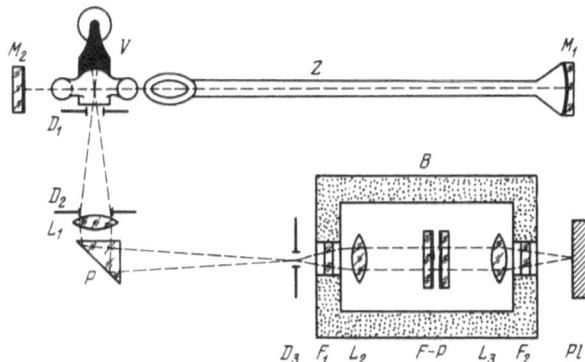

Fig. 4. Arrangement of the apparatus for studying the fine structure of the Rayleigh line in a liquid. Z) Laser tube; M_1,M_2) dielectric mirrors; F-P) Fabry—Perot interferometer; L_1,L_2,L_3) objectives; Pl) photographic plate; F_1,F_2) filters; B) isothermal chamber.

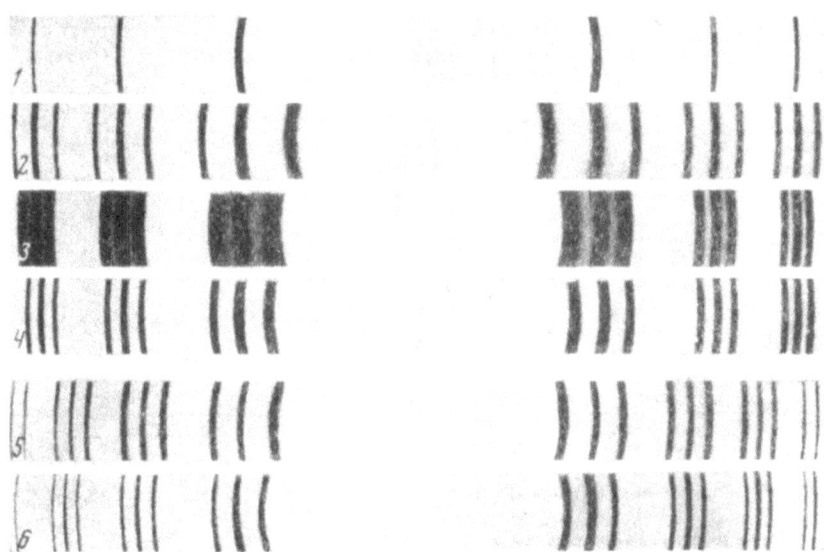

Fig. 5. Fine structure of the line representing the Rayleigh scattering of light in a liquid. 1) Exciting line, $\lambda = 6328$ Å; 2) benzene; 3) carbon tetrachloride; 4) chloroform; 5) toluene; 6) methylene chloride.

(Fig. 4) and the aperture angle was $2\delta\vartheta = 0.03$. The divergence of the exciting laser beam may be neglected.

In order to calculate the velocity of hypersound by reference to the distance between the fine-structure components, we must know the scattering angle ϑ. It is not hard to estimate from (III.3) that in order to measure the velocity of hypersound to an accuracy of $\pm 0.5\%$ the scattering angle $\vartheta = 90°$ must be established to an accuracy of $\pm 0.5°$. This requirement was satisfied in our apparatus. The scattering angle was established each time to the required accuracy before taking each picture of the fine-structure spectrum. The scattering spectrum of each liquid was photographed at least six times and the final result was obtained by averaging the

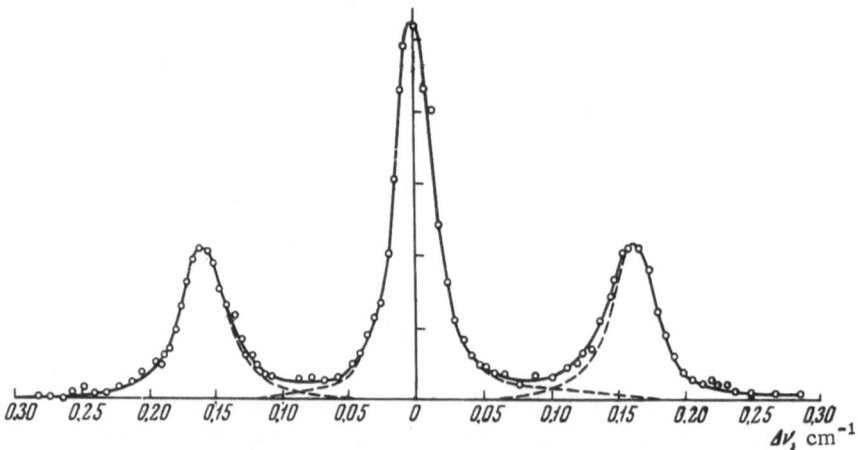

Fig. 6. Intensity distribution in the fine-structure components of the scattering line of benzene. Solid line, experimental contour; broken line, result of resolution.

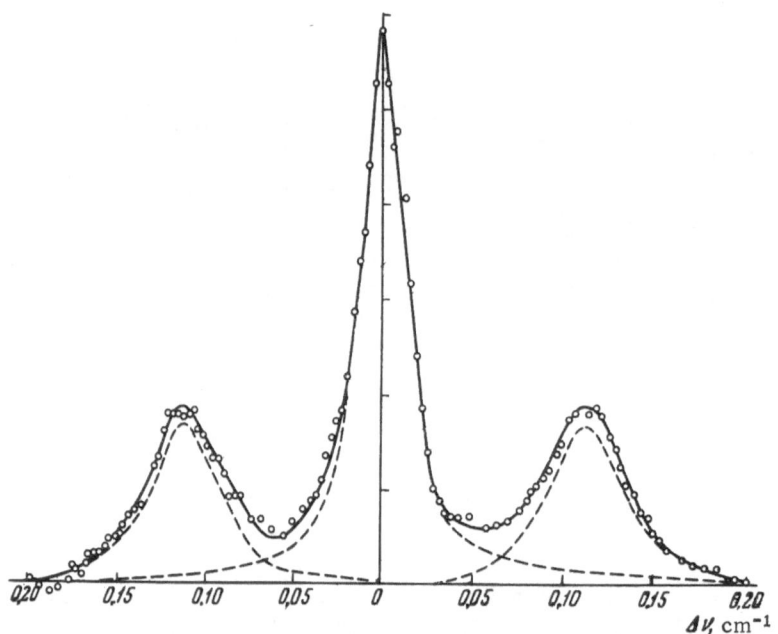

Fig. 7. Intensity distribution in the fine-structure components of the scattering line of carbon tetrachloride. Notation as in Fig. 6.

results of measurements based on all the fine-structure photographs. Thus, random error in determining the velocity of hypersound due to inaccuracy in setting the angle was reduced by a factor of several times and equalled ~0.2%.

The visible contour of the Mandelshtam—Brillouin components $F(x)$ is a convolution (III.1) of the true and apparatus contours (together with the contour of the exciting light).

As apparatus contour we used the central component of fine structure, the intrinsic width of which was 10^{-4} cm^{-1} and lay far beyond the limits of accuracy of our measurements. The intensity distributions in the contours of the central component and the Mandelshtam—Brillouin components were close to the dispersion form. The width of the central component (~ 0.02 cm^{-1}) was mainly made up of the width of the apparatus contour, the width of the exciting line, and a slight broadening associated with the influence of changes in atmospheric pressure and the temperature of the surrounding air on the interferometer during the period of exposure [70]. This total width we had to subtract (for a dispersion-type intensity distribution in the contours) from the width of the visible contour of the Mandelshtam—Brillouin components in order to obtain the intrinsic width associated with the attenuation of the hypersound. Also important was the fact that the apparatus contour and the contours of the Mandelshtam—Brillouin components had to be photographed simultaneously under identical conditions. Figure 5 shows photographs of the fine-structure components of the light scattered in our liquids. The distance between the fine-structure components was measured with an IZA-2 comparator. Analysis of the measurements in order to determine the extent of the displacement of the components in cm^{-1} was carried out by the eccentric method [70].

In photometering the fine-structure photographs with an MF-2 microphotometer, the width of the slit used corresponded to a spectral range of $\sim (3-5) \cdot 10^{-3}$ cm^{-1} on the photographic film, so that there was no need to make any allowance for the finite slit width of the microphotometer. The degree of blackening (photometric density) recorded by the microphotometer was then converted into intensity. For this purpose, we used a photometric density scale obtained by means of a seven-step platinum attenuator. The distribution of the intensity in the fine-structure components was close to a dispersion type; the results for C_6H_6 and CCl_4 are shown in Figs. 6 and 7.

Very important when preparing to measure the width of the fine-structure components (and hence the absorption coefficient of hypersound) is the purification of the liquids. It is well known [17, 71] that even slight impurities greatly alter the absorption coefficient of hypersound. For this reason we paid special attention to the careful purification of the sample liquids and the removal of dust. Chemically pure liquids were used for the work. The purified and dust-freed liquids after repeated distillation were poured into the test vessel. Before this the vessel, already cleaned and degreased, was washed several times with the same liquid. The liquid poured into the vessel was frozen and the vessel was evacuated with a backing pump through a filter. In one of the side tubes of the vessel a drop of unfrozen liquid remained. When the main part of the liquid was frozen, the vessel was sealed off and the liquid was thus kept free from possible contaminants.

The construction of the vessel provided for the possibility of further purifying the liquid and cleaning the walls of the vessel after sealing off (Martin's method).

The accuracy of measuring the width of the Mandelshtam—Brillouin components was $\sim 20\%$, the accuracy of determining the velocity of hypersound was $\sim 0.5\%$, and the accuracy of determining the ratio of the intensity of the central component to the intensity of the Mandelshtam—Brillouin components was $\sim 10-15\%$.

3. Method of Studying the Stimulated
Scattering of Light

The stimulated Mandelshtam—Brillouin scattering and the stimulated scattering of the light on the wing of the Rayleigh line were studied with the apparatus shown schematically in Fig. 8.

As source of exciting light we used a composite ruby laser with modulated Q. The ruby rods R_1 and R_2, each 120 mm long, were placed in a resonator consisting of a mirror with a dielectric coating and a reflection coefficient of $\sim 100\%$ on one side and two glass plane-parallel

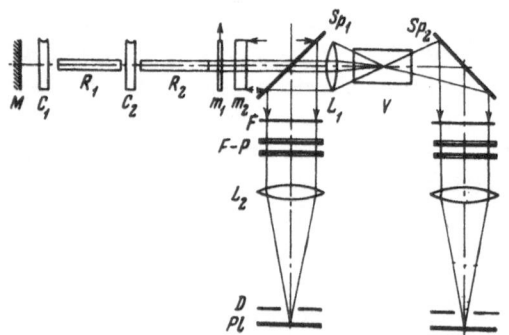

Fig. 8. Arrangement of the apparatus for observing stimulated scattering of the light on the Rayleigh-line wing. M) Mirror (R ~ 100%); C_1, C_2) cuvettes; R_1, R_2) rubies 12 cm long; m_1, m_2) plane-parallel plates; Sp_1, Sp_2) separating glass plates; L_1) lens (f = 120 cm); V) cuvette for the scattering liquid; F-P) Fabry—Perot interferometer; Pl) photographic plate.

plates m_1 and m_2 on the other. The Q modulation of the resonator was effected by means of two glass cuvettes with plane-parallel windows, filled with a solution of cryptocyanin in methyl alcohol [72]. One of the cuvettes C_1 was placed between the mirror with the 100% reflection coefficient and the ruby R_1, and the other between the two rubies. The thickness of the cuvettes was 10 mm; the concentration of the cryptocyanin in the methyl alcohol was $1.8 \cdot 10^{-6}$ M.

Our apparatus incorporated ruby rods with 0.05% Cr^{3+} ions and a 90° orientation. Each of the ruby rods was in an elliptical illuminating system with two IFP-2000 lamps. The ruby crystals and lamps were cooled with running water. A voltage of ~1200–1500 V from a battery of condensers was applied to each of the lamps. The ruby crystals were arranged with their optic axes parallel. The laser radiation was polarized in a vertical plane.

The ruby optical generator (laser) used for studying stimulated scattering gave a giant pulse 10–12 nsec long with a power of ~100 MW.

Mode selection may be effected by inclining one of the plane-parallel plates m_1 and m_2. The thickness of these plates differed considerably: that of m_1 being 2.5 and that of m_2 being 13 mm. The pair of plates operated as a composite multiplex interferometer. This multiplex interferometer selected only the particular mode of radiation which coincided simultaneously with the characteristic modes of resonators m_1 and m_2. By slightly inclining one of these plates the degree of coincidence between the modes of the resonators m_1 and m_2 could be regulated. Sometimes for good selection we also had to introduce a slight inclination of ruby R_2 with respect to ruby R_1 in the vertical plane.

Since there were many reflecting surfaces inside the laser, forming resonators with different ranges of dispersion, several modes corresponding to some of these resonators were observed in the radiation spectrum. By slightly inclining one of the plates m_1 and m_2 (and sometimes ruby R_2) we were able, after several attempts, to obtain a "single-mode" picture within the limits of a spectral width of ~0.03 cm^{-1}. This spectral width might contain the modes of resonators formed either by the surfaces M and m_1 or by the surfaces M and m_2 (the distance between the modes of these resonators was ~0.01 cm^{-1}). Hence, the term "single-mode" picture is here used in an arbitrary sense, simply meaning that the spectral width of the laser radiation was ~0.03 cm^{-1}.

The radiation of the ruby laser was focused with a lens inside the vessel V filled with the sample liquid or inside the glass sample. In the first case we used a lens with a focal length of f = 3 cm and in the second one with f = 1.5 cm. Light scattered at an angle of 0° to the direction of the exciting light and light scattered backward, passing through the optical generator and returning to the vessel after reflection from the mirror M, were both able to fall into the optical system. In addition to this, some of the laser's own radiation passed into the optical system. After passing the vessel V the radiation was directed by means of a glass separating plate Sp_2 through the light filter F onto the Fabry—Perot interferometer F-P, and the spectrum of this radiation was photographed on the photoplate Pl. The focal length of the camera objective L_2 was 60 or 120 cm.

As indicated in Fig. 8, we were also able to photograph the light scattered backward by the sample liquid on its own, using the separating plate Sp_1. In this case, however, the intensity of the initial exciting radiation fell owing to losses in reflection from the plate Sp_1. Hence, the majority of the experiments were carried out without Sp_1, using the radiation reflected from plate Sp_2.

The stimulated Mandelshtam—Brillouin scattering was studied using interferometers with a range of dispersion of 1.5, 8.3, and 16 cm^{-1}. When studying the stimulated scattering of light on the wing of the Rayleigh line we also used an interferometer with a dispersion range of 50 cm^{-1}.

The power of the exciting light pulse was attenuated by introducing a pile of glass plates in front of the vessel (or other sample).

In adjusting the laser, the mirror M, the glass plates m_1 and m_2, and the ends of the rubies were first set strictly parallel to each other. Then the laser radiation spectrum was photographed by means of the optical system and interferometer. Only after satisfactory mode selection had been achieved by the means indicated in the foregoing were the stimulated-scattering spectra photographed. For control purposes we usually photographed the laser spectrum before and after recording the stimulated scattering.

RESULTS OF AN EXPERIMENTAL STUDY
OF THE SPECTRUM OF DEPOLARIZED SCATTERED LIGHT
IN LOW-VISCOSITY LIQUIDS

1. General Characteristics of the Spectrum
of the Depolarized Scattering of Light.
Diffusion Wing of the Rayleigh Line

We studied the spectral intensity distribution of the light scattered by dipolar liquids (toluene, chloroform, acetic acid) and nondipolar liquids (carbon disulfide, benzene, and carbon tetrachloride) characterized by different physicochemical properties, the molecules of which differed from each other in shape, size, and moment of inertia. The character of the intensity distribution in the wing was the same for all these, although the width and extent of the wing differed for each liquid.

In studying the wing of the Rayleigh line we set up a relationship between the reciprocal of the intensity $1/\mathcal{J}$ and the square of the frequency, reckoned from the undisplaced line. The intensity distributions plotted in this way for the wings of carbon disulfide and chloroform are shown in Figs. 9 and 10. Three parts may be distinguished in the contour of the Rayleigh-line wing of each of the liquids studied. The first part AB (Figs. 9 and 10) corresponds to a dispersion type of intensity—frequency relationship. In the graphs illustrated, the value of $1/\mathcal{J}$ varies approximately linearly with the square of the frequency in the section AB. After this the intensity starts falling less rapidly with frequency (BC), and later still the intensity again drops sharply (CD). The rate of fall of intensity in this part of the wing is much greater than in the first (dispersion) section. In all the low-viscosity liquids studied the part of the wing AB near the undisplaced line is, according to the interpretation given in Chapter II, due to the Brownian rotational motion of the molecules. Since for these liquids condition (II.16) is adequately satisfied, we may make a direct comparison between the anisotropy relaxation times found experimentally (from the width of the near section of wing) and those calculated from relation (I.4).

Table 1 shows the anisotropy relaxation times measured by the author and those determined from Fabelinskii's [6] interferometric measurements. The agreement between the two sets of results is excellent. Comparison of these experimental data with the relaxation times calculated from relation (I.4) shows that in the cases studied agreement here is equally good. This agreement serves to confirm the model selected, i.e., the one based on the assumption that the rotational Brownian motion of the molecules in the liquid was of a diffusion nature. It should be emphasized that our conclusion regarding the validity of relation (I.4) is only applicable to low-viscosity liquids. In the case of liquids with high viscosity (glycerin, triacetin, Salol, benzophenol, phenol, etc.), the part of the wing corresponding to the diffusive rotational motion of the molecules lies in a spectral range inaccessible to interference technology, so that as yet this has not been studied. In these liquids this part of the wing lies at 10^{-2}-10^{-4} cm^{-1} from the undisplaced line.

Table 1. Measured and Calculated Anisotropy Relaxation Times

Substance	Viscosity $\eta \cdot 10^3$ (20° C)	$\tau_a \cdot 10^{12}$, sec, accord. to [6]	$\tau_a \cdot 10^{12}$, sec, accord. to author's data	$\tau_a \cdot 10^{12}$, sec, calc.
Carbon disulfide	3.7	2.4	2.3	2.8 (2.1)
Chloroform	5.7	—	5.9	5.8
Toluene	5.9	5.3	4.1	8.6
Benzene	6.5	3.3	3.8	7.5 (5.3)
Acetic acid	12	17.5	14	13
Carbon tetrachloride	9.5	—	20	14

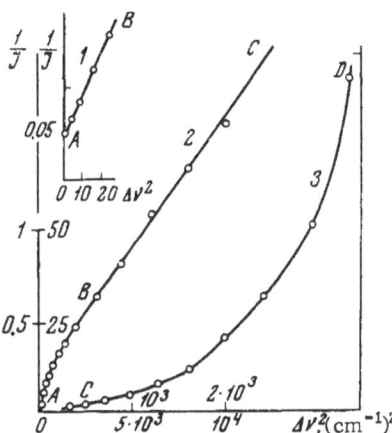

Fig. 9. Reciprocal of the intensity of the visible line contour as a function of the square of the frequency for carbon disulfide. 1) Part of the wing close to the undisplaced line; 2) close and medium-distance parts of the wing; 3) remote part of the wing (as compared with curve 2 the scale is here compressed five times along the axis of abscissas and 50 times along the ordinate axis).

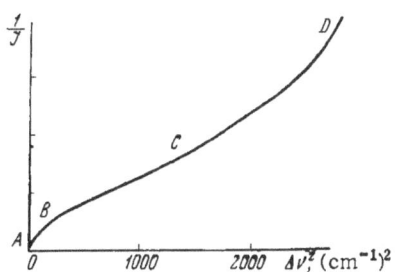

Fig. 10. Reciprocal of intensity in the wing as a function of the square of the frequency for chloroform.

As regards the calculation of the values of τ_a (Table 1) it should be noted that there is a certain indeterminacy in the choice of the radius of the molecule r in relation (I.4) (r^3 comes into this relation). Thus, for example, the brackets in Table 1 indicate values of τ_a for several liquids, calculated on the basis of ultrasonic data [71] regarding the dimensions of the molecules. These values of τ_a differ by 20-30% from the values calculated from critical data [73].

As mentioned earlier in Section 1 of Chapter I, relation (I.4) has been repeatedly verified by a large number of authors (see, for example, [6, 30, 35, 44]) for various liquids. By these the photographic method of studying the wing of the Rayleigh line has been employed. For the majority of the liquids studied there is also satisfactory agreement between the calculated and measured values of τ_a. The same agreement is exhibited by experiments on the width of the depolarized Raman [38] and nuclear magnetic-resonance lines. All this leads to the conclusion that relation (I.4) [and so much the more (II.19)] is applicable to a large number of low-viscosity liquids.

The remote section of the wing (BCD in Figs. 9 and 10) is interpreted as the result of the modulation of the scattered light by aperiodic rotational oscillations of the molecules (see Chapter II). On this basis the transitional region (the point B in Figs. 9 and 10) should correspond to the frequency with which the molecules pass from one temporarily occupied position of equilibrium orientation into another. This frequency should satisfy the relation $\omega_l = 1/\tau_l$, where τ_l is the mean time between two jumps of orientation in the molecules. Since the transitional region (corresponding to ω_l) on the experimental curve is quite wide, we could only roughly estimate the values of τ_l. These were several times smaller than the anisotropy relaxation time τ_a, the greatest value being that of acetic acid ($\tau_l \sim 2 \cdot 10^{-12}$ sec) and

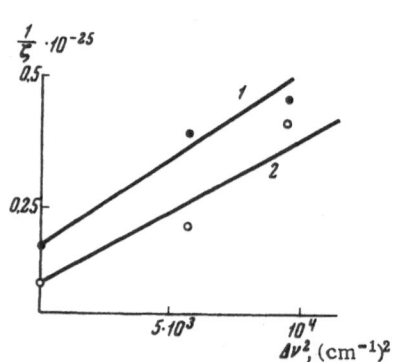

Fig. 11. Reciprocal of internal friction as a function of the square of the frequency for: 1) CS_2; 2) benzene.

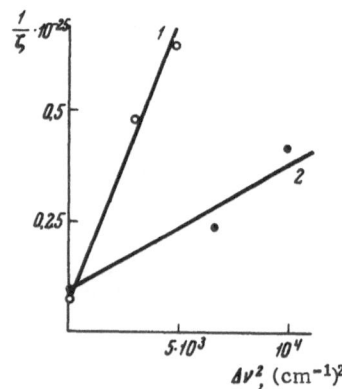

Fig. 12. Reciprocal of internal friction as a function of the square of the frequency for: 1) chloroform; 2) toluene.

the smallest that of carbon disulfide ($\tau_l \sim 5 \cdot 10^{-13}$ sec). Subsequently we shall set out the results of our experimental investigation into the remoter parts of the wing in more detail.

2. Results of a Study of the High-Frequency Part of the Rayleigh-Line Wing in Low-Viscosity Liquids at Room Temperature

In order to compare the experimental intensity distribution in the remote region of the wing (section BCD) with the distribution calculated theoretically (II.23) [with parameters satisfying relation (II.25)], we require to know two quantities for each liquid: the internal-friction relaxation time τ_ζ and the limiting parameter g_0. We regard the static value of the internal-friction coefficient ζ_0 and the moment of inertia I as known. The coefficient ζ_0 is determined from the anisotropy relaxation times τ_a measured in the near part of the wing, using the relation

$$\zeta_0 = 6kT\tau_a. \tag{IV.1}$$

If we know τ_ζ and g_0 and match the distribution (II.23) with the experimental curve at one point, we may determine how well the theory is confirmed by experiment. However, we have no information regarding the values of τ_ζ and g_0, and we must therefore determine these quantities also, from the intensity distribution in the wing. The internal-friction relaxation time for each of the sample liquids was determined as follows.

The high-frequency part of the contour, starting at a certain distance from the turning point B (Fig. 10) (at which we may entirely neglect the effect of molecules jumping from one potential well to another on the wing spectrum), was divided into two parts. In each of these fairly small sections (20-40 cm^{-1}) the elastic constant g and the internal-friction coefficient ζ were regarded as independent of frequency, so that ζ/g and I/g could be found for each section by means of Eq. (II.23) [in order to do this, three intensity values taken for three different frequencies from the part of the spectrum under examination had to be substituted into (II.23)]. Knowing the moment of inertia I, we may find ζ. This latter was regarded as different for the different sections and was also distinguished from ζ_0 found by analyzing the first dispersion section of the wing on the basis of formula (IV.1).

We plotted the relationship between 1/ζ and the square of the frequency (taking the mean frequency for each section for a frequency of $\omega = 0$; we used the value of ζ_0 determined by

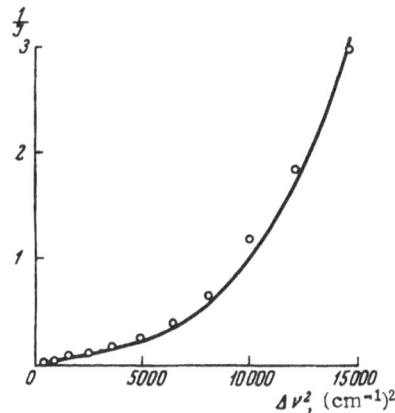

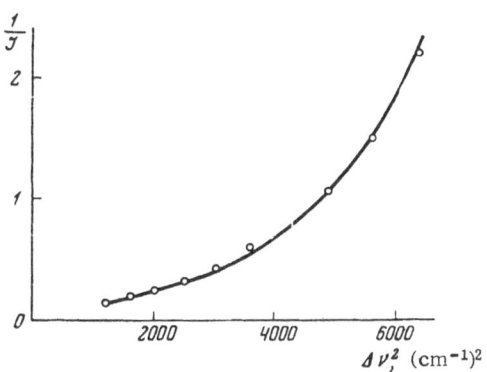

Fig. 13. Reciprocal of the intensity in the wing as a function of the square of the frequency for carbon disulfide. Solid line, theoretical; circles, experimental data.

Fig. 14. Reciprocal of the intensity in the wing as a function of the square of the frequency for chloroform. Notation as in Fig. 13.

means of formula (IV.1) from the experimental results of Table 1. This relationship was linear for all cases studied (Figs. 11 and 12), which may be regarded as a confirmation of the dispersion law chosen (II.25).

The internal-friction relaxation time τ_ζ was determined from the figures for each liquid studied. For CS_2 we obtained $\tau_\zeta = 7.4 \cdot 10^{-14}$ sec, for benzene and toluene $\tau_\zeta = 8.5 \cdot 10^{-14}$ sec, and for chloroform $\tau_\zeta = 2.4 \cdot 10^{-13}$ sec.

The method just described also enables us to estimate the value of g_0 for any liquid. In fact, the quantity $I/g(\omega)$ is determined in each of the sections into which the remoter region of the wing was divided. Hence, if we know $g(\omega)$ at the two frequencies corresponding to the two sections we may estimate g_0 by extrapolating to zero frequency.

The values of τ_ζ and g_0 (mainly g_0) thus found were refined by making slight variations in these parameters so that the experimental intensity distribution should agree more closely with expressions (II.23) and (II.25). As a result of this analysis we found that for carbon disulfide $g_0 = -1.2 \cdot 10^{-12}$ dyn $\cdot$ cm, and for chloroform, $g_0 = -4.5 \cdot 10^{-12}$ dyn $\cdot$ cm.

The intensity distribution in the remote region of the wing plotted on the basis of the resultant values of τ_ζ and g_0 is shown in Figs. 13 and 14 for carbon disulfide and chloroform (continuous lines). The points show the experimental values of $1/\mathcal{J}$. We see from the figures that formulas (II.23) and (II.25) describe the experimental intensity distribution in the remote region of the Rayleigh line wing very well.

It should be noted that formulas (II.23) and (II.25) were calculated for low-viscosity liquids consisting of molecules with a single moment of inertia. For this reason carbon disulfide and chloroform, which satisfy this requirement most closely, were studied more carefully and completely than the others. In carbon disulfide the Rayleigh-line wing was of low intensity, and under our conditions it was therefore almost impossible to study the remote region. As regards the rest of the liquids studied (those with molecules having two or three different moments of inertia), in these cases we confined ourselves to estimating the values of τ_ζ by the method just described, since formulas (II.23) and (II.25) cannot claim to give any quantitative description of the intensity distribution in the wing.

Table 2. Comparison of the Frequencies ν_C in the Liquid and the
Low-Frequency Spectrum of the Corresponding Crystals

Substance	ν_C, author's data	Low-frequency spectrum of the crystals [74-78]
Carbon disulfide	71	69, 80
Benzene	80	35, 55, 65, 105
Toluene	84	48, 70, 88, 100, 128
Chloroform	60	75, 94
Acetic acid	49 (120)*	49 (119)*

*For the explanation of the frequencies given in brackets, see [80].

As indicated in the foregoing, the ratio g/I could be determined from the most distant part of the wing. This region corresponds to frequencies for which we may consider $g(\omega)$ as close to its limiting value $g(\omega \to \infty) = g_\infty$. The quantity

$$\nu_c = \frac{1}{2\pi c} \sqrt{\frac{g_\infty}{I}}, \tag{IV.2}$$

characterizing the limiting (highest) frequency of the elastic oscillations of the molecules in the liquid is interesting to compare with the low-frequency spectrum of the corresponding molecular crystals. This comparison is made in Table 2 for the liquids studied.

This comparison shows that the agreement between the limiting frequencies of the elastic oscillations of the molecules in the liquids and the frequencies determined for the crystal spectra can hardly be considered a matter of chance.

Thus, for high frequencies (CD part of the wing) the liquid appears to have a high degree of "solidity." In this range of frequencies the viscosity has already largely relaxed and the coefficient of elasticity g is close to its limiting value. At these frequencies the elastic properties of the liquids and the elastic properties of the corresponding molecular crystals differ very little.

We note that the values of the frequencies ν_C (Table 2) may be a little too low, since they were obtained from a part of the spectrum corresponding to high but still finite frequency values. For some liquids (benzene, toluene) the analysis of the experimental results could only be approximate, since in these cases we were only able to take an average moment of inertia and obtain a single frequency ν_C. In molecular crystals consisting of molecules with two or three moments of inertia there are several frequencies corresponding to differently oriented waves [79].

The fact that, on comparing the experimental contour with the theoretical expression (II.23) [together with (II.25)], g_0 is negative must to a large extent be considered as an experimental fact. The qualitative explanation of this fact (see Section 3 of Chapter II) is neither exhaustive nor final. Nevertheless, the fact that the molecule executes one reorientation on average in a time τ_l should affect its motion for periods of time $t < \tau_l$ and this effect should therefore appear in the Rayleigh-line wing for frequencies of $\omega > \omega_l$. The reorientation of the molecules is caused by the combined action of the thermal motion of neighboring molecules. This action is such that it leads to reorientation of the molecules on average in a time τ_l, while for periods of $t < \tau_l$ it weakens the coupling between the neighboring molecules. In theory (see Section 3 of Chapter II) this weakening of the coupling appears in that the elastic constant g in (II.26) is smaller by an amount g_0 over short periods of time $(t < \tau_l)$ than it would have been if no such weakening had taken place.

Theory gives the relation (II.26) between $g_0 g_\infty$ and τ_ζ, which we write in the following form:

$$\tau_\zeta = \frac{\zeta_0}{g_\infty - g_0}. \tag{IV.3}$$

For carbon disulfide and chloroform we know ζ_0 [from (IV.1) and the data in Table 1], g_∞ [from (IV.2) and the data in Table 2], and g_0. Hence we may calculate τ_ζ from these data and formula (IV.3) and compare the results with the values obtained directly from experiment and presented above.

Calculations based on (IV.3) give $\tau_\zeta = 7.3 \cdot 10^{-14}$ sec for carbon disulfide, and $\tau_\zeta = 2 \cdot 10^{-13}$ sec for chloroform, which agree closely with the earlier values of $7.4 \cdot 10^{-14}$ and $2.4 \cdot 10^{-13}$ sec.

Thus the parameters obtained experimentally (g_0, g_∞, and τ_ζ) obey the relation (IV.3). This again tends to favor the original premises of the theory.

We note that expression (II.23) [together with (II.25)] for the intensity distribution in the remote region of the Rayleigh-line wing is only applicable for the frequency range in which $g(\omega) > 0$, i.e., at frequencies $\omega > \omega_l = 1/\tau_l$.

3. Study of the Remote Region of the Rayleigh-Line Wing in Liquids at Various Temperatures [80]

A study of the temperature dependence of the spectral distribution in the remote parts of the wing and a comparison of this relationship with the predictions of theory, in the author's opinion subjects the theory to its most stringent test. The difficulty of such an investigation lies in the fact that the liquids (low-viscosity, consisting of molecules with a single moment of inertia) to which relation (II.23) is applicable [together with (II.25)] only change their viscosity by a factor of 2 or 3 as temperature varies by 80 to 100°C. Such a change in viscosity (as confirmed by experiment) does not lead to any substantial change in the intensity distribution over the remoter parts of the Rayleigh-line wing. Further, in temperature experiments the vessel with the scattering liquid is placed in a glass Dewar (see Chapter III) and this leads to a weakening of the intensity of the exciting and hence of the scattered light. If we add to this the necessity of keeping the temperature of the vessel with the scattering liquid constant and the danger of the windows of the vessel becoming misty (on cooling), it becomes clear how great are the difficulties of conducting such a temperature experiment.

In order to study the temperature changes in the remote region of the Rayleigh-line wing we chose chloroform and carbon disulfide.

Chloroform was studied at −34, 20, and 45°C. On changing the temperature from −34 to 45°C the viscosity changed by a factor of 2.5. Carbon disulfide was studied at −65 and 20°C. In this temperature range the viscosity changed by a factor of 3. Figures 15 and 16 show the intensity distributions with respect to frequency [in coordinates of $1/\mathcal{J}(\Delta\nu^2)$] in the Rayleigh-line wing for carbon disulfide and chloroform at various temperatures. The values of the quantities in question were made to coincide for the different temperatures at a point corresponding to a frequency of 30 cm^{-1} for chloroform and 35 cm^{-1} for carbon disulfide. Since the matching points were chosen arbitrarily, the curves presented in Figs. 15 and 16 only enable us to draw conclusions regarding the way in which intensity varies with frequency at various temperatures; we cannot make any judgment as to whether the absolute intensity of the scattered light increases or decreases in any particular range of frequencies as the temperature of the liquid changes. This is because the Rayleigh-line wing cannot be recorded simultaneously at different temperatures.

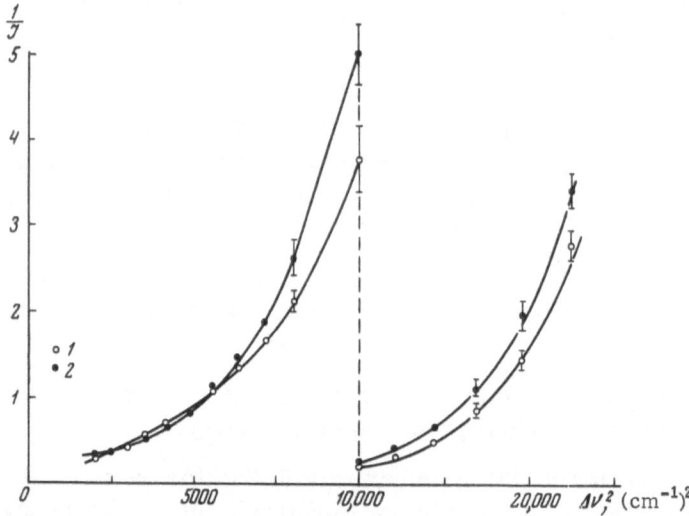

Fig. 15. Reciprocal of the intensity in the wing as a function of the
square of the frequency for carbon disulfide at temperatures of:
1) 20°C; 2) −65°C.

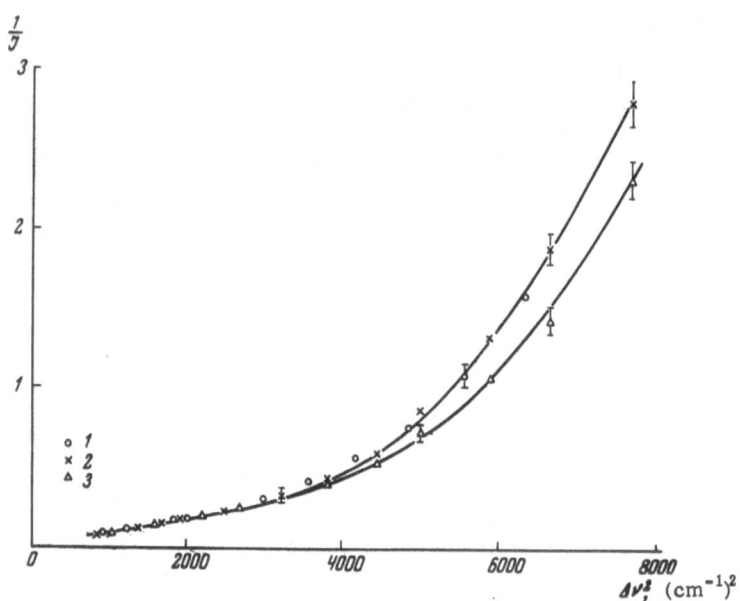

Fig. 16. Reciprocal of the intensity in the wing as a function of the
square of the frequency for chloroform at temperatures of: 1) 20°C;
2) 45°C; 3) −34°C.

On analyzing the experimental data relating to the temperature dependence of the Rayleigh-
line wing we found τ_ζ by comparing the experimental curve at a given temperature with the
theoretical relations (II.23) and (II.25) as in the previous section. The quantity τ_ζ is a fundamental parameter of the theory and the way in which it varies with temperature may either confirm or refute this. The internal-friction relaxation time τ_ζ was determined experimentally at
various temperatures in the following way. For two different frequencies ω_1 and ω_2 of the remote part of the wing we determined the values of $\zeta_1(\omega_1)$ and $\zeta_2(\omega_2)$ by the means described in

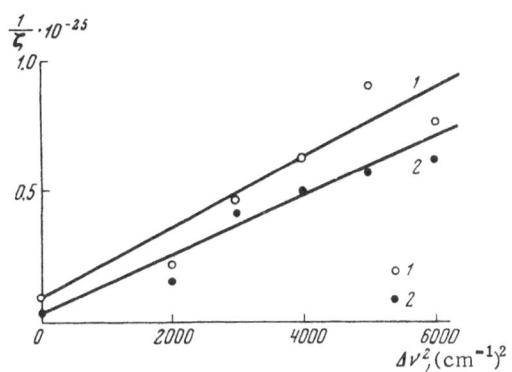

Fig. 17. Reciprocal of the internal friction as a function of the square of the frequency for chloroform at: 1) 45°C; 2) −34°C.

the previous section. Then we found the static (corresponding to $\omega = 0$) internal friction ζ_0. The value of ζ_0 at room temperature may easily be determined from relation (IV.1) and Table 1 for the anisotropy relaxation times. According to (I.11) the values at various temperatures for the same liquid are related in the same way as the viscosities at these temperatures:

$$\frac{\zeta_0(T_1)}{\zeta_0(T_2)} = \frac{\eta(T_1)}{\eta(T_2)}. \qquad (IV.4)$$

The ratio of the viscosities of the liquids at different temperatures T_1 and T_2 we already knew (these were given above); we also knew the value of ζ_0 at room temperature T_1, and could therefore find ζ_0 at temperature T_2.

In this way we found the values of ζ (ζ_0, ζ_1, ζ_2) for three frequencies at a given temperature. Then we plotted the relation between $1/\zeta$ and the square of the frequency. This relationship is shown in Fig. 17 for chloroform at 45 and −34°C. Considering the difficulty of the experiment indicated earlier (due to the large error in determining $1/\zeta$, an error possibly reaching 30%), it is reasonable to suppose that the relation $1/\zeta (\Delta\nu^2)$ may be described by a straight line at both temperatures in accordance with the theoretical expression (II.25). The internal-friction relaxation times for chloroform have the following values: at 45°C, $\tau_\zeta = 2.1 \cdot 10^{-13}$ sec; at −34°C, $\tau_\zeta = 3.3 \cdot 10^{-13}$ sec. In the same way we found the internal-friction relaxation time for carbon disulfide at −65°C. This equalled $\tau_\zeta = 19 \cdot 10^{-14}$ sec (we remember that at room temperature for CS_2, $\tau_\zeta = 7.3 \cdot 10^{-14}$ sec).

It follows from relations (IV.3) and (IV.4) that, if we consider g_0 as weakly temperature-dependent, then the internal-friction relaxation time should vary in direct proportion to the shear viscosity. The values given above for τ_ζ in the case of the two liquids show that this theory is confirmed by experiment to an accuracy of 30%. Considering the difficulty of the experiment and the probable dependence of g_0 on temperature, we must accept the agreement between theory and experiment as quite satisfactory.

Regarding the physical nature of τ_ζ we may make the following comments. The internal-friction relaxation time τ_ζ is determined by the condition that for a time $t < \tau_\zeta$ the molecule undergoing rotational motion (oscillations or vibrations) is unable to transfer moment of momentum to the surrounding medium. Hence, for short intervals of time the rotational motion of the molecule is executed without friction. However, if the motion of the molecule is considered for time intervals of $t > \tau_\zeta$, when the molecule is in fact able to exchange moment of momentum with its neighbors, then ordinary internal friction starts acting. It is quite clear that the condition $\tau_\zeta \geq T$ must be satisfied, where T is the period of elastic oscillations of the molecules.

Thus we consider the rotational motion of the molecules and internal friction during this motion in a manner analogous to that in which Andrade [81] considered the translational motion of molecules and the viscosity associated with this motion. The internal-friction relaxation time τ_ζ for the rotational motion of the molecules and the period spent by the molecules in the potential well τ_l should not be identified with the analogous times in the case of translational motion and in translational motion the relaxation times are proportional to the vis- both cases these parameters have many common features (for example, both in orientational motion and in the case of translational motion the relaxation times are proportional to the viscosity), they correspond to different forms of motion and have different numerical values. As regards the shear viscosity for translational motion, it may be noted that Andrade [81] in his theory considers a viscosity relaxation time analogous to our τ_ζ for the case of orientational motion, while Frenkel' [3] considers the time spent by the molecule in a resting position between two translational jumps, analogous to our time τ_l (time between two jumps of orientation). Andrade considers the time (or frequency ν) necessary for the molecule to reach a specific (extremal) position at which it can transfer momentum to its neighbors. The molecule reaches such a position in the course of vibrational motion around a position of equilibrium as it approaches a neighboring molecule, also vibrating. It is clear that analogous considerations may be extended to rotational motion.

It follows from the foregoing that translational and orientational motion have many common features and the study of one form of motion may help in understanding the chief characteristics of the other. However, for studying the translational motion at frequencies of $\sim 10^{11}$ to 10^{13} cps there is no such convenient method as that of the Rayleigh-line wing for studying the orientational motion of molecules in a liquid.

DETERMINATION OF THE VELOCITY AND ABSORPTION OF HYPERSOUND IN LIQUIDS FROM THE WIDTH OF AND SPACING BETWEEN THE MANDELSHTAM -BRILLOUIN COMPONENTS

1. Measured Values of the Absorption and Velocity of Hypersound

The new source of light provided by the gas laser offered the possibility of greatly improving the technology of studying the spectrum of scattered light. Not only did it become possible to measure the velocity of hypersound more accurately, but even the absorption of hypersound, hitherto impossible to measure, became susceptible to measurement.

For studying the velocity and absorption of hypersound we chose liquids in which the volume viscosity greatly exceeded the shear viscosity, as indicated by ultrasonic measurements.

We studied the fine structure of the scattered light in benzene, carbon tetrachloride, toluene, carbon disulfide, chloroform, and methylene chloride. In all these liquids the absorption coefficient and velocity of sound had been measured earlier in the ultrasonic range. In some of these liquids (carbon disulfide [82], methylene chloride [83, 84]) a relaxation of the volume viscosity coefficient had already been observed at ultrasonic frequencies. Stewart and Stewart [48], using an ultrasonic technique, observed a dispersion of the velocity of sound in carbon tetrachloride at a frequency of $3 \cdot 10^9$ cps. Earlier the dispersion of the velocity of sound had been determined in certain liquids from measurements of the distance between the Mandelshtam—Brillouin components by Fabelinskii [6], Pesin [11], and Tunin and Shakhparanov [12]. The accuracy of these measurements, however, was not very high.

Our measurements of the velocity of hypersound by an optical method gave an accuracy an order higher than in previous measurements. This increased accuracy was obtained by using a gas laser working with a mixture of Ne and He as the exciting light source ($\lambda = 6328$ Å).

Table 3 shows the frequency and velocity of hypersound at this frequency for each liquid. The velocity of hypersound was calculated for all the liquids from formula (I.13). The error in measuring the velocity of hypersound ($\sim \pm 0.5\%$) is made up of random errors committed in determining the position of the fine-structure components and in establishing the scattering angle. The velocity of hypersound in benzene and carbon tetrachloride agrees closely with the previous measurements of Fabelinskii [6] and subsequent data [11, 85].

Our own data regarding the dispersion of the velocity of sound in carbon tetrachloride confirm the earlier measurements of Fabelinskii [6] but differ from those of Stewart and Stewart [48], who give a value of $\sim 4\%$ for the dispersion of the velocity of sound at a frequency of $3 \cdot 10^9$ cps. It is possible that this fact is due to the difficulty of measuring the velocity of

Table 3. Velocity of Hypersound Measured in Liquids

Substance	$f \cdot 10^{-9}$, cps	Hyper-sound velocity V_f, m/sec	$\frac{\Delta v}{V_f} \cdot 10^2$	ρ, g/cm³	$\eta \cdot 10^3$, P	η'_0, P	$\tau \cdot 10^{10}$, sec	$\alpha \cdot 10^{-3}$, cm⁻¹	V_∞, m/sec
Benzene...	4.94 ± 0.02	1471 ± 8	10	0.878	6.5	0.921	2.5 ± 0.1	5.5 ± 0.4	1474 ± 8
Carbon tetrachloride.....	3.31 ± 0.01 (3,54) *	1015 ± 6 (1022) *	10	1.595	9.7	0.339	0.93 ± 0.1 (0.82) *	14.4 ± 2.3 (18.1) *	1036 ± 13 (1043) *
Chloroform	3.4 ± 0.01	1055 ± 5	5	1.498	5.8	0.306	1.68 ± 0.2	4.52 ± 1.1	1096 ± 14
Methylene chloride··	3.53 ± 0.02	1113 ± 6	2	1.336	4.3	1.071	17 ± 6	0.91 ± 0.1	1113 ± 8

* This value of the velocity of hypersound in CCl₄ is obtained if we consider the effect of the absorption of the hypersound on the position of the maximum of the Mandelstam—Brillouin component. Thus [1], $\Delta v_{max} \approx (1 - .25\delta v^2 / \Delta v^2)$, where Δv is determined by (I.13). In the rest of the liquids studied this correction was unimportant.

Table 4. Measured Widths of the Mandelshtam—Brillouin Components in Liquids

Substance	$\delta v \cdot 10^3$, cm⁻¹	$\alpha_{meas} \cdot 10^{-3}$, cm⁻¹	$\alpha_\eta \cdot 10^{-3}$, cm⁻¹	$\alpha_\eta \cdot 10^{-3}$, cm⁻¹	$\tau \cdot 10^{10}$, sec	$\alpha/f^2 \cdot 10^{17}$
Benzene	7 ± 2	4.5 ± 1.3	2.0	2.5	3.03 ± 1.0	18.4 ± 0.5
Carbon tetra-chloride ...	17 ± 3	16 ± 3	2.0	14	0.83 ± 0.1	146 ± 27
Chloroform	11 ± 2	10 ± 2	1.15	9	0.96 ± 0.1	87 ± 17
Toluene	10 ± 2	7 ± 1.5	1.49	5.6	0.46 ± 0.1	36 ± 8
Methylene chloride	~ 2	—	—	—	—	—
Carbon disulfide ··	~ 3	—	—	—	—	—

sound at such high frequencies by means of ordinary ultrasonic techniques. It should be mentioned that our data relating to the velocity of hypersound in carbon tetrachloride agree within measuring error with those of Chiao and Stoicheff [86], in which the velocity of hypersound was determined by reference to the distance between the fine-structure components, the source of exciting light being an Ne—He laser.

As already mentioned (see Section 2 of Chapter I), if the absorption of the hypersound is very high, then for calculating the velocity of hypersound from the distance between the Mandelshtam—Brillouin components we must use formula (I.15) instead of (I.13). Estimates show that it is only necessary to allow for the effect of the absorption of the hypersound on the position of the Mandelshtam—Brillouin maxima in the cases under consideration for the case of carbon tetrachloride. In the remaining cases this correction is negligibly small. For CCl₄, allowance for absorption gives a value of the velocity of hypersound at our frequency 0.7% greater than that calculated by formula (I.13). Table 3 shows the velocity of hypersound and the corresponding relaxation parameters obtained after allowing for the effect on hypersound absorption on the position of the fine-structure maxima in brackets.

In the case of chloroform and methylene chloride, the accuracy of the previous measurements [11] was apparently rather overestimated and therefore the new data (Table 3) disagree with the previous hypersound—velocity measurements.

Table 4 gives the measured widths $\delta\nu$ of the Mandelshtam—Brillouin components and the corresponding values of the absorption coefficient of hypersound calculated from the formula

$$\alpha = \pi \frac{c}{V} \delta\nu. \qquad (V.1)$$

A particularly large number of photographs (several tens) was analyzed in studying the width of the fine-structure components in benzene and carbon tetrachloride. In these liquids we studied the widths of the Mandelshtam—Brillouin components as functions of the aperture of the scattered light with the intention of extrapolating the resultant values of $\delta\nu$ to zero aperture (see Section 2 of Chapter III). This extrapolation gives component widths differing very little from those obtained at an aperture of $2\delta\vartheta \sim 0.03$ for both liquids. For this reason the rest of the liquids were simply studied with an aperture angle of the scattered light equal to $2\delta\vartheta \sim 0.03$.

The width of the Mandelshtam—Brillouin components in benzene, carbon tetrachloride, chloroform, and toluene was also studied in [86]. The authors of [86] gave a value of $\tau \geq 4 \cdot 10^{-10}$ sec for all these liquids, which corresponds to a component width of $\delta\nu \leq 5 \cdot 10^{-3}$ cm^{-1}. This result contradicts our own measurements (see Table 4) and may be a result, for example, of insufficient purity of the liquids studied in [86]. In addition to this, the interferometer used in [86] had a low range of dispersion and this led to overlapping of the components belonging to different orders of interference. Hence, in deciphering the contours of these overlapping components of fine structure, results differing sharply from the truth may well have been obtained. It is possible that there may also have been other as yet uncertain reasons for the difference between our own results and those of [86].

2. Analysis of the Results from the Point of View of the Phenomenological Relaxation Theory

The simultaneous measurement of the velocity and absorption coefficients of hypersound for the first time offered the possibility of directly verifying the applicability of the version of the relaxation theory given in [16] (with one relaxation time) to any particular liquid and finding the value of the corresponding relaxation time.

We set about this verification in the following way. After measuring $\delta\nu$ and using formula (I.16"), we determine α. Since α_η is easily calculated, it is not difficult to find $\alpha_{\eta'} = \alpha - \alpha_\eta$. Then, knowing η_0', it is easy to find the relaxation time of volume viscosity τ from formula (I.19"). The values of τ determined in this way for the liquids under consideration are given in Table 4. Table 3 gives the values of the same relaxation time, but determined from the dispersion of the velocity of hypersound and formula (I.21). Comparison of the values of τ obtained from the dispersion of the velocity of hypersound (Table 3) and those obtained by measuring the absorption coefficient of hypersound (Table 4) enables us to judge the applicability of the one-relaxation-time relaxation theory to any given liquid. The values of τ in both tables agree for benzene and carbon tetrachloride. Hence, we may conclude that these liquids obey the relaxation theory up to frequencies of $\sim 5 \cdot 10^9$ cps.

In the case of toluene we were able to measure the width of the Mandelshtam—Brillouin components and hence determine τ from these measurements (Table 4). From the resultant value of τ it is not difficult to estimate the expected dispersion of $\Delta V_f / V_f \sim 3\%$ at a frequency of $f = 4.4 \cdot 10^9$ cps.

For methylene chloride we found the dispersion of the velocity of sound and were able to determine the relaxation time τ from this measurement. Measurement of the width of the

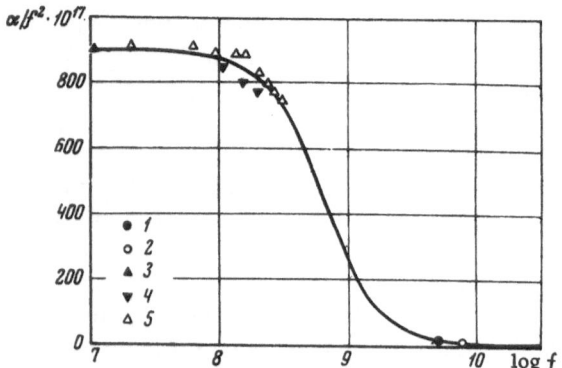

Fig. 18. Dependence of α/f^2 on $\log f$ for benzene. Results of: 1) the author; 2) [6]; 3) [71]; 4) [84]; 5) [87].

components $\delta\nu$ showed that this value lay within the limits of measuring error ($\delta\nu \sim 2 \cdot 10^{-3}$ cm^{-1}). We should expect this result from the determination of the dispersion of the velocity of sound.

Measurements in carbon disulfide gave the width of the Mandelshtam—Brillouin components as $\delta\nu \sim 3 \cdot 10^{-3}$ cm^{-1}, also within the limits of measuring error and agreeing with the expected values based on earlier measurements of the dispersion of the velocity of sound [6, 11].

For chloroform the values of τ calculated from the dispersion of the velocity of sound (Table 3) and from the absorption of hypersound (Table 4) differ considerably. This clearly means that the relaxation of volume viscosity in chloroform cannot be described by the formulas of the simplified one-relaxation-time relaxation theory.

According to relaxation theory [16] we may write

$$\frac{\alpha}{f^2} = A + \frac{B}{1 + (f/f_c)^2}, \qquad\qquad (V.2)$$

where

$$A = \frac{\alpha_\eta}{f^2} = \frac{2}{3}\frac{\eta}{V_0^3\rho}, \qquad\qquad B = \frac{\alpha_{\eta_0'}}{f^2} = \frac{\eta_0'}{2V_0^3\rho}$$

(f_c is the critical frequency equal to $f_c = .5\pi\tau$). It follows from this expression that at low frequencies ($f \ll f_c$), α/f^2 has a constant value of $\alpha/f^2 = A + B$, and at high frequencies ($f \to \infty$) a constant value of $\alpha/f^2 = A$.

Using expression (V.2) we plotted the relation between α/f^2 and $\log f$ for all the substances studied. For ultrasonic frequencies we used data obtained by ultrasonic measurements, and in the hypersonic range we used values of α/f^2 obtained from the results of Tables 3 and 4. Figures 18-22 show the dependence of α/f^2 on $\log f$ as obtained from [6, 11, 71, 84, 87-89]. These graphs and all that has been said up to now lead to the conclusion that in benzene (Fig. 18), carbon tetrachloride (Fig. 19), and methylene chloride (Fig. 20), the relaxation of volume viscosity is properly described by the one-relaxation-time relaxation theory. Particularly noteworthy is the curve of α/f^2 as a function of $\log f$ for methylene chloride (Fig. 20), where the various ultrasonic measurements and our own hypersonic data lie on a single relaxation curve. This experimental result may be explained by the molecular theory of sound absorption, which is used for describing the results of ultrasonic investigations [90]. We note that direct measurements of the absorption of hypersound could not be carried out in methylene chloride, since this quantity lay beyond the limits of sensitivity of our method. Absorption in this case is calculated from the dispersion of the velocity of sound.

The dependence of α/f^2 on $\log f$ for toluene (Fig. 21) is based on ultrasonic data and our own results obtained for the absorption of hypersound on the basis of the measured widths of the Mandelshtam—Brillouin components. In the case of toluene the extrapolation of the absorption coefficient measured at low frequency into the hypersonic region, using formula (I.18) (on the assumption that there is no relaxation of volume viscosity), gives a value of $\alpha = 16 \cdot 10^3$

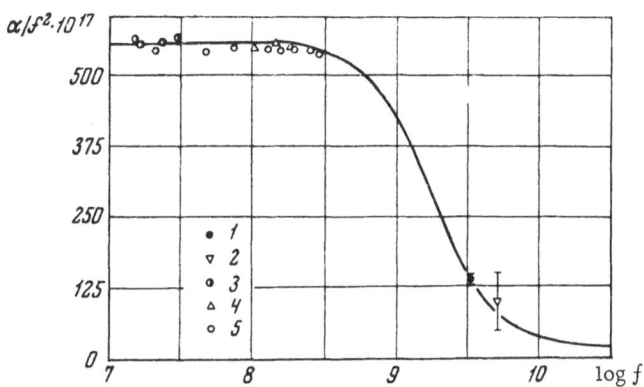

Fig. 19. Dependence of α/f^2 on $\log f$ for carbon tetrachloride.
Results of: 1) the author; 2) [6]; 3)[88]; 4) [84]; 5) [89].

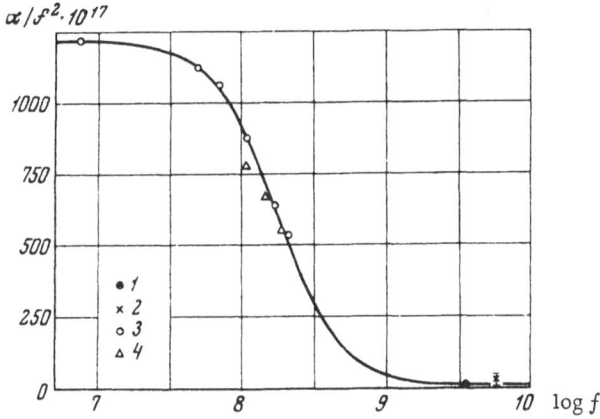

Fig. 20. Dependence of α/f^2 on $\log f$ for methylene chloride.
Results of: 1) the author; 2) [11]; 3) [83]; 4) [84].

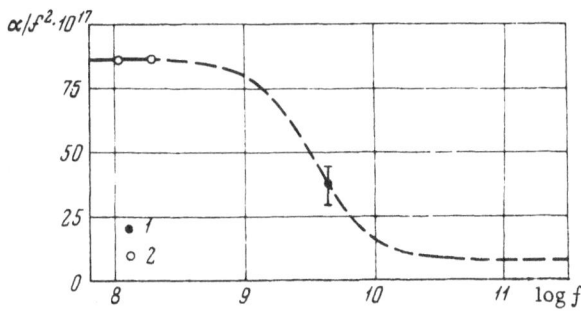

Fig. 21. Dependence of α/f^2 on $\log f$ for toluene. Results
of: 1) the author; 2) [84].

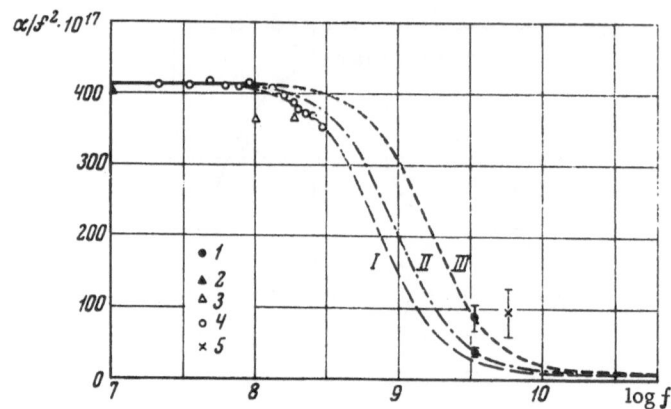

Fig. 22. Dependence of α/f^2 on $\log f$ for chloroform. I) Curve based on the results of Lezhnev [89]; II) curve based on measurements of the velocity of sound; III) curve based on absorption measurements. Results of: 1) the author; 2) [89]; 3) [84]; 4) [71]; 5) [11].

cm^{-1}. Measurements of the width of the Mandelshtam—Brillouin components, however, give $\alpha = (7 \pm 1.5) \cdot 10^3$ cm^{-1}. This disagreement suggests that relaxation of the volume viscosity occurs in toluene with a relaxation time of $\tau = 5 \cdot 10^9$ sec.

Figure 22 shows the relaxation graph for chloroform. The three different curves are based on: 1) ultrasonic measurements [89]; 2) ultrasonic measurements [89] and our own hypersonic data for absorption, calculated from the dispersion of the velocity of sound; 3) ultrasonic data [89] and hypersonic data obtained from the width of the Mandelshtam—Brillouin components. The disagreement between these three curves lies far beyond possible experimental errors.

Thus, we may conclude from our experiments that the relaxation of volume viscosity in benzene, carbon tetrachloride, and apparently methylene chloride is adequately described by the formulas of the one-relaxation-time relaxation theory. As regards toluene, as yet we have insufficient data for a conclusive decision. The graph (Fig. 21) of α/f^2 as a function of $\log f$ should in this case be considered as a working hypothesis reflecting the possible picture of volume-viscosity relaxation if in fact only one relaxation time exists.

3. Analysis of the Results from the Point of View of the Molecular Theory of Relaxation

The phenomenological theory of Mandelshtam and Leontovich does not specify the molecular mechanisms of the hyper-Stokes part of the absorption of sound in liquids [see formula (I.19')]. It is well known (see, for example, [17, 71]) that in gases the redistribution of energy between the internal and external degrees of freedom of the molecule has a considerable effect on sound propagation (this is the so-called Kneser relaxation process).

Even before the appearance of the phenomenological relaxation theory [16], Kneser [91] suggested that in liquids the transfer of energy to the internal degrees of freedom of the molecule during the propagation of a sound wave took place in a manner analogous to that holding in the case of gases. The extension of the Kneser theory to liquids was effected by Herzfeld in 1941 [92].

On considering the absorption of sound due to the exchange of energy between internal and external degrees of freedom, it is convenient to use the concept of the effective complex specific heat

$$(C_v)_{\text{eff}} = C_v - \frac{C'i\omega\tau'}{1 + i\omega\tau'}, \tag{V.3}$$

where C_V is the specific heat at constant volume, C' is the specific heat of the internal degrees of freedom of the molecule, and τ' is the relaxation time. It is assumed in the Herzfeld theory that all the degrees of freedom of the molecule relax with the same relaxation time τ'. As we shall see later, for some liquids this assumption is by no means valid. For the quantity $S = (\rho - \rho_0)/\rho$ we may write the expression

$$\frac{\partial^2 S}{\partial t^2} + \Gamma' \frac{\partial}{\partial t} \Delta S - \left(\frac{1}{\rho_0} \frac{\partial p}{\partial S}\right) \Delta S = 0. \tag{V.4}$$

Here

$$\Gamma' = \frac{\eta'}{\rho} = \frac{2\alpha_{\eta'} V^3}{\omega^2}. \tag{V.5}$$

In (V.4) the extra term (extra in relation to the ordinary version of [93]) containing Γ' compensates the attenuation contained in the last term, and hence the solution of this equation will be a nonattenuating harmonic wave.

The solution of Eq. (V.4) may be sought in the form

$$S = \sum_n S_n(\omega_n, k_n) e^{i(\omega_n t - \mathbf{k}_n \mathbf{r})}. \tag{V.6}$$

Substituting (V.6) into (V.4) and considering that $\mathbf{k}^2 = \omega^2/V^2$ (replacing ω_n by ω and $\mathbf{k}_n$ by $\mathbf{k}$), we obtain

$$\frac{1}{V} + \frac{2i\alpha_{\eta'}}{\omega} = \frac{1}{V^3}\left(\frac{1}{\rho_0}\frac{\partial p}{\partial S}\right). \tag{V.7}$$

It is not hard to show [93] that $(1/\rho_0)(\partial p/\partial S)$ may be expressed in the following way:

$$\frac{1}{\rho_0}\frac{\partial p}{\partial S} = \frac{1}{\rho_0 k_T}\frac{(C_v)_{\text{eff}} + \Delta}{(C_v)_{\text{eff}}}, \tag{V.8}$$

where k_T is the isothermal compressibility of the liquid, and $\Delta = C_p - C_v$, C_p is the specific heat at constant pressure. Putting (V.8) into (V.7) and separating the real and imaginary parts, we obtain

$$\frac{\alpha_{\eta'}}{\omega^2} = \frac{V_0^2}{2V^3}\frac{C'\Delta}{C_p(C_v - C')}\frac{\tau}{1 + \omega^2\tau^2}, \tag{V.9}$$

$$\frac{V^2}{V_0^2} = 1 + \frac{C'\Delta}{C_p(C_v - C')}\frac{\omega^2\tau^2}{1 + \omega^2\tau^2}. \tag{V.10}$$

Here $\tau = [(C_v - C')/C_v]\tau'$. In deriving (V.9) and (V.10) we have assumed that the absorption of sound due to volume viscosity may be considered independently of the absorption due to shear viscosity, and thus in (V.4) and (V.7) we should write $\alpha_{\eta'}$ and not $\alpha = \alpha_\eta + \alpha_{\eta'}$.

From (V.9) with $\omega\tau \ll 1$ we find an expression for the relaxation time τ:

$$\tau = \frac{2\alpha_{\eta_0'} V_0}{\omega^2}\frac{C_p(C_v - C')}{C'\Delta}. \tag{V.11}$$

Putting $\omega \rightarrow \infty$ in (V.10), we obtain for the dispersion

$$\frac{V_\infty^2}{V_0^2} - 1 = \frac{\Delta C'}{C_p (C_v - C')},$$ (V.12)

where V_∞ is the limiting value of the velocity of sound as $\omega \rightarrow \infty$.

It follows from (V.9) and (V.11) that the relaxation time of volume viscosity τ differs from the relaxation time of the specific heat τ', which is equal to

$$\tau' = \frac{2\alpha_{\eta_0}' V_0}{\omega^2} \frac{C_p C_v}{C' \Delta}.$$ (V.13)

Nevertheless, attempts are sometimes made [51, 94] to compare the experimental volume-viscosity relaxation time with the calculated values of τ', naturally leading to false conclusions. In [11], the following expression, given in [93], was used for τ instead of expression (V.11):

$$\tau'' = \frac{2\alpha_{\eta_0}' V_0}{\omega^2} \frac{C_v (C_p - C')}{C' \Delta}.$$ (V.14)

This formula was obtained by a simplified derivation [93]. Its inaccuracy may be demonstrated in the following way. From the general phenomenological theory (see Section 2 of Chapter I) we may easily find the following relation between τ and V_∞^2 / V_0^2:

$$\tau = \frac{2\alpha_{\eta_0}'}{\omega^2} \frac{V_0^3}{V_\infty^2 - V_0^2}.$$ (V.15)

It is not hard to see that from (V.11) and (V.12) we obtain the general relation (V.15), whereas, from (V.12) and (V.14) we obtain τ'', larger by a factor of V_∞^2 / V_0^2.

In calculating the relaxation time of the volume-viscosity coefficient and the dispersion of the velocity of sound from the formulas of the molecular theory, it is usually assumed that all the degrees of freedom of the molecule relax with the same relaxation time. The specific heat of the i-th vibrational degree of freedom of the molecule is calculated from the Einstein formula

$$C_i = d_i \frac{R x^2}{e^x (1 - e^{-x})^2},$$ (V.16)

where d_i is the multiplicity of the degeneracy of the frequency ν_i, $x = h\nu_i/kT$, and $R = 1.986$ cal/g-mole.

Table 5 gives the calculated and experimental values (the latter determined from the width and spacing of the fine-structure components) of the relaxation times of volume viscosity and the dispersion of the velocity of sound for five liquids, as well as some data for carbon disulfide, a liquid already thoroughly studied both in the ultrasonic range [82] and by the molecular scattering of light [6, 11]. This liquid is a classical example of a "Kneser liquid," since here the relaxation of volume viscosity is excellently described by the formulas of the theory of thermal relaxation. However, none of the other liquids in Table 5 are described by the theory of thermal relaxation if all the degrees of freedom of the molecules of each of these liquids have the same relaxation time.

In methylene chloride we may obtain a good agreement between theory and experiment if we omit the specific heat of the lowest vibrational frequency $\nu_1 = 283$ cm^{-1} from the calculation. Thus we assume that, whereas the remaining eight frequencies [11, 93] relax with the same relaxation time, given in Table 1, the first frequency relaxes with a different, much shorter time. On this assumption, the calculated value of $\tau = 1 \cdot 10^{-9}$ sec and our value measured from the dispersion of the velocity of sound $\tau = (1.7 \pm 0.6) \cdot 10^{-9}$ sec agree quite satisfactorily. This assumption is also confirmed by a special experiment [90] on the measurement of the absorption of sound in methylene chloride at low temperatures, and by the fact that in the vapor ν_1 relaxes separately, at higher frequencies, whereas all the other eight frequencies have

Table 5. Calculated and Measured Values of τ and $D = (V_\infty - V_0)/V_0$

Substance	$\dfrac{\alpha_{\tau_0}'}{f^2} \cdot 10^{15}$	C_p, cal/ g-mole	C_v, cal/ g-mole	C', cal/ g-mole	Calculated		Measured	
					$\tau \cdot 10^{10}$, sec	D	$\tau \cdot 10^{10}$, sec	D
Carbon disulfide . . .	60	18.2	11.7	3.9	20	0.09	22	0.09
Benzene	8.9	32.2	22.2	12.6	1.5	0.19	2.5±0.1	0.11
Carbon tetra- chloride	5.3	30.8	20.8	11.8	0.57	0.20	0.9±0.1	0.13
Methylene chloride .	12	24.1	15.6	2.57 *	9.5 *	0.03 *	17±6(10)	0.02
Chloroform	4.1	28.0	18.5	7.8	0.82	0.12	0.96±0.1 1.7 ±0.2	0.09
Toluene	0.75	37	27	17	0.11	0.15	0.46±0.1	0.007

*In calculating τ and D in methylene chloride we reject the specific heat corresponding to the lowest vibrational frequency $\nu_1 = 283$ cm^{-1}. Hence, the above value of $C' =$

$$C_{2-9}' = \sum_{i=2}^{9} c_i \quad \text{(more details in text)}.$$

one and the same relaxation time. Finally, the absorption of ultrasound in methylene chloride was studied in [49] by ultrasonic techniques in the frequency range 30-510 Mc/sec. The resultant value of $\tau = 1 \cdot 10^{-9}$ sec (given in brackets in Table 5) also agrees with the calculated value.

In order to explain the difference between theory and experiment in the case of benzene and carbon tetrachloride, Pesin [11] suggested that in these liquids, as in methylene chloride, not all the degrees of freedom of the molecules relaxed with the same time.

In benzene satisfactory agreement between the calculated and measured values of τ and $D = (V_\infty - V_0)/V_0$ may be secured if we suppose that all the frequencies belonging to the E type of symmetry relax with the same relaxation time, while the rest do so at considerably higher frequencies, as yet inaccessible to study. In this case, the relaxing part of the vibrational specific heat equals $C' = 8.9$ cal/g-mole and calculation gives $\tau = 2.8 \cdot 10^{-10}$ sec, $D = 0.10$, which agree satisfactorily with the experimental data of Table 5.

In the case of carbon tetrachloride, if of the four vibrational frequencies we reject just the lowest, then C' becomes 8.25 cal/g-mole, which give values of $\tau = 1.1 \cdot 10^{-10}$ sec and $D = 0.11$, also satisfactorily agreeing with experiment. It should be noted that in CCl_4 there may be Fermi resonance, since the sum of the two frequencies $\nu_2 = 314$ cm^{-1} and $\nu_2 = 458$ cm^{-1} has a value (772 cm^{-1}) close to $\nu_4 = 776$ cm^{-1}, and in addition to this the frequency ν_4 and the composite frequency $\nu_2 + \nu_3$ relate to the same F type of symmetry. Hence, the assumption as to the possibility of frequencies ν_2, ν_3, and ν_4 having the same relaxation time may be fully justified.

Thus, in benzene and carbon tetrachloride, as in methylene chloride, the molecular theory satisfactorily describes the experimental results if we assume the existence of at least two processes in these liquids with different relaxation times. However, as noted in [11], there are no well-founded principles which would enable us in advance to select a particular group of frequencies relaxing with the same time. In each specific case we must seek such a group of frequencies on the basis of agreement between the calculated and measured relaxation parameters. It may well prove to be the case that it is not a matter of different groups of frequencies relaxing with different values of τ so much as an inadequacy in the basic principles underlying the theory. From this point of view we shall not require more of this molecular theory than agreement with experiment on order of magnitude, and as indicated in Table 7, such an agreement does exist.

On comparing the molecular theory with experiment in the case of chloroform and toluene, we cannot draw any definite conclusion regarding the contribution of any particular groups of frequencies in the course of thermal relaxation.

The values of τ given in Table 5 for chloroform were found both from the width of the Mandelshtam—Brillouin components and from the dispersion of the velocity of sound. These values of τ differ by a factor of 2. The τ values determined from the width of the Mandel-shtam—Brillouin components agree closely with the calculated values.

The calculated and measured values of τ for toluene differ very severely. This may be because, in the case of toluene (as in chloroform) different degrees of freedom of the molecules relax with different relaxation times. We should then have to suppose in the case of toluene that several internal degrees of freedom relaxed at much higher frequencies than that at which measurement was made (f = 4.4 · 10^9 cps). However, we cannot at the present moment draw any definite conclusions regarding such a complicated liquid as toluene.

The application of the theory of thermal relaxation to liquids can only be justified by practical success. In liquids there may also be relaxation processes of another type, such as, for example, structural relaxation, the disruption of the equilibrium distribution of the rotational isomers of the molecules by the sound wave, and so on. Frequently we cannot say in advance which process will predominate in any particular liquid.

In the cases here studied, the Kneser nature of the volume viscosity raises no doubts. However, as indicated earlier, for liquids the theory itself is based on assumptions which do raise serious doubts. This is true in particular of the assumption that the mechanism underlying the exchange of energy between the translational and rotational degrees of freedom in the liquid takes place by way of pair collisions.*

4. Intensity Ratio of the Fine-Structure Components

The sharp, fine structure arising on excitation by the stimulated emission of a gas laser enables us to measure the integral intensities of all three components of fine structure, comparing the results with the results of Fabelinskii's calculation [1, 6, 95]:

$$\frac{\mathscr{I}_c}{2\mathscr{I}_{M\text{-}B}} = K \frac{\sigma^2 T}{C_p \rho \beta_s} , \tag{V.17}$$

where

$$K = \frac{\left(\dfrac{1}{\sigma} \dfrac{\partial \varepsilon}{\partial T}\right)^2_p}{\left(\rho \dfrac{\partial \varepsilon}{\partial \rho}\right)^2}. \tag{V.18}$$

Here σ is the coefficient of volume expansion, β_S is the adiabatic compressibility, and β_S = $1/\rho V^2$.

If we neglect the difference between $\left(\rho \dfrac{\partial \varepsilon}{\partial \rho}\right)_s$ and $\left(\dfrac{1}{\sigma} \dfrac{\partial \varepsilon}{\partial T}\right)_p$ and take no account of the dispersion of the velocity of sound, formula (V.17) transforms into the well-known Landau—Placek formula [96]

$$\frac{\mathscr{I}_c}{2\mathscr{I}_{M\text{-}B}} = \gamma - 1, \tag{V.19}$$

where γ = C_p/C_v.

---

*For a discussion of this question see, for example, [17].

We see from (V.17) that in liquids for which there is a considerable dispersion of the velocity of sound, the ratio of the integral intensities calculated from (V.17) will differ considerably from that calculated by means of (V.19).

Thus, a measurement of the ratio $\mathcal{I}_c/2\mathcal{I}_{M-B}$ and a comparison of this with the calculated value based on formula (V.17) may serve as a further criterion for the correctness of the sound-velocity measurement.

When studying the fine-structure components of benzene we analyzed 15 different photographs and found that the intensity distribution of all three components was nearly of the dispersion type.

We regarded the intensity between the n-th order of the Stokes component and the (n − 1)-th order of the anti-Stokes component as equal to the intensity of the continuous spectrum created by the wing of the Rayleigh line.* The intensity at the same point due to the difference of the reflection coefficient of the interferometer mirrors used in our experiments from unity was negligibly small.

The means used for analyzing the fine-structure spectra may introduce considerable errors into the "tails" of the intensity distribution of all three components. It therefore seemed desirable to determine the integral intensities of the components, not from direct measurement of the areas formed by the contours of the components, but from a measurement of the maximum intensities of the components and their visible half-widths (see Chapter III). The results of experimental data for $\mathcal{I}_c/2\mathcal{I}_{M-B}$, and also previous measurements made with a mercury arc as exciting light source are presented in Table 6.

We note that although the previous data practically overlapped the contemporary results (within the limits of a rather large error), the later results were systematically lower than the earlier.

Rank et al. [99] also measured the intensity ratios in the fine-structure components for a number of liquids, including carbon tetrachloride and toluene. As exciting light source these authors used an Ne—He gas laser (λ = 6328 Å). For CCl_4 the value obtained was $\mathcal{I}_c/2\mathcal{I}_{M-B}$ = 0.75, and for toluene, $\mathcal{I}_c/2\mathcal{I}_{M-B}$ = 0.41, which agrees closely with our own results (see Table 6).

There is no difficulty in comparing the experimental data with calculations based on formula (V.19). Such a comparison (Table 6) shows that agreement with experiment is satisfactory for the majority of liquids here studied.

Comparison of the measured values of $\mathcal{I}_c/2\mathcal{I}_{M-B}$ with the results of a calculation based on formula (V.17) encounters purely technical difficulties in that the quantities in (V.17) are either quite unknown or only known very inexactly. These relate in particular to such quantities as $\left(\rho \frac{\partial \varepsilon}{\partial \rho}\right)_s$ and $\left(\frac{1}{\sigma}\frac{\partial \varepsilon}{\partial T}\right)_p$. Table 7 presents some data from which the calculations summarized in Table 6 were carried out.

---

*In the present investigation the scattered light was excited by the light of a laser polarized in a plane perpendicular to the plane of scattering. On the photograph of the scattered-light spectrum, in addition to the polarized discrete components there was also some continuous depolarized light (Rayleigh-line wing). The scattered light was not photographed in two mutually perpendicular polarizations; hence the analysis described in [6] was not carried out. The range of dispersion of the interferometer was 0.625 cm^{-1}, and the half-width of the narrowest part of the wing ~2 cm^{-1}. We therefore considered that in our spectra the intensity distribution in the continuous spectrum was uniform.

Table 6. Intensity Ratios in the Fine-Structure Components

Substance	Experiment $\frac{\mathscr{I}_c}{2\mathscr{I}_{M-B}}$		Calc. from (V.19) $\frac{\mathscr{I}_c}{2\mathscr{I}_{M-B}}$	Calc. from (V.17), values of K	Dynamic values $\frac{\rho\frac{\partial\varepsilon}{\partial\rho}}{\frac{\mathscr{I}_c}{2\mathscr{I}_{M-B}}}$	Calculation from (V.17)*			
						"low" values of R		"high" values of R	
	present work	previous meas.				K	$\frac{\mathscr{I}_c}{2\mathscr{I}_{M-B}}$	K	$\frac{\mathscr{I}_c}{2\mathscr{I}_{M-B}}$
Benzene	0.72±0.1	0.97 [97] 0.98±0.25 [6]	0.46	1.10	0.60	1.11	0.60	0.62	0.34
Carbon tetra- chloride . . .	0.65±0.15	0.84 [87] 1.10±0.3 [6]	0.46	—	—	1.04	0.56	0.65	0.35
Chloroform . .	0.58±0.1	—	0.51	1.08	0.58	1.41	0.75	0.79	0.42
Toluene	0.45±0.06	0.40 [98]	0.36	1.24	0.43	1.74	0.60	0.93	0.32
Methylene chloride . . .	0.72±0.1	—	0.54	—	—	—	—	—	—

* Values of $\rho(\partial\varepsilon/\partial\rho)$ calculated from the light-scattering coefficient $R_{\pi/2}$

Table 7. Parameters Used in Calculating the Intensity Ratios of the Fine-Structure Components

Substance	ρ g/cm³	n	$\sigma\cdot10^5$, deg⁻¹	$C_p\cdot10^{-7}$, erg/deg	$\frac{dn}{dT}\cdot10^5$	Scattering coeff., $R\cdot10^6$		Dynamic values of $\left(\rho\frac{\partial\varepsilon}{\partial\rho}\right)_s$	Depolarization coefficient $\Delta\cdot100$
						"low" value of $R_{\pi/2}$	"high" value of $R_{\pi/2}$		
Benzene . . .	0.879	1.499	121	1.70	64	32	48	1.56	42
Carbon tetra- chloride	1.595	1.462	122	0.84	56	10.4	14	—	6
Chloroform . . .	1.489	1.445	126	0.966	61	12.8	18.6	1.36	20
Toluene . . . . ,	0.865	1.494	106	1.68	61	35.8	55	1.6	48

In principle, the best method of finding $\left(\rho\frac{\partial\varepsilon}{\partial\rho}\right)_s$ is to calculate this quantity from the experimentally determined scattering coefficient and the general formula for scattered light, from which it follows that

$$\left(\rho\frac{\partial\varepsilon}{\partial\rho}\right)_s^2 = \rho V^2\left[\frac{2\lambda^4 R_{\pi/2}}{\pi^2 kTf(\Delta)} - \left(\frac{1}{\sigma}\frac{\partial\varepsilon}{\partial T}\right)_p^2\frac{\sigma^2 T}{\rho C_p}\right], \qquad (V.20)$$

where $f(\Delta) = \frac{1+\Delta}{1-\frac{7}{6}\Delta}$, and Δ is the depolarization coefficient of the scattered light.

As yet, however, the accuracy of measuring $R_{\pi/2}$ is low. Moreover there are two groups of scattering-coefficient measurements: the so-called "high" and "low" values [1, 6]. Inside each group we find a good agreement between the results of measurements, but between the groups the measurements may disagree by, for example, 40% in benzene.

We therefore determined $\left(\rho\frac{\partial\varepsilon}{\partial\rho}\right)_s$ from the "low" and "high" values of $R_{\pi/2}$ separately. In addition to this we calculated the value of K from dynamic measurements of $\left(\rho\frac{\partial\varepsilon}{\partial\rho}\right)_s$.

According to thermodynamic calculations, K should be equal to or greater than unity [6]. This is in fact found from the dynamic values of $\left(\rho\frac{\partial\varepsilon}{\partial\rho}\right)_s$ and the "low" values of $R_{\pi/2}$, with formula (V.20).

From the "high" values of $R_{\pi/2}$ and formula (V.20) we systematically find $K < 1$; this suggests that the "high" values are wrong.

There is no doubt that the use of gas lasers for studying the intensities of the fine-structure components will enable us to measure the quantity $\dfrac{\mathcal{I}_c}{2\mathcal{I}_{\mathbf{M-B}}}$ with still greater accuracy than now, and then it will be desirable to determine the quantity $\left(\rho\dfrac{\partial \varepsilon}{\partial \rho}\right)_s$ from these measurements and formula (V.17).

Thus the use of the radiation from a gas laser as source of exciting light has for the first time made it possible, by studying the fine-structure spectrum of the molecular scattering of light, to measure the absorption of sound in the frequency range 10^9-10^{10} cps in certain liquids and to determine the velocity of hypersound and the ratio of the integral intensities of the fine-structure components much more accurately than heretofore. This kind of complex study of the characteristics of the molecular scattering of light on the same samples under the same conditions should enable us to secure mutually consistent and self-checking results.

The results obtained for the absorption and velocity dispersion of sound are of independent interest and enable us to draw certain conclusions regarding the applicability of the general phenomenological theory and the molecular theory of relaxation processes to the liquids studied. The extension of these studies will doubtless be very important in developing the theory of liquids in general and of molecular acoustics in particular.

CHAPTER VI

STIMULATED MOLECULAR SCATTERING OF LIGHT

1. Stimulated Mandelshtam—Brillouin Scattering in Liquids and Glasses

We have observed stimulated Mandelshtam—Brillouin scattering [22] in three sorts of optical glasses, in fused quartz, and in seven different liquids, by focusing the light of a giant pulse, obtained from a ruby laser with an output power of 80-100 MW inside these media. We studied the following liquids: carbon disulfide, benzene, toluene, water, acetic acid, nitrobenzene, and Salol, the latter two being studied over a wide temperature range (20-180°C).

The liquids were placed in a cylindrical metal vessel 10 cm long with two glass windows 3 cm in diameter. The radiation from the ruby laser was focused inside the vessel by means of a lens with a focal length of $f = 3$ cm.

In the majority of the liquids studied we observed several Stokes and anti-Stokes Mandelshtam—Brillouin components. The second, third, and so on components clearly appeared as a result of successive scattering* (see Section 3 of Chapter I). The largest number of components (up to 17) was observed in carbon disulfide. It should be noted that the anti-Stokes components were not much weaker than the Stokes components in the majority of cases. From the point of view of theory, the intensity of the anti-Stokes components may be much lower than that of the Stokes components. Why there is this difference between theory and experiment is not yet quite clear. It may be, as indicated by Brewer [101], that owing to the asymmetry of the K_1 line of the ruby the anti-Stokes components are intensified more than the Stokes components. We notice that in viscous Salol, glasses, and crystals, only Stokes components are observed. In nitrobenzene the anti-Stokes components are much weaker than the Stokes components, if found at all. Figure 23 shows, for example, some photographs of stimulated Mandelshtam—Brillouin scattering spectra in carbon disulfide at room temperature and in nitrobenzene at three different temperatures.

Table 8 presents the results of some of our measurements of the frequency of the Mandelshtam—Brillouin components and the velocity of hypersound. This latter is compared with the velocity found from thermal scattering and direct ultrasonic measurements.

---

*The authors of [21] claimed to have observed components belonging to harmonics of the frequencies 2Ω and 3Ω caused by the nature of the nonlinearity of the stimulated scattering. In order to solve the problem as to the feasibility of such a process with laser powers of the type now available it is desirable to study the time sequence of the appearance of the components. Mandelshtam—Brillouin thermal scattering in the second harmonic of a hypersonic wave due to stimulated scattering was observed in [100].

Table 8. Velocities of Hypersound in Several Substances Obtained from the Stimulated and Thermal Scattering of Light

Substance	Stimulated scattering		Thermal scattering, V, m/sec	Ultrasound, V, m/sec
	Δν, cm⁻¹	V, m/sec		
Benzene	0.203±0.002	1434±15	1471±8 [14]	1324
Carbon disulfide.*	0.181±0.002 *	1162±15 *	1265±22 [6]	1158
	0.192±0.002 **	1232±15 **		
Nitrobenzene	0.232±0.002	1546±15	—	1473
Acetic acid	0.145±0.002	1105±20	1140±35 [6]	1144
Salol	0.232±0.002 (20° C)	1544±15	—	—
	0.106±0.02 (180° C)	740±20		
Fused quartz	0.811±0.004	5804±30	5590 5840 [102]	5970
Crown glass (K-8) . . .	0.856±0.005	5906±40	—	—

*For photographs of stimulated scattering, when about 10 fine-structure components were observed.

**For photographs in which two components were observed.

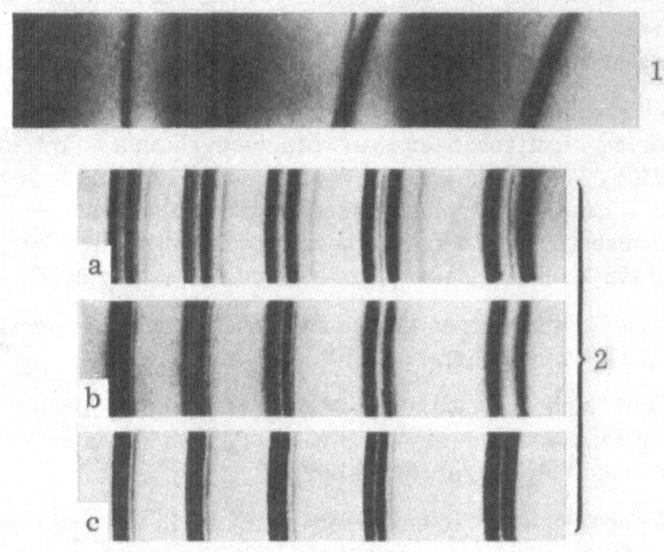

Fig. 23. Spectra of stimulated Mandelshtam—Brillouin scattering in carbon disulfide and nitrobenzene. 1) Spectrum of the light scattered in carbon disulfide. Range of dispersion of the interferometer 8.33 cm⁻¹, 17 components observed; 2) spectrum of the light scattered in nitrobenzene at temperatures of: a) 20; b) 65; c) 180°C.

Attention should be drawn to a certain systematic difference between the velocities of hypersound obtained in the stimulated and thermal scattering of light. This difference is sometimes no greater than the experimental error, but in the case of carbon disulfide it considerably exceeds this limit.

The velocity of hypersound in carbon disulfide is smaller, the greater the number of Mandelshtam—Brillouin components there are in the scattered light; the number of these increases with increasing intensity of the light pulse.

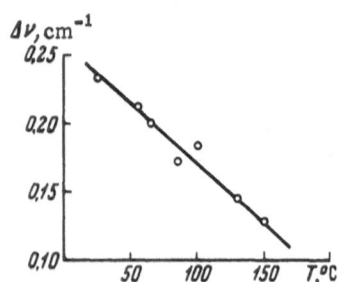

$\Delta\nu$, cm^{-1}

Fig. 24. Displacement of the Mandelshtam—Brillouin components as a function of temperature in Salol.

The observed dependence of the velocity (or $\Delta\nu$) on the number of components is presumably due to the heating of the scattering medium by the absorption of energy from the light pulse. In the case of carbon disulfide it is sufficient to suppose, in order to explain the observed difference in velocities, that the space in which the stimulated scattering takes place is heated* by about 30°C. According to theory [1, 54], energy some (ω_0/Ω) times less than the amount scattered (stimulated scattering) is transformed into hypersound. If we consider that about 10^{-5} (for one Mandelshtam—Brillouin component) of the energy of the giant pulse (1-1.5 J) is converted into hypersound and then absorbed, then altogether (if there are 10-15 components) some $(1-2) \cdot 10^{-4}$ J are converted into hypersound and then heat the scattering volume. This means that, in order to secure a heating of 30°C, it would be necessary for the scattering volume to be 10^{-5}-10^{-6} cm^{-3}, which is quite reasonable.

It is possible that in certain other cases also the reduction in the velocity of hypersound in stimulated scattering is explained by the heating of the scattering volume. If this is so, then hopes [20] of obtaining considerably greater accuracy in determining the velocity of hypersound from the distances between a large number of stimulated Mandelshtam—Brillouin scattering components can hardly be realized quickly.

Despite the reduced values of the velocity of hypersound obtained from stimulated Mandelshtam—Brillouin scattering, in nitrobenzene we were nevertheless able to observe a dispersion of the velocity of sound $\Delta V/V \sim 5\%$ (t = 20°C), which enabled us to use formula (I.22) in order to estimate the relaxation time of the volume viscosity coefficient as $\tau \sim (4-5) \cdot 10^{-11}$ sec. Considering that up to the present time there have been no data regarding the relaxation time of the second viscosity-coefficient in nitrobenzene, this estimate of τ may really prove useful.

As already mentioned, we observed [22] stimulated Mandelshtam—Brillouin scattering in supercooled viscous liquids, fused quartz, and silicate glasses.

Figure 24 shows the results of measuring the $\Delta\nu$ of the components of stimulated Mandelshtam—Brillouin scattering at various temperatures in Salol. The first point on the graph of Fig. 24 corresponds to liquid Salol supercooled by 22°C.

A study of the fine structure of the scattered-light line in viscous liquids and glasses is undoubtedly of great interest. It is well known that for a long time no one was able to observe fine structure in ordinary thermal scattering, not only in glasses but also in viscous liquids. Fairly recently thermal-scattering fine structure was observed at last in viscous liquids [102] and fused quartz [103]. In optical glasses, however, no fine structure could be found, although repeated attempts were made to do so [104-107]. This led certain authors to assert [106] that in view of the high static viscosity of glasses no fine structure should ever be observed, since according to (I.16) and (I.18) the width of the Mandelshtam—Brillouin components is proportional to the viscosity, and if the latter is large then the fine structure will not be resolved. This point of view is refuted by the theory of Isakovich [108], which referred the concept of relaxation processes in liquids to the shear viscosity. According to this theory a large static viscosity should fall considerably at high frequencies owing to relaxation, allowing fine structure to be observed. Our observation of well-resolved stimulated Mandelshtam—Brillouin scattering components in fused quartz and three kinds of optical glass supports this point of view.

―――――――

*The temperature coefficient of the velocity of sound in carbon disulfide $\Delta V/\Delta t = -3.2$ m/sec·deg.

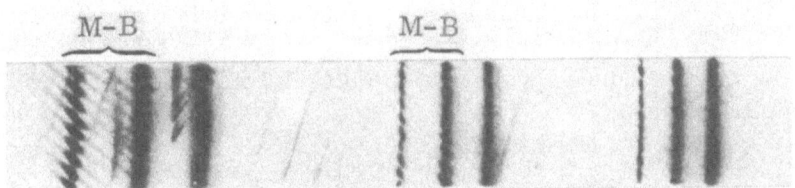

Fig. 25. Mandelshtam—Brillouin stimulated scattering spectrum in fused quartz. Interferometer dispersion range 5 cm^{-1}.

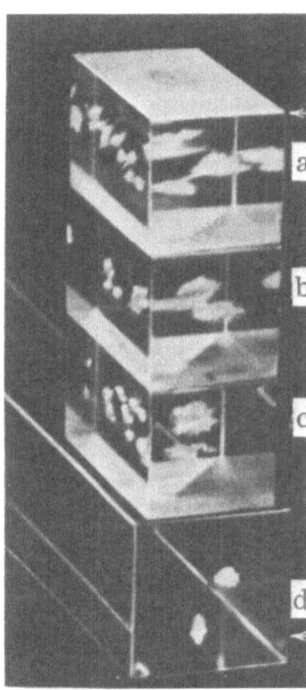

Fig. 26. Picture of damage in various kinds of glass and fused quartz. Arrows indicate the direction of propagation of the light from the giant pulse of the ruby laser. a) Heavy flint glass TF-3; b) light flint glass LF-5; c) crown glass K-8; d) fused quartz

The Mandelshtam—Brillouin stimulated scattering spectrum in fused quartz is presented in Fig. 25.

When studying the stimulated scattering in fused quartz and glasses, the light from the giant pulse of the ruby laser was focused inside these media by a lens with a focal length of $f = 1.5$ cm. The intensity of the exciting light at the lens focus was such as to break up the solid material in this region.

Damage at the focus was also observed earlier for various crystals and glasses [19, 109-111]. We must draw attention to the different nature of the damage in the three sorts of glass and the fused quartz.

Figure 26 presents a photograph of such damage. In fused quartz and in optical glass K-8 the internal damage resembled ellipsoids of rotation almost spherical in shape. In glass LF-5 these ellipsoids were severely drawn out, while in glass TF-3 they were drawn out still more, having a length of ~1 cm. The hardest medium was fused quartz and the softest TF-3. The nature of the damage to the glass samples differed considerably from that recently observed in ruby [111].

There are various points of view regarding the reasons for the damage [109-111] and this question cannot yet be regarded as settled. There is no doubt that for damage to occur to transparent dielectrics a high proportion of the energy of the light pulse must be absorbed in the region of the focus. The mechanism responsible for this absorption is not yet clear.

Direct optical absorption in a transparent dielectric can hardly be so intense as to create an explosive wave or severe thermal expansion. Giuliano [109] set up a number of interesting experiments on the damaging of transparent dielectrics and came to the conclusion that the reason lay in the generation of an intense hypersonic wave accompanying stimulated Mandelshtam—Brillouin scattering. Harper [110] considers this hypersonic wave as the reason for the overheating and softening of the glass. This causes strong absorption of light, as a result of which an explosive wave develops.

However, whatever damage mechanism may in fact act,* it is clear that the phenomenon of stimulated Mandelshtam—Brillouin scattering is able to develop before the damage takes place. This is all the more so since we observe two Stokes components arising as a result of successive stimulated scattering.

Further study of the stimulated Mandelshtam—Brillouin scattering in glasses and super-cooled liquids is of interest not only because this is a very simple (and often the only) method of studying the kinetics of hypersound propagation at a frequency of $\sim 10^{10}$ cps, but also because a powerful sound wave is developed, and we are therefore presented with a powerful generator of hypersound. With the contemporary power of ruby lasers, hypersound of several kilowatts or even several tens of kilowatts may be achieved.

Of no less importance, of course, is the fact that, if the medium is nonlinear, when a powerful light-wave field passes through it stimulated scattering will occur; we thus have the possibility of studying nonlinear optical effects in the medium by reference to the stimulated scattering.

The phenomenon of stimulated scattering was observed just a year ago, and there are still many questions awaiting solution.

2. Results of an Investigation into the Stimulated Scattering of Light on the Wing of the Rayleigh Line

It follows from simple theory (Section 4 of Chapter II) that the stimulated scattering spectrum of the light on the Rayleigh-line wing should differ considerably from the thermal-scattering spectrum. Instead of a monotonic fall of intensity on both sides of the exciting line such as occurs in thermal scattering, in stimulated scattering there should be only a Stokes wing, and (most important) in the wing there should be a maximum at a frequency displaced from the frequency of the exciting line by an amount $\Omega_{max} \sim 1/\tau_a$ (τ_a is the anisotropy relaxation time). In other words, whereas the half-width of the Rayleigh-line wing in thermal scattering is $\Delta\nu = 1/\pi c \tau_a$ (cm^{-1}), the lowest threshold value of the power of the exciting light will occur at precisely the half-width of the thermal wing, i.e., at a frequency of $\frac{1}{2}(\Delta\nu) = 1/2\pi c\tau_a$.

Naturally, τ_a means the effective value, since in reality in thermal scattering the wing has a very complicated spectral distribution and not a dispersion-type one, as proposed in Chapter II when deriving the relation $\Delta\nu_{max} = 1/2\pi c\tau$.

Of the seven liquids mentioned in the preceding section, we only observed stimulated scattering in the wing of the Rayleigh line for three: carbon disulfide and nitrobenzene at room temperature and Salol at 170°C. In the remaining liquids the threshold for the development of this phenomenon was apparently above the power of our ruby laser.

In Fig. 27b the arrow shows the stimulated scattering of light on the Rayleigh-line wing in nitrobenzene. For comparison (Fig. 27a) the spectrum of the ruby laser is also given. For the laser power used the stimulated Stokes wing in nitrobenzene has a range of ~ 1 cm^{-1} and its distribution shows a maximum in the region of 0.5 cm^{-1}. However, it is hard to determine the position of this maximum exactly, since the wing is quite wide, and a Mandelshtam—Brillouin component also lies in the region of the wing at a distance of ~ 0.23 cm^{-1} from the exciting line. The 0.5 cm^{-1} maximum in the stimulated wing of nitrobenzene agrees with the data relating to the half-width of the thermal wing of the Rayleigh line in this liquid ($\tau_a = 10^{-11}$ sec) [112].

*We note that the damage cannot be explained by light pressure or optical phonons arising as a result of stimulated Raman scattering [109].

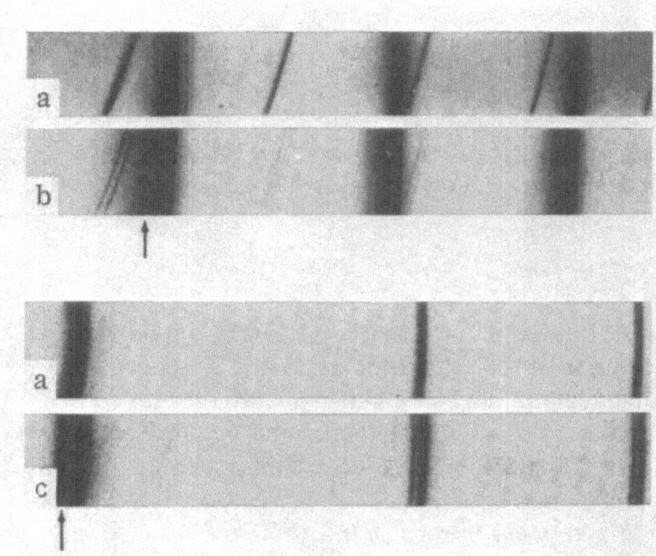

Fig. 27. Stimulated light-scattering spectra on the Rayleigh-line wing for nitrobenzene and Salol (interferometer dispersion range 5 cm^{-1}). a) Ruby laser emission-spectrum; b) stimulated scattering spectrum for the Rayleigh-line wing in nitrobenzene at 20°C; c) stimulated scattering spectrum for the Rayleigh-line wing in Salol at 170°C.

It follows from (II.56) and (II.59) that, for a given power of the exciting light, the maximum in the stimulated wing of the Rayleigh line will be narrower, the greater the τ_a, and hence the closer to the exciting line this maximum lies. In Salol (Fig. 27c) the stimulated wing at a temperature of 170°C constitutes a clearly visible band with a maximum at $\Delta\nu_{max} = 0.16$ cm^{-1}. It should be noted that, as indicated by the results of the previous section (Fig. 26), the Mandelshtam—Brillouin component should occur at a distance of $\Delta\nu = 0.11$ cm^{-1} (t = 170°C). In Fig. 27c this component is not to be seen. In fact, in all cases studied it is hard to discover the components of stimulated Mandelshtam—Brillouin scattering when stimulated scattering of the Rayleigh-line wing occurs. These components either fail to appear at all or appear in much smaller numbers than usual. This is apparently connected with competition between the processes in question, so that when stimulated scattering of the Rayleigh-line wing occurs most of the available energy goes preferentially into this process.

As we should expect, in carbon disulfide there is a wide stimulated Rayleigh-line wing (Fig. 28), the width of which in some pictures reaches 15 cm^{-1}. On photographing the stimulated wing spectrum of carbon disulfide with an interferometer having a 5 cm^{-1} dispersion range, an intense continuous background was observed between the orders, comparable in intensity with the exciting line and the Mandelshtam—Brillouin components. The dispersion of the interferometer used in obtaining the photographs of Fig. 28 was 50 cm^{-1}. The width of the apparatus function in this interferometer has such a considerable value that the Mandelshtam—Brillouin components are unresolved and merge into a single wide line. This may be seen from Fig. 28a, in which we show a photograph of the scattered light for a ruby laser power 10–15% greater than in the case of Fig. 28b. All this indicates that the minimum in the intensity which should occur between the exciting line and the stimulated wing is partly filled with unresolved Mandelshtam—Brillouin components. Considering also that the value of τ in carbon disulfide is small, and hence the maximum in the stimulated wing for the power used has a considerable width, we can

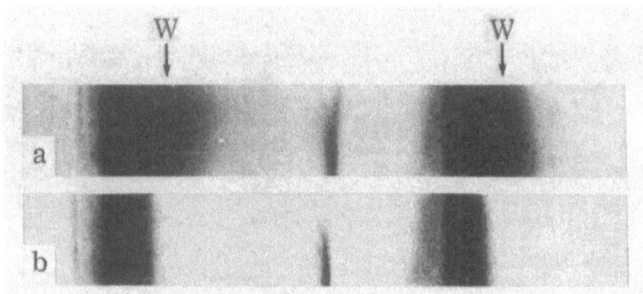

Fig. 28. Spectrum of stimulated scattering on the Rayleigh-
line wing in carbon disulfide at 20°C. a) Stimulated scattering spectrum in carbon disulfide obtained with an exciting
light pulse of ~100 MW; interferometer dispersion range 50
cm^{-1}; b) the same, with 10-15% less power.

understand why no sharp maximum occurs in the photographs (Fig. 28) of the stimulated Rayleigh-line wing scattering in carbon disulfide.*

In all three cases studied, the stimulated Rayleigh-line wing has a sharp threshold. The
threshold value of the intensity of the exciting light for the development of stimulated scattering
of the light in the Rayleigh-line wing was ~10^5 MW/cm^2 in our experiments.

The existence of the threshold, the intensity of the scattering (comparable with the intensity of the exciting light), the presence of the maximum in the wing, and the absence of an anti-
Stokes wing — all this indicates that, in all three cases described in the foregoing, we are observing stimulated scattering of the light in the Rayleigh-line wing.

Calculations based on formula (II.59) give an intensity $S_{thr} = 10^{11}k_\omega$ W/cm^2 for the
threshold value of the intensity of the exciting light giving stimulated scattering on the Rayleigh-
line wing in carbon disulfide. For a well-purified liquid and low light intensities, the coefficient of optical losses [113] is $k_\omega \sim 10^{-3}$ cm^{-1} and the threshold value of the intensity is $S_{thr} \sim$
100 MW/cm^2. However, in our experiments, when studying the stimulated scattering of light,
the liquid was not subjected to special purification, and hence the value of k_ω should be considerably greater. Further, the threshold value of the intensity of the exciting light associated
with the development of stimulated Mandelshtam—Brillouin scattering in carbon disulfide is, according to [20], 30 MW/cm^2. Hence, stimulated Mandelshtam—Brillouin scattering should occur
(as indeed found experimentally) earlier than the stimulated scattering of light on the Rayleigh-
line wing. The amount of exciting light flux lost to stimulated Mandelshtam—Brillouin scattering should very greatly increase the value of k_ω in formula (II.59).

The stimulated Raman scattering of light, clearly, should have an even lower threshold
than stimulated Mandelshtam—Brillouin scattering, which also leads to a rise in k_ω.

Thus we cannot at the moment compare the threshold value of the exciting light power calculated from formula (II.59) with that determined experimentally. This is because the loss coefficient k_ω is not known even approximately. The value of the coefficient k_ω is determined by
the losses associated with stimulated Mandelshtam—Brillouin and Raman scattering. This
means that in carbon disulfide, as the flux of exciting light approaches the lens focus, stimulated

---

*According to our data regarding the width of the Rayleigh-line wing due to thermal scattering
(see Chapter IV), the maximum in the stimulated wing of carbon disulfide should occur at a
distance of ~2 cm^{-1} from the exciting line (at room temperature).

Mandelshtam—Brillouin scattering arises first and the energy of the exciting light is initially expended in precisely this process. Whether the energy in the region of the lens focus will be sufficient to produce stimulated Rayleigh-line wing scattering depends both on the initial power of the exciting light and on the losses sustained by this power in producing stimulated Raman and Mandelshtam—Brillouin scattering.

However, as soon as the power becomes sufficient to produce stimulated scattering on the Rayleigh-line wing, this process may continue to progress as a result of the self-focusing of the laser beam. In fact, if stimulated scattering of the light on the Rayleigh-line wing has begun, for example, in some part of the lens focus, this means that in this region the molecules are oriented in a specific direction by the electric field of the light wave. As a result of this, in the region in question the refractive index increases, which should lead to self-focusing of the laser beam, and hence to an increase in energy density in this region and an intensification of the process of stimulated light scattering. It follows from the foregoing that the threshold of stimulated light scattering on the Rayleigh-line wing should be quite sharp, as indeed we have seen experimentally.

Thus, the stimulated Raman-scattering of light, the stimulated Mandelshtam—Brillouin scattering, the stimulated scattering of light on the wing of the Rayleigh line, and the phenomenon of laser-beam self-focusing should all be closely connected. There is no doubt that the theoretical and experimental study of the propagation of powerful light fluxes through matter should take account of this relationship.

Finally, let us turn our attention to the possibility of using the stimulated scattering of light on the Rayleigh-line wing in order to determine the anisotropy relaxation time τ_a by measuring the position of the maximum in the stimulated Rayleigh-scattering spectrum. This method of measuring τ_a may be particularly effective when studying viscous liquids, where τ_a is large so that the stimulated wing should be quite narrow. The usual methods of determining τ_a from the width of the thermal-scattering wing encounters great difficulties for viscous liquids, since the width of the thermal-scattering wing is often much smaller than the width of the apparatus contour of the spectral apparatus employed in the case of viscous liquids.

In conclusion, the author wishes to express his sincere thanks to his director, I. L. Fabelinskii for proposing the subject and for constant guidance in the course of the work, to V. L. Ginzburg, D. I. Mash, and I. I. Sobel'man for discussing the results, and to M. A. Vysotskaya, V. P. Zaitsev, G. I. Zaitsev, V. V. Morozov, and E. V. Tiganov for help in carrying out the experiments.

LITERATURE CITED

1. I. L. Fabelinskii, Molecular Scattering of Light, "Nauka," Moscow (1965).
2. I. Z. Fisher, Statistical Theory of Liquids, Fizmatgiz, Moscow (1961).
3. Ya. I. Frenkel', Kinetic Theory of Liquids, Izd. Akad. Nauk SSSR, Moscow (1959).
4. M. V. Vol'kenshtein, Molecular Optics, Gostekhizdat, Moscow (1951).
5. M. I. Shakhparonov, Methods of Studying the Thermal Motion of Molecules and the Structure of Liquids, Izd. Mosk. Univ. (1963).
6. I. L. Fabelinskii, Tr. Fiz. Inst. Akad. Nauk SSSR, 9:181 (1957).
7. L. D. Landau and E. M. Lifshits, Electrodynamics of Continuous Media, Fizmatgiz, Moscow (1959).
8. M. A. Leontovich, J. Phys., 4:449 (1941).
9. S. M. Rytov, Zh. Eksp. i Teor. Fiz., 33:514, 669 (1957).
10. M. S. Pesin and I. L. Fabelinskii, Dokl. Akad. Nauk SSSR, 122:575 (1958).
11. M. S. Pesin, Tr. Fiz. Inst. Akad. Nauk SSSR, 30:158 (1964).
12. M. S. Tunin and M. I. Shakhparanov, in collection: Use of Ultrasonics in Studying Matter, No. 14, Izd. MOPI (1961).
13. D. I. Mash, V. S. Starunov, and I. L. Fabelinskii, Zh. Eksper. i Teor. Fiz., 47:783 (1964).
14. D. I. Mash, V. S. Starunov, E. V. Tiganov, and I. L. Fabelinskii, Zh. Eksper. i Teor. Fiz., 49:1764 (1965).
15. Kh. E. Sterin, Dokl. Akad. Nauk SSSR, 62:219 (1948).
16. L. I. Mandelshtam and M. A. Leontovich, Zh. Eksper. i Teor. Fiz., 7:438 (1936).
17. I. G. Mikhailov, V. A. Solov'ev, and Yu. P. Syrnikov, Fundamentals of Molecular Acoustics, "Nauka," Moscow (1964).
18. C. H. Townes, Enrico Fermi International School of Physics, XXXI Course, August 19-31, 1963.
19. R. Y. Chiao, C. H. Townes, and B. P. Stoicheff, Phys. Rev. Letters, 12:552 (1964).
20. E. Garmire and C. H. Townes, Appl. Phys. Letters, 5:84 (1964).
21. R. G. Brewer and K. E. Eieckhoff, Phys. Rev. Letters, 13:334 (1964).
22. D. I. Mash, V. V. Morozov, V. S. Starunov, E. V. Tiganov, and I. L. Fabelinskii, Zh. Eksper. i Teor. Fiz., Pis'ma, 2:246 (1965).
23. D. I. Mash, V. V. Morozov, V. S. Starunov, and I. L. Fabelinskii, Zh. Eksper. i Teor. Fiz., Pis'ma, 1:41 (1965).
24. V. S. Starunov, Dokl. Akad. Nauk SSSR, 153:1055 (1963).
25. V. S. Starunov, Opt. i Spektroskopiya, 18:300 (1965).
26. J. Cabannes and R. Daure, Compt. Rend. Acad. Sci., Vol. 186 (1928).
27. C. V. Raman and R. S. Krishnan, Nature, 122:278, 882 (1928).
28. J. Brandmüller, Z. Phys., 140:75 (1955).
29. K. Venkateswarlu and G. Thyagarajan, Z. Phys., 154:81 (1959).
30. I. L. Fabelinskii, Opt. i Spektroskopiya, 2:510 (1957).
31. Yu. N. Zhivlyuk, Diploma Work, Moscow Physico-Technical Institute, Physical Institute of the USSR Academy of Sciences, Moscow (1960).

32. K. A. Valiev and L. D. Éskin, Opt. i Spektroskopiya, 18:300 (1965).
33. M. F. Vuks and V. L. Litvinov, Dokl. Akad. Nauk SSSR, 105:696 (1955).
34. M. F. Vuks, Opt. i Spektroskopiya, 9:92 (1960).
35. A. K. Atakhodzhaev, Izv. Akad. Nauk UzbSSR, Ser. Fiz.-Matem., No. 1, p. 86 (1962).
36. I. I. Sobel'man, Izv. Akad. Nauk SSSR, Ser. Fiz., 17:554 (1953).
37. K. A. Valiev, Zh. Eksper. i Teor. Fiz., 40:1832 (1961).
38. A. V. Rakov, Tr. Fiz. Inst. Akad. Nauk, 27:111 (1962).
39. V. L. Ginzburg, Zh. Eksper. i Teor. Fiz., 34:246 (1957).
40. D. V. Sivukhin, Zh. Eksper. i Teor. Fiz., 9:1258 (1939).
41. H. A. Kramers, Physica, 7:284 (1940).
42. R. A. Sack, Proc. Phys. Soc., B70:414 (1957).
43. E. F. Gross, Nature, 126:201, 400 (1930); Izv. Akad. Nauk SSSR, Ser. Fiz., 5:19 (1941).
43a. E. F. Gross and M. F. Vuks, J. Phys., 6:457 (1936); 7:113 (1936).
44. N. A. Chernyavskaya and G. P. Roshchina, Vestn. Leningr. Univ., 16:26 (1964).
45. L. I. Mandelshtam, Zh. Ross. Fiz.-Khim. Obshch., 58:381 (1926).
46. L. Brillouin, Ann. Phys., (9)17:88 (1922).
47. E. F. Gross, Z. Phys., 63:685 (1930); Nature, 125:603 (1930).
48. E. S. Stewart and J. L. Stewart, Phys. Rev. Letters, 13:437 (1964).
49. J. L. Hunder and H. D. Dardy, J. Chem. Phys., 42:2961 (1965).
50. L. D. Landau and E. M. Lifshits, Mechanics of a Continuous Medium, Gostekhizdat, Moscow (1953).
51. A. A. Berdyev, Izv. Akad. Nauk TadzhSSR, Ser. Fiz.-Tekh., Khim. i Geol. Nauk, No. 3, p. 16 (1965).
52. L. I. Komarov and I. Z. Fisher, Zh. Eksper. i Teor. Fiz., 43:1927 (1962).
53. M. A. Isakovich and I. Chaban, Dokl. Akad. Nauk SSSR, Vol. 65, No. 2 (1965).
54. S. A. Akhmanov and Chin Dong Ah, Zh. Eksper. i Teor. Fiz. (in press).
55. N. Bloemberger, Nonlinear Optics, W. A. Benjamin, Inc., New York (1965).
56. G. Stewart, Structure of Molecules [Russian translation], ONTI, Khar'kov-Kiev (1937).
57. V. V. Vladimirskii, Zh. Eksper. i Teor. Fiz., 12:199 (1942).
58. L. D. Landau and E. M. Lifshits, Statistical Physics, Gostekhizdat, Moscow (1961).
59. V. P. Milant'ev, Vestn. Mosk. Univ., Ser. Fiz., No. 5, p. 71 (1960).
60. G. M. Panchenko, Theory of the Viscosity of Liquids, Gostekhizdat, Moscow (1947).
61. K. A. Valiev and A. Sh. Agishev, Opt. i Spektroskopiya, 16:881 (1964).
62. P. Debye, Polar Molecules [Russian translation], GNTI, Moscow (1931). [Dover, New York.]
63. S. Bhagavantam, Indian J. Phys., 6:319 (1931).
64. S. A. Akhmanov and R. V. Khokhlov, Problems of Nonlinear Optics 1962-1963, Izd. Akad. Nauk SSSR (1964).
65. A. A. Andronov and S. É. Khaiken, Theory of Vibrations, Izd. NKTP SSSR, Moscow (1937).
66. M. L. Sosinskii, Izv. Akad. Nauk SSSR, Ser. Fiz., 17:621 (1953).
67. A. Rousset, Ann. Phys., 6:1 (1936).
68. J. Weiler, Z. Phys., 68:782 (1931).
69. H. C. Van de Hulst and J. J. Reesink, Astrophys. J., 106:121 (1947).
70. S. Tolansky, High-Resolution Spectroscopy [Russian translation], IL, Moscow (1955).
71. L. Bergmann, Ultrasound [Russian translation], IL, Moscow (1956).
72. P. Kafalas, J. I. Masters, and E. M. E. Murray, J. Appl. Phys., 35:2349 (1964).
73. Charles D. Hodgman et al. (editors), Handbook of Chemistry and Physics, 1955-1956 edition, Chemical Rubber, Cleveland, Ohio.
74. N. C. Majumdar, Indian J. Phys., 23:253 (1949).
75. A. Kastler and A. Fruhling, Compt. Rend. Acad. Sci., 218:997 (1944).

76.   A. K. Ray, Indian J. Phys., 24:111 (1950).
77.   S. C. Sirkar, Indian J. Phys., 10:189 (1936).
78.   A. J. Sidorowa (Sidorova), Acta Physico-Chim., URSS, 7:193 (1937).
79.   A. I. Ansel'm and N. N. Porfir'eva, Zh. Eksper. i Teor. Fiz., 19:438 (1949).
80.   G. I. Zaitsev and V. S. Starunov, Opt. i Spektroskopiya (in press).
81.   E. N. Andrade, Phil. Mag., 17:497, 698 (1934).
82.   J. H. Andreae, E. L. Hessel, and J. Lamb, Proc. Phys. Soc., B69:625 (1956).
83.   J. H. Andreae, Proc. Phys. Soc., B70:71 (1957).
84.   R. L. Hessel and J. Lamb, Proc. Phys. Soc., B69:869 (1956).
85.   G. B. Benedek, J. B. Lastovka, K. Fritsch., and T. Greytak, J. Opt. Soc. Am., 54:1284 (1964).
86.   R. Y. Chiao and B. P. Stoicheff, J. Opt. Soc. Am., 54:1286 (1964).
87.   A. A. Berdyev and I. B. Lezhnev, Izv. Akad. Nauk TadzhSSR, Ser. Fiz.-Tekh., Khim. i Geol. Nauk, 3:104 (1963).
88.   P. A. Bazhulin, Tr. Fiz. Inst. Akad. Nauk, 5:261 (1950).
89.   I. B. Lezhnev, Dissertation, Physico-Technical Institute of the TadzhikSSR Academy of Sciences, Ashkhabad (1963).
90.   J. H. Andreae, P. L. Joyce, and R. J. Oliver, Proc. Phys. Soc., B75:82 (1960).
91.   H. O. Kneser, Ann. Phys., 32(5):277 (1938).
92.   K. F. Herzfeld, JASA, 13:33 (1941).
93.   K. F. Herzfeld and T. A. Litovitz, Absorption and Dispersion of Ultrasonic Waves, Academic Press, New York (1959).
94.   A. A. Berdyev, Author's Abstract of Dissertation, Moscow (1965).
95.   I. L. Fabelinskii, Usp. Fiz. Nauk, 63:355 (1957).
96.   L. D. Landau and G. Placek, Phys. Zeit. Sowjetunion, 5:172 (1934).
97.   C. S. Venkateswaran, Proc. Indian Acad. Sci., 15:322 (1942).
98.   K. Birus, Phys. Z., 39:80 (1938).
99.   D. H. Rak, E. M. Kiess, U. Fink, and T. A. Wiggins, J. Opt. Soc. Am., 55:925 (1965).
100. R. G. Brewer, Appl. Phys. Letters, 6:165, 230 (1965).
101. R. G. Brewer, Appl. Phys. Letters, 5:127 (1964).
102. M. S. Pesin and I. L. Fabelinskii, Dokl. Akad. Nauk SSSR, 129:299 (1959); 135:1114 (1960).
103. P. F. Flubacher, A. L. Leodbetter, J. A. Morrison, and B. P. Stoicheff, Intern. J. Phys. Chem. Solids, 12:53 (1960).
104. W. Ramm, Phys. Z., 35:756 (1934).
105. E. Gross, Z. Phys., 63:685 (1930).
106. D. H. Rank and A. E. Douglas, J. Opt. Soc. Am., 38:966 (1948).
107. T. S. Velichkina, Tr. Fiz. Inst. Akad. Nauk, 9:59 (1958).
108. M. A. Isakovich, Dokl. Akad. Nauk SSSR, 23:782 (1939).
109. C. R. Giuliano, Appl. Phys. Letters, 5:137 (1964).
110. D. W. Harper, Brit. J. Appl. Phys., 16:751 (1965).
111. T. P. Belikova and É. A. Sviridenkov, Zh. Eksper. i Teor. Fiz., Vol. 1 (1965) (letter).
112. A. K. Atakhodzhaev, M. F. Vuks, and V. L. Litvinov, Transactions of the Tenth Conference on Spectroscopy, Izd. L'vov Univ. (1957), p. 117.
113. R. C. C. Leite, R. S. Moore, and J. R. Whinnery, Appl. Phys. Letters, 5:141 (1964).

GAMMA- AND PHOTOLUMINESCENCE
OF ALKALI IODIDES*

*Dissertation in pursuit of the degree of Candidate of Physicomathematical Sciences. Defended
May 17, 1965, at the P. N. Lebedev Physical Institute of the Academy of Sciences of the USSR.
Scientific Director: Senior Research Fellow Z. L. Morgenshtern.

CHAPTER I

PRESENTATION OF THE PROBLEMS

Single crystals of alkali halides have recently attracted more and more scientific and practical interest. These compounds have a number of valuable properties. They are transparent over a wide spectral range, including the whole visible and near-infrared regions as well as a considerable part of the ultraviolet spectrum, and they have lattices of the purely ionic type, crystallizing in the cubic system. Methods of growing very large alkali halide single single crystals are now known; also phosphors may be prepared from alkali halides without melting (and thus disrupting the crystal lattice both spectrally and chemically). All this has made alkali-halide crystals subjects for study, both in connection with discovering the nature of the luminescence of crystalline substances and in studying a wide variety of processes taking place in dielectrics and semiconductors [1].

Thanks to their high yield under hard excitation, alkali halides have found an important application in scintillation counters and spectrometers widely used in experimental techniques. The scintillation method was first used in 1908 for the visual recording of nuclear particles, but later, with the invention of Geiger counters at the beginning of the thirties, the use of scintillation counters diminished. However, with the appearance of photomultipliers the scintillation method was resuscitated, and beginning from 1949 the history of the development of scintillation counters was closely connected with experimental nuclear physics; the scintillation method became one of the most important and universal methods of investigation.

Originally the visual method was used: α particles and protons, which have a strongly ionizing effect on matter, were counted in thin layers of material, but β particles and γ rays could not be observed because of their weaker ionizing effects in such layers. However, in 1947, Kallmann [2] showed that the scintillations produced by β and γ rays in a large transparent piece of naphthalene could be recorded with the help of a photomultiplier. Later, Bell [3] observed that it was better to work with anthracene, which in scintillating properties was five times better than naphthalene; with anthracene even neutrons could be recorded on the basis of their ionizing effect produced in the phosphor by recoil protons. A little later, in 1948, Hofstadter [4] found that the amplitude of the light pulses in NaI—Tl was greater than in anthracene owing to the strong photoelectric absorption in iodine; he proposed using these crystals and other alkali iodides for the γ spectroscopy of very weak sources.

More and more alkali halide crystals found application as scintillation materials, and this encouraged an all-around study of their properties. The spectral characteristics of alkali halides were studied in [4-10], the dependence of the light-pulse amplitude and radiation yield on the energy of the incident particles in [11-15], the dependence of these factors on the density of ionization in [16-18], excitation by hard particles and x rays in [19], and photoexcitation in [10, 20]; the temperature [21-27] and concentration [9, 28-30] characteristics of phosphors were investigated, the kinetics of scintillation processes were studied [24, 31-33], and the influence on these of such phenomena as plastic deformation [34] and electric fields [35] was considered. All this helped to establish that the energy of the incident particles absorbed in the main substance was efficiently transferred to the emitting centers of the activator; the intensity of the scintillations was proportional to the energy of the incident particles up to a certain limit, after which saturation set in, and depended on the ionizing capacity of the different particles. Both the energy yield and the time characteristics depended greatly on temperature, and for different scintillators there existed certain optimum temperatures at which their use was most efficient.

The study of the concentration dependences of the scintillation characteristics led to the observation of several forms of luminescence (fluorescence) in the same phosphors. It was reported that pure salts of various alkali halides scintillated in hard excitation [4, 21, 22, 31, 36-39] and in some cases had a greater yield than activated phosphors. Thus, according to Hahn and Rossel [40] the absolute yield of pure CsI on excitation by γ quanta with energies between 40 keV and 1.3 MeV was about 40% and at 100°K the height of the light pulse was twice that of such scintillators as NaI—Tl. A high intensity of scintillation pulses in pure alkali iodides was observed by Bonanomi and Rossel [11]; thus, if we take the emission of an NaI—Tl phosphor at room temperature to represent 100, the intensity of the emission from pure KI at 77°K is 180, pure CsI at the same temperature, 700, pure NaI at −100°C, 55, and unactivated RbI at room temperature, 35.

All this inspired a further all-around study of pure salts. Thus, Van-Sciver and Hofstadter [4, 22, 38] showed that pure NaI crystals at low temperatures gave a band of luminescence in the ultraviolet part of the spectrum with $\lambda_{max} \sim 300$ mμ both under hard radiation and photoexcitation. Teegarden [41] observed luminescence from apparently unactivated KI at low temperatures. Knöpfel, Loepfe, and Stoll [39] studied the luminescence of unactivated CsI mainly under α radiation as a function of temperature. Morgenshtern [42] studied the emission spectra, the duration of the luminescence, the spectral distribution of the absorption coefficient, and the relative yield of unactivated luminescence from CsI crystals subjected to excitation by γ rays, and also on excitation in a narrow band lying near the self-absorption edge of the crystals.

A number of authors, seeking to explain the nature of the luminescence from pure salts, have associated this with exciton processes [22, 31, 39, 41-47]. Thus, apart from the development of luminescence, the external [48-50] and internal [51-53] photoeffects and various photochemical transformations [47, 54-56] caused by hard excitation are associated with excitons. Although, according to Seitz [48], the exciton emits on annihilation, it may also dissipate its energy by migrating about the crystal and interacting with impurities or lattice defects. For example, in interaction with phonons the exciton may decompose into a free electron and a free hole [57], or radiationless recombination may occur, the energy of the exciton being dissipated in heat [58], or if the energy of the excitons is transferred to lattice defects various photochemical reactions may take place [59]. Diemer and Hoogenstraaten [60], using the probe method for studying photoconductivity in CdS, observed the development of a photocurrent at large distances from the point of excitation and explained this as being due to the migration of excitons. The diffusion of excitons was also studied by Vorob'ev and Karkhanin [61] for the case of Cu_2O, where the displacement equalled about 5 μ.

We thought it would be interesting to determine the impurity-ion concentration in the crystals of alkali iodides for which excitons were annihilated earlier by the impurities than by the spontaneous annihilation associated with the luminescence of the pure salts. However, in order to study the effects of the impurities it was essential first to study the luminescence of the unactivated salts, particularly as their luminescent properties had not yet been examined fully. By no means enough attention has been paid to the luminescence of alkali iodides arising at low temperatures. The study of this luminescence is essential in order to understand the mechanisms responsible for the transfer of energy from the main substance to the activator; this is also of practical interest, since scintillation counters are also used in some cases at low temperatures.

A study of the scintillation growth time as a function of concentration might help in determining the diffusion coefficient of the excitons in various crystals, if we consider that the growth time of the scintillations incorporates both the lifetime of the activator in the excited state, and the time required to transfer the energy absorbed by the base to the emitting activator ion.

Considering all the foregoing, we set ourselves the problem of studying the unactivated luminescence of alkali halides under both γ- and photoexcitation, and also the effect of foreign impurities and lattice defects on this, as well as the concentration dependence of the time characteristics of the scintillations of phosphors based on alkali iodides.

STUDY OF γ-LUMINESCENCE IN ALKALI IODIDES

1. Review of Published Data Relating to the
Spectral Characteristics of Alkali Iodides

A study of the luminescence of pure salts is of great scientific and practical interest; however, the literature contains insufficient information regarding their luminescence and scintillation properties. A little information exists for the luminescence of pure NaI and CsI, but very little is known of other alkali iodides.

Eby and Jentschke [29] studied scintillations for a small activator concentration (c_{Tl} = $6.3 \cdot 10^{-5}$ mole Tl/mole NaI) and found an emission band at room temperature with $\lambda_{max} \sim$ 410 mμ and $\Delta\lambda$ = 50 mμ, as well as a short-wave band with $\lambda_{max} \sim$ 320 mμ and $\Delta\lambda$ = 45 mμ, the intensity of which fell with increasing activator concentration. The luminescence of NaI crystals was studied in more detail by Van Sciver and Hofstadter [21, 22, 38]. These authors studied pure NaI crystals under γ and α excitations and found an emission band in the ultraviolet part of the spectrum with λ_{max} near 310 mμ at room temperature. With falling temperature, the maximum of the band shifted in the short-wave direction and at $-190°C$, λ_{max} = 303 mμ (or 295 mμ according to [38]). The intensity of the emission meanwhile increased over 10 times, and as the temperature fell an additional band appeared with a maximum at about 410 mμ (or 425 mμ [38]). Regarding the short-wave band, it is known that this is excited by light with λ < 225 mμ and not by the absorption of light in the activator band ($\lambda_{max} \sim$ 290 mμ). According to [38], the 295-mμ band observed in all samples with a small activator content vanishes for a Tl concentration of 0.1 mol.%. Enz and Rossel [31], after special purification of the original material, obtained results agreeing with those of Van Sciver and Hofstadter for NaI.

Thus, it was found by studying pure NaI that there were at least two emission bands: one lying in the ultraviolet part of the spectrum with a maximum at 295-303 mμ at a temperature of $-190°C$ and the other (not appearing in all crystals) in the blue part of the spectrum with $\lambda_{max} \sim$ 410 mμ.

On irradiating pure CsI crystals with α particles, Hahn [36] observed that these luminesced with a blue light. Knöpfel, Loepfe, and Stoll [39], studying the temperature dependence of the yield and τ for CsI and CsI—Tl, found that the component ascribed to the emission of the pure material, having a wide band at $-180°C$ with $\lambda_{max} \sim$ 460 mμ and $\Delta\lambda$ = 120 mμ, vanished with increasing temperature and increasing activator concentration. These authors also noted a change in the luminescent yield with heat treatment of the crystals. After the pure CsI sample had been heated in a sealed tube at about 900°C, there was a considerable iodine vapor pressure inside the tube, and the luminescent yield on α excitation doubled.

Enz and Rossel [31], studying the same cold component of the luminescence of CsI, showed that the luminescence died out in accordance with an exponential law. Studying the spectral characteristics of these scintillations, the authors determined the emission spectrum

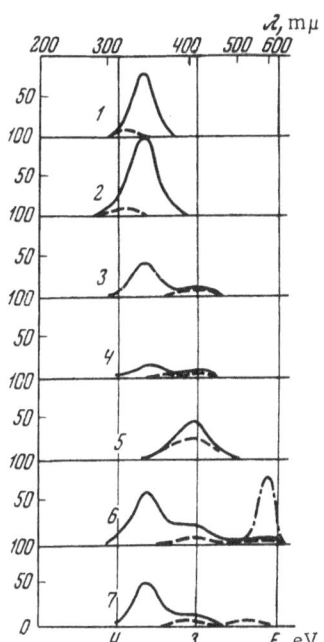

Fig. 1. Emission spectra of various CsI crystals grown by various methods (taken from [42]): 1) from solution; 2) by zone refining; 3,4,5) by the Stockbarger method; 6) CsI—Tl · (10^{-6} mole Tl/mole CsI); 7) CsI—In · (10^{-5} mole In/mole CsI). Continuous curves correspond to liquid nitrogen temperature; broken curves to room temperature, and the dot—dash curve to the emission of CsI—Tl (large amount of Tl) at room temperature.

at 77°K, consisting of two bands with $\lambda_{max} \sim$ 325 and 356 mμ, diffuse on the long-wave side, where the emission spectrum broke off at about 430 mμ. Raising the temperature led to a severe weakening of the emission from both bands. The intensity of the band at 325 mμ fell 15 times and that of the 350 mμ band 150 times as temperature was increased to 200°K. The authors explained a slight difference relative to the emission spectra of [39] as being due to differences in the components studied.

A clearer picture of the luminescence of CsI crystals, either unactivated or activated with small quantities of Tl and In, was obtained by Morgenshtern [42], who studied in great detail the luminescence of CsI at room temperature and the temperature of liquid nitrogen. Figure 1 shows the emission spectra of various CsI crystals, pure and activated; at low temperatures all crystals grown from solution and many crystals grown from the melt showed an emission band with a maximum at about 340 mμ and a halfwidth of about 0.4 eV. In addition to this, both in unactivated crystals and in those activated with small quantities of metal there was another component of luminescence, a blue band with $\lambda_{max} \sim$ 405 mμ and a half-width of 0.45 eV. However, in crystals grown from solution this band failed to appear; it also failed to occur in the spectrum of a crystal grown by zone-refining the original material. This band occupies an intermediate position between the ultraviolet and the activator luminescence; hence we shall subsequently call all analogous bands in the emissions of alkali iodides "intermediate emission bands."

The introduction of an activator in small quantities does not lead to the elimination of the blue or ultraviolet bands, but increasing the activator concentration does weaken these; the addition of a large amount of activator destroys both kinds of unactivated luminescence.

The ultraviolet band vanishes in almost all crystals with increasing temperature, while the blue and activator luminescences remain even at room temperature. However, in crystals grown from solution or obtained by zone melting, in which the intensity of the ultraviolet band is considerable, the latter even appears in the luminescence at room temperature, although its maximum moves 20 mμ in the short-wave direction.

Thus we see that in CsI-base phosphors there are three forms of luminescence: the ultraviolet, the intermediate, and that of the activator. We also see that breaks in crystal structure caused by the introduction of foreign atoms and ordinary lattice defects have a harmful effect on the unactivated luminescence of CsI.

Chanvy and Rossel [62] also noted a change in the spectral composition of the emission of CsI, according to the method used for obtaining the crystals; thus, in samples obtained from aqueous solution, only the ultraviolet emission occurred, while in those obtained from the melt, the ultraviolet band with $\lambda_{max} \sim 347$ mμ was accompanied by a blue band with $\lambda_{max} \sim 437$ mμ.

2. Choice of Subjects for Study

As subjects for study we chose alkali-iodide single crystals, both pure and activated with traces of Tl or In, small enough to allow the unactivated luminescence to appear in the spectrum of the samples. We used crystals already available in the laboratory, grown either by the Stockbarger method (obtained from L. M. Shamovskii's laboratory of the VIMS) or by the Kiropoulos method (obtained from L. M. Belyaev's laboratory of the Institute of Crystallography, Academy of Sciences of the USSR), or grown by zone melting of the original raw material (obtained from the Khar'kov Institute of Single Crystals).* For obtaining the most perfect samples the single crystals were grown from aqueous solution. The activator concentration in the crystals was determined either polarographically or from the absorption in the first activator band.

It was harder to grow the crystals from solution than from the melt. However, the advantage of the former method was that the crystal lattice was more perfect, the equilibrium number of lattice defects being many times smaller than in crystals obtained from the melt; this held particularly for iodides, since iodine is very volatile and the number of anion vacancies in the crystal depends strongly on temperature and on the time spent by the sample at high temperatures.

In growing single crystals from solution, specially purified raw material subjected to double recrystallization was used. The crystals were grown by evaporation from aqueous solution at room temperature.

In studying series of crystals with varying concentrations, the crystals used were grown by the Stockbarger method and as far as possible were equal in size and shape.

3. Method of Studying the γ-Luminescence

Spectra of Alkali Iodides

The emission spectra were studied after γ excitation. After hard excitation, the emission spectrum of the phosphor contains all forms of luminescence characterizing the particular sample, and there is no difficulty in separating the exciting radiation from the sample emission, such as occurs in photoexcitation. As source of γ radiation we used Co^{60}. The radiation was recorded by means of a quartz spectrograph on photographic films with a sensitivity of 1200 GOST units.

The source was placed as near as possible to the sample in order to increase the intensity of luminescence and hence reduce the exposure time. The minimum distance between sample and source was set by the dimensions of the Dewar containing the sample. A schematic picture of the Dewar for studying the excitation and emission spectra appears in Fig. 2. When studying an emission spectrum it was desirable to make a semi-quantitative estimate of the effect of activator concentration on the luminescence of the pure salt. In order to make the measurements as reliable as possible, all the parameters of the apparatus in which the emission was recorded were kept constant, and the samples were made as far as possible similar in size and shape.

Since it was hard to make the dimensions of different samples agree exactly, cuvettes with specified window sizes were made. For a known crystal thickness, the volume of luminescing

---

*The author takes this opportunity to thank A. A. Dunin, G. F. Dobrzhanskii, and L. M. Soifer for kindly presenting these crystals.

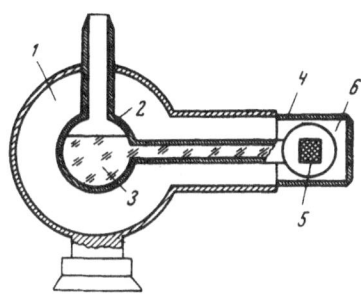

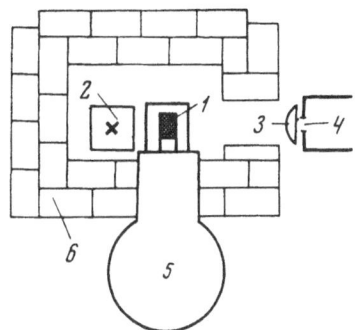

Fig. 2. Arrangement of the Dewar for studying crystals at different temperatures. 1) Volume evacuated; 2) inner vessel filled with liquid nitrogen; 3) liquid nitrogen; 4) quartz window holder; 5) sample; 6) quartz window.

Fig. 3. Arrangement of the apparatus for recording γ-luminescence spectra. 1) Sample; 2) Co^{60} excitation source; 3) lens; 4) entrance slit of the spectrograph; 5) Dewar; 6) lead barrier furnace.

material emitting the radiation being recorded was calculated; reabsorption of the luminescence in the crystal was neglected, and the density of the luminescence was regarded as uniform throughout the volume of the crystal. The luminescing material being recorded was regarded as constituting a cylinder with a base equal to the area of the cuvette window and a height equal to the thickness of the crystal. It would be more accurate to consider the luminescing material in the form of a truncated cone, but, for the existing apparatus parameters, the cylindrical assumption introduced very little error into the results of the measurements; for an average crystal thickness of 5 mm and a crystal—spectrograph slit distance of 50-60 mm, the error introduced equalled 7-9%; this lies within the limits of experimental error for photographic recording methods.

In order to secure better heat transfer and also in order to fix the crystals firmly in the cuvette, the samples were sealed with Wood's alloy (melting point 57°C).

A schematic representation of the apparatus appears in Fig. 3. The source of γ radiation together with the Dewar containing the crystal were placed in a lead barrier furnace. The emission from the sample fell on the entrance slit of the spectrograph via a lens ensuring uniform illumination over the height of the slit [63, 63a, 64], which was essential for the microphotometer measurement of the spectrograms. As source of γ radiation we used a radioactive sample of Co^{60}, the energy of the γ quanta from which was 1.17 and 1.33 MeV. The exposure time was varied, according to the intensity of the luminescence in question, between 5-10 and 80 h. The resultant spectrograms were measured on an MF-4 microphotometer [64].

In order to simplify the problem of explaining the laws governing the behavior of the unactivated bands of luminescence, the main factors affecting these were deliberately separated, i.e., we either studied the effect of impurities, keeping the other parameters unaltered, or else studied the effect of the structural defects by themselves on the unactivated luminescence.

4. Effect of Foreign Impurities on the Unactivated Fluorescence of Alkali Iodides

a) Study of CsI Crystals

In order to study the effect of foreign impurities on the unactivated luminescence of CsI we chose a CsI—Tl concentration series (CsI containing various amounts of Tl). The crystals

Table 1. Concentration of Tl in the CsI—Tl Crystal Studied

Crystal No.	Conc. in original charge, %	True Tl conc. (g/g) deter-mined polaro-graphically	No. of Tl ions per cm³ 1/cm³	$c_{Tl} \frac{\text{mole Tl}}{\text{mole CsI}}$	Mean distance between Tl ions	
					in lattice constants d	in Å
1	0	$1.1 \cdot 10^{-5}$	$1.50 \cdot 10^{17}$	$1.40 \cdot 10^{-5}$	42	191
2	0	$4.6 \cdot 10^{-5}$	$6.25 \cdot 10^{17}$	$5.84 \cdot 10^{-5}$	26	117
3	0.1	$5.6 \cdot 10^{-5}$	$7.76 \cdot 10^{17}$	$7.26 \cdot 10^{-5}$	24	108
4	—	$1.23 \cdot 10^{-5}$	$1.67 \cdot 10^{18}$	$1.56 \cdot 10^{-4}$	19	85
5	0.6	$2.2 \cdot 10^{-4}$	$3.09 \cdot 10^{18}$	$2.88 \cdot 10^{-4}$	15	69
6	0.5	$4.8 \cdot 10^{-4}$	$6.57 \cdot 10^{18}$	$6.17 \cdot 10^{-4}$	12	54
7	1.0	$1.24 \cdot 10^{-3}$	$1.69 \cdot 10^{19}$	$1.57 \cdot 10^{-3}$	6	27

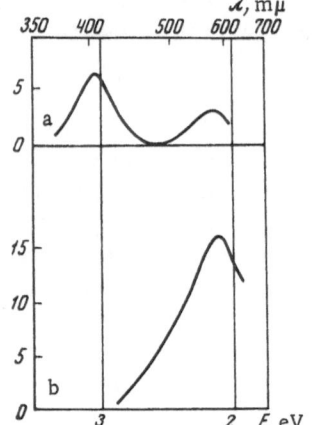

Fig. 4. Emission spectra of CsI—TlNO$_3$ crystals grown from the melt. a) At liquid-nitrogen temperature; b) at room temperature.

were obtained by the Stockbarger method; the activator concentration varied from $1.40 \cdot 10^{-5}$ mole Tl/mole CsI to $1.575 \cdot 10^{-3}$ mole Tl/mole CsI. The concentration characteristics of the crystal in question are given in Table 1.

It was found in [65], as in [42], that the intensity of the luminescence in the short-wave bands fell with increasing activator concentration, vanishing entirely for high concentrations. The short-wave bands were only observed in the first four samples and not in the last three; thus, in the emission spectrum of crystal No. 4 at liquid nitrogen temperature all three forms of luminescence appeared, while in crystal No. 5 there was only the activated type. The activator concentrations in these crystals were, respectively, $c_4 = 1.56 \cdot 10^4$ mole Tl/mole CsI and $c_5 = 2.88 \cdot 10^{-4}$ mole Tl/mole CsI, which correspond to a mean distance between the activator ions of 19d (lattice constant) in the first case and 15d in the second.

Usually on growing from solution (as indicated in [42]), the luminescence of CsI contains no intermediate band at λ max ~ 405 mμ. On growing activated CsI—Tl phosphors from solution, however, adding the activator not in the form of its iodide, which is the usual way, but as a nitrate TlNO$_3$, the luminescence of the crystals obtained at liquid nitrogen temperature includes the intermediate band as well as the activator band, but has no ultraviolet component.

The intermediate band, usually absent in crystals grown from solution, is quite strong in this case (Fig. 4).

Thus, the creation of defects associated with different types of cations or anions in the crystal eliminates the ultraviolet luminescence, while anion impurities may intensify the intermediate luminescence at low temperatures.

b) Study of RbI Crystals

The least studied of the alkali iodides is RbI.* In contrast to CsI, which has a bcc lattice, RbI has a fcc lattice like the other alkali iodides. We expected to observe luminescence in the activator-free salts in this case also.

*Data relating to the luminescence of lithium iodide occur in the literature [11, 66-68]; however,

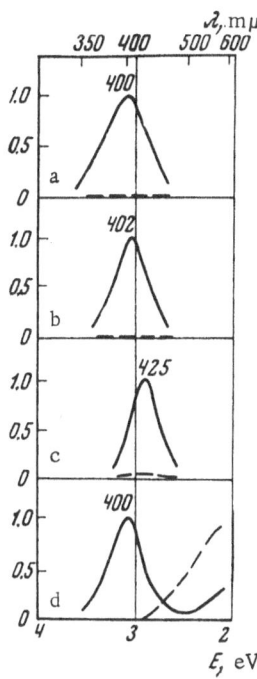

Fig. 5. Emission spectra of RbI-base phosphors grown in various ways. a) RbI from solution; b) RbI by the Stockbarger method; c) RbI—TlI from solution; d) RbI—InI by the Stockbarger method. Continuous curves at liquid nitrogen temperature, broken curves at room temperature.

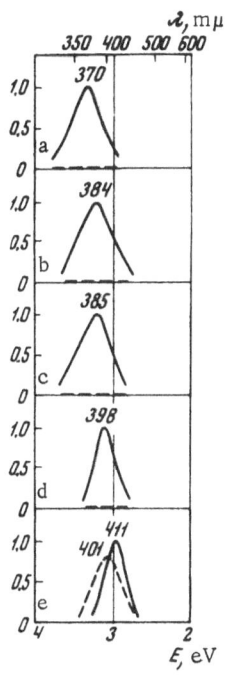

Fig. 6. Emission spectra of KI single crystals grown by different methods. a) From solution; b) by the Stockbarger method; c,d) by the Kiropoulos method; e) from solution but with a very imperfect structure (crystal cloudy).Continuous lines at temperature of liquid nitrogen; broken lines at room temperature.

In RbI crystals [65, 69] grown from solution we found one emission band with a maximum at about 400 mμ and a half-width of 0.54 eV (Fig. 5a).

The intensity of the emission from these crystals falls sharply with increasing temperature. In the case of room temperature, with exposures double the length of those employed for photographs taken at liquid nitrogen temperature, the emission could not be recorded.

A structural band was also observed in RbI. This was a band with a maximum at about 402 mμ and a half-width of 0.42 eV (Fig. 5b). At room temperature the intensity of the luminescence in question fell very sharply. The band observed was rather narrower than the short-wave band and displaced very slightly from the latter, so that the two bands could hardly be resolved when both occurred in the spectrum of pure RbI. However, a separation of this kind could be effected by reference to the photoexcitation spectra; the short-wave band had a complicated excitation spectrum, situated in the characteristic absorption region of the crystal, while the longer-wave band was excited in a narrow range of frequencies on the falling part of the characteristic absorption curve of RbI. If the emission spectrum of the crystal contained both bands, the photoexcitation spectrum was correspondingly complicated (more will be said of the excitation spectra in Chapter III).

the study of this substance is very difficult in view of its considerable hygroscopic properties. We therefore refrained from studying LiI.

In determining the influence of foreign impurities on the unactivated luminescence of RbI we used traces of Tl and In. We had no RbI—TlI crystals capable of revealing unactivated luminescence (Fig. 5c). The emission band of Tl in RbI is very close to the bands of unactivated luminescence ($\lambda_{max} \sim 425$ mμ) and it might be expected that the activator band would overlap the short-wave bands, giving unresolved luminescence from several kinds of centers in the emission spectrum. However, unactivated luminescence failed to appear, even in the photoexcitation spectrum of the phosphor.

On activation with indium, however, we were unable to obtain samples having activator luminescence only; although the unactivated luminescence became weaker with increasing activator content, it never quite fell to zero. This may be explained by the fact that the indium atoms fitted the lattice badly, so that the impurity concentration at which the unactivated luminescence should vanish entirely was never achieved in the crystal.* With increasing temperature the activator luminescence increased at the expense of the unactivated component, and at room temperature only the activator luminescence was visible.

Thus, RbI-base phosphors also manifest the three forms of luminescence in their spectra: that of the activator type and two forms of unactivated luminescence with severely overlapping bands.

c) Study of KI Crystals

In studying the γ luminescence of KI single crystals we had the same problems in mind as in the case of the other alkali iodides. Depending on the prehistory of the crystals, a number of different emission spectra were obtained [65, 69-71]. On growing crystals from solution, one band with $\lambda_{max} \sim 370$ mμ and a half-width of 0.45 eV is obtained in the emission spectrum (Fig. 6a) at liquid nitrogen temperature; at room temperature we found no luminescence.

In other pure KI crystals grown either by the Stockbarger or Kiropoulos methods we also found one band, but this was displaced slightly in the long-wave direction from the luminescence of KI samples grown from solution (Figs. 6b, c); the displacement varied slightly from sample to sample. Such behavior of the emission band may easily be explained if we suppose that it has a complex structure and consists of two elementary bands: a band with $\lambda_{max} \sim 370$ mμ, corresponding to the shortest-wave emission observed in the luminescence spectra of crystals grown from solution, and a band analogous to the intermediate band of luminescence in pure CsI and RbI salts, situated in the blue part of the spectrum. Then the overall maximum lie somewhere between the maxima of the component luminescence bands, the precise value depending on the relative intensities of these.

In order to confirm this idea, it was essential to obtain crystals emitting only the intermediate band. Such crystals were sometimes obtained by growing from solution, but they had an imperfect structure and the samples were cloudy. The luminescence spectrum of such crystals (Fig. 6e) showed one band with $\lambda_{max} \sim 410$-413 mμ and a half-width of 0.35 eV. The excitation spectrum of this band, as in the case of analogous alkali iodides, was situated at the characteristic absorption edge of the crystals and constituted a narrow band with a maximum at around 230 mμ. Thus, the unactivated luminescence of KI, observed in the form of a wide band occupying an intermediate position between 370 and 410 mμ, may be regarded as the superposition of two elementary bands, a short-wave one and an intermediate one.

---

*According to A. A. Dunina, who kindly gave us the RbI—In crystals, larger amounts of In could not be introduced into the crystal; the In precipitated in atomic form on the crystal surface or block boundaries.

Table 2. Concentration of Indium in the KI—In Phosphors Studied

Crystal No.	Conc. in original charge, %	True In conc. determined from absorption coeff. in first activator band, $\frac{\text{mole In}}{\text{mole KI}}$	No. of In ions per cm³, 1/cm³	Av. distance between In ions	
				in lattice constants d	Å
1	0.001	—	—	—	—
2	0.01	—	—	—	—
3	0.1	$8.7 \cdot 10^{-6}$	$0.99 \cdot 10^{17}$	62	218
4	>0.1	$2.54 \cdot 10^{-5}$	$2.89 \cdot 10^{17}$	43	152

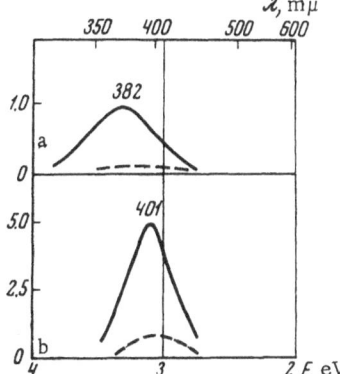

Fig. 7. Emission spectra of KI—Tl with different nominal concentrations of Tl. a) 0.05%; b) 2%. Continuous curves at liquid-nitrogen temperature, broken curves at room temperature.

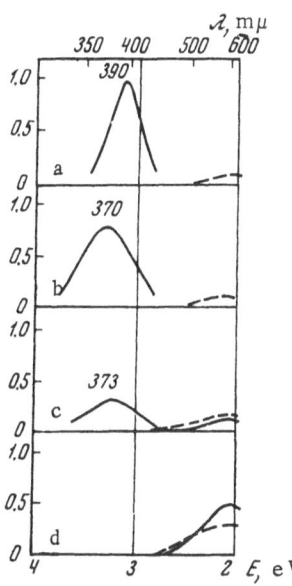

Fig. 8. Emission spectra of KI—In with different nominal concentrations of In. a) 0.001%; b) 0.01%; c) 0.1%; d) >0.1%. Continuous curves at liquid-nitrogen temperature; broken curves at room temperature.

Among the various activated KI-base phosphors we studied KI—Tl and KI—In. In KI—Tl the activated luminescence severely overlapped the unactivated band, and separation could only be achieved at room temperature, when the luminescence of pure KI vanished (Fig. 7a). It was also possible to achieve separation with high activator concentrations, when the unactivated luminescence was extinguished by the presence of foreign impurities in the crystal, in the present case Tl (Fig. 7b); in this case only the Tl luminescence was observed, and this changed position very little on passing from liquid-nitrogen to room temperature. Owing to the severe overlapping of the emission bands of the activated and unactivated luminescence in the KI—Tl phosphor, it was inconvenient to study the effect of impurities on the luminescence of the pure material in this case. In order to make a complete study of the effect of the activator on the unactivated luminescence of pure KI we therefore studied a number of KI—In phosphors (Table 2). The results of these investigations are shown in Fig. 8. In this case, as in all the previous cases, the intensity of the unactivated luminescence fell with increasing number of foreign atoms and finally vanished altogether.

Since the intensity ratio of the unactivated luminescence bands changes from sample to sample, the position of the overall emission spectrum differs for different samples, i.e., we have the same phenomenon as in pure KI samples, with two forms of unactivated luminescence.

Using KI—In as an example, we estimated the activator concentrations at which the luminescence of the pure base would still be just visible and also the concentration which would be sufficient to extinguish (quench) this luminescence completely. In the emission of crystal No. 3 (Table 2) unactivated luminescence (Fig. 8c) was still clearly visible, while in the emission of crystal No. 4 there was no such luminescence at all (Fig. 8d). The true indium concentration in these crystals was determined from the absorption coefficient in the first activator band, using Shamovskii and Zhvanko's data [72] on the relation between the absorption coefficient in this band and the In concentration in the KI—In crystals.

For the first of the two crystals indicated, the In concentration was $0.87 \cdot 10^{-5}$ mole In/mole KI, which corresponds to a mean distance between the activator ions of ~62d or 218 Å; for the second crystal the In concentration was $2.54 \cdot 10^{-5}$ mole In/mole KI, i.e., the average distance between the activator ions was about 43d or 152 Å. Thus, in all our alkali halide-base phosphors the activator emission is accompanied by two kinds of unactivated luminescence; the latter vanish in the presence of foreign impurities, and this may well be why the luminescence of the pure salts remained so long undiscovered. The observation of unactivated luminescence was made more difficult because it usually occurred at low temperatures, and also because the bands of unactivated luminescence often overlapped the activator absorption bands; all this promoted the reabsorption of the unactivated luminescence, with subsequent emission of the absorbed energy in the activator center. Thus, CsI—Tl crystals with a Tl concentration of 0.002 mol.% and a thickness of 0.2 cm absorb 50% of the light in the region of 330 mμ, while the same crystals with a Tl concentration of 0.17 mol.% absorb over 99% in the same region [73]. These circumstances are exacerbated by the fact that, with increasing temperature, both the unactivated bands of luminescence and the activator absorption bands widen, so that they overlap over a wider optical range.

5. Effect of Structural Defects in the Crystals on the Unactivated Luminescence

As indicated earlier, our study of the quenching of unactivated luminescence was divided into two stages: the effect of impurities (treated in the preceding section) and the effect of structural defects deliberately created in the crystals by special heat treatment.

We chose crystals grown from solution or obtained by zone refining of the original material; the emission spectrum of these showed just one short-wave band. These crystals we subjected to heat treatment, holding them for a long period at a temperature close to the melting point but low enough to prevent melting of the material. The crystals were heated in a platinum crucible or in quart ampoules and then quenched (cooled rapidly).

In studying CsI, crystals and samples were heated to 590°C (melting point of CsI, 621°C). Even after 4 h heating the intensity of the ultraviolet band fell sharply and the emission spectrum developed an intermediate band (Fig. 9). The band so formed did not vanish with increasing temperature, i.e., it behaved just as in the case in which the centers responsible for it were obtained in the course of crystal formation. Then the same sample was subjected to further heat treatment under the same conditions as in the previous experiment. Further heating for 54 h completely annihilated the ultraviolet emission, while the intensity of the intermediate band increased.

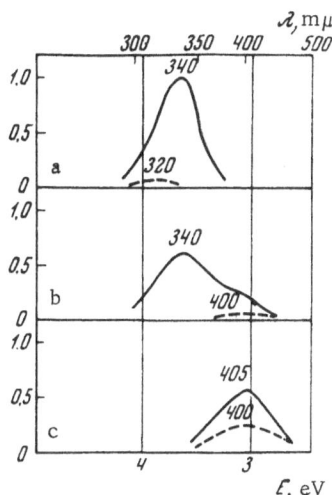

Fig. 9. Emission spectra of CsI single crystals. a) Before heat treatment; b) after 4 h annealing at 590°C; c) after 54 h annealing at the same temperature followed by quenching. Solid lines at liquid-nitrogen temperature; broken lines at room temperature.

In order to confirm the fact that the band thus arising really corresponded to the intermediate band in the cases considered earlier, we measured its photoexcitation spectrum, which coincided with that of the blue band in CsI. (More will be said of the excitation spectra in Chapter III.)

On raising the temperature, iodine evolves from the alkali iodides; this may clearly be seen on heating the samples in ampoules by the formation of a brown deposit on the sides. The number of defects in the crystal increase accordingly and this annihilates the ultraviolet emission, but at the same time forms luminescent centers which give rise to the intermediate band.

In studying the effect of structural imperfections on the unactivated luminescence of RbI, the crystals were heated to about 500°C (melting point of RbI, 642°C). The heating time varied from 4 to 25 h. The emission spectra of the crystals broadened after heat treatment and their intensity fell slightly, without any marked displacement to either side. It is well known that any disruption of the structure of crystals tends to broaden both their absorption and their emission bands [74, 75]. We may, on the one hand, explain the observed broadening of the emission bands of RbI crystals after heat treatment in this way. On the other hand, the fact that the emission band broadened after the deliberate creation of structural defects in the crystal may be explained by supposing that an intermediate luminescence band has appeared in the emission spectrum. This band is situated too close to the short-wave band and cannot be resolved from it, so that the wide band observed constitutes the composite emission of both bands of unactivated luminescence. The presence of the intermediate band may be established by studying the photoexcitation spectra of the crystals, and this will be considered in more detail in Chapter III.

In order to discover the effect of structural defects on the unactivated luminescence of KI we carried out experiments analogous to those relating to CsI and RbI. Heat treatment of the KI crystals was carried out at about 500°C (melting point of KI, 723°C), the heating time being the same as in the case of RbI, varying from 7 to 25 h. On the basis of the same considerations as those employed in the case of RbI (mutual proximity of the short-wave and intermediate bands), the resultant effects may be distinguished by analyzing the excitation spectra.

In this way we found that both bands of unactivated luminescence were very sensitive to disruptions of the crystal structure, but this sensitivity operated in different senses. The ultraviolet band vanished as the crystal structure of the sample was damaged, while the intermediate band became stronger.

6. Temperature Behavior of the Activator
Emission Bands of Phosphors

a) Thallium Phosphors

As already indicated, we chose crystals with small activator concentrations for our investigations. On studying Tl-containing phosphors we observed thermal quenching of the activator luminescence. Morgenshtern observed a similar phenomenon in CsI—Tl [42]. The Tl luminescence became stronger with increasing activator concentration; this may be simply explained by the increasing number of emitting centers per unit volume.

A similar behavior of the Tl emission bands occurs in other alkali iodides. Thus, in the RbI—Tl phosphor the intensity of the activator luminescence falls more than 15 times as temperature rises from that of liquid nitrogen to room temperature (see Fig. 5c).

In the case of KI—Tl the same picture holds as in the other Tl-activated alkali iodides (see Fig. 7): The intensity of the activator luminescence falls with increasing temperature and rises with increasing Tl content.

Thus the behavior of the thallium emission bands may be easily explained by considering the thermal quenching of activator luminescence and the rise in the number of emitting centers with increasing activator content. On studying the activator luminescence of indium phosphors, however, there is no such simple temperature dependence, and we feel that a study of the latter is important in order to understand the mechanism underlying the transfer of energy from the base to the activator under hard excitation.

b) Indium Phosphors

It was noted earlier in [42] that in a CsI—In phosphor with a low activator content, for which the unactivated luminescence had not yet vanished, the indium emission band failed to appear at liquid-nitrogen temperature and only emerged on heating, while the ultraviolet emission correspondingly weakened.

It was interesting to trace the behavior of the indium luminescence in other alkali iodides as a function of temperature and activator concentration. For this purpose we chose a concentration series of KI—In with a nominal indium concentration of between 0.001 and ~0.5 wt.%, which in the latter case corresponded to an In concentration of $2.54 \cdot 10^{-5}$ mole In/mole KI. The emission spectrum of KI—In in our case is a band lying in the yellow part of the spectrum with λ_{max} about 565 mμ, which agrees with the results of [76]. We found (Fig. 8) that the activator band was absent from samples having unactivated luminescence at liquid-nitrogen temperature. On raising the temperature, however, so that the unactivated luminescence diminished, the activator band began to appear. Then, with increasing activator content (but still low enough to allow unactivated luminescence to appear), the luminescence spectrum of the phosphor started showing an activator band even at 77°K. As before, the intensity of this band continued rising with increasing temperature; the weaker the unactivated luminescence, the smaller was the difference in the intensity of the activator luminescence at room temperature and the temperature of liquid nitrogen, respectively. When the activator concentration was so great that the unactivated emission vanished completely, however, the behavior of the indium luminescence changed, and the activator emission fell with increasing temperature.

The same behavior of the activator emission was observed in the phosphor RbI—In. As temperature increased, the activator emission increased at the expense of the unactivated luminescence (Fig. 5d).

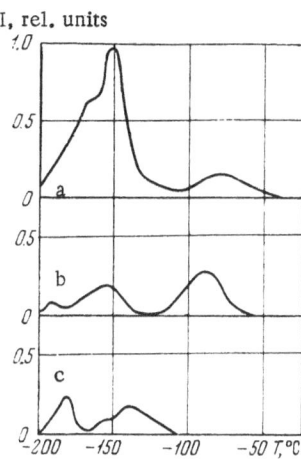

I, rel. units

Fig. 10. Thermoluminescence curves of alkali iodides subjected to heat treatment. a) CsI heated for 54 h; b) KI heated for 25 h; c) RbI heated for 25 h.

Thus, we get the impression that, in alkali iodides activated with In, the excitation energy obtained by the absorption of a γ quantum is emitted by the base at liquid-nitrogen temperature to the detriment of the activator emission; however, with increasing temperature the energy received is, as it were, "thrown over" to the activator, which emits it in its own band. When the unactivated luminescence fails to appear altogether at 77°K, i.e., when the "reservoir" from which the activator drew its extra energy is exhausted, then the activator luminescence loses intensity with increasing temperature as a result of thermal quenching.

7. Thermoluminescence of Alkali Iodides

In order to obtain more complete data regarding the nature of the luminescence of pure alkali iodides, we studied their thermoluminescence. Regarding the storage of light (light sum) in CsI, it was shown in [42] that in the ultraviolet band the storage is insignificant, while the blue band stores a considerable supply of light, and its thermoluminescence curve has a number of maxima, the largest at about 130°K and the next at about 190°K.

In studying thermoluminescence we used existing laboratory apparatus* to ensure uniform heating at a rate of 8 deg/min over the range in question. The excitation of the crystals was carried out at liquid-nitrogen temperature for 20 min, the γ source being Co^{60} with an activity of 3.3 mCi.

Crystals grown from solution stored very little light, whereas crystals subjected to heat treatment had a considerable storage capacity (Fig. 10). The thermoluminescence curve of CsI samples subjected to heat treatment has a complicated form: a sharp maximum at about −150°C and a wide, diffuse band at −70°C; this agrees closely with the results of [42] regarding the thermoluminescence of the blue band of CsI. The thermoluminescence curve of RbI subjected to heat treatment at 500°C for 25 h also has a complicated form with maxima at about −180 and −140°C and a small maximum in an intermediate position. The thermoluminescence curve of KI crystals also heat-treated at about 500°C for 25 h shows three maxima at about −180, −150, and 95°C.

Thus we have established that in all the alkali iodides studied there is very little light storage in the ultraviolet emission band, whereas crystals subjected to heat treatment, having an intermediate band in their thermoluminescence characteristic, do store light. According to [22], there is no light storage in the ultraviolet emission band of NaI with $\lambda_{max} \sim 303$ mμ, whereas storage does occur in the 410-mμ band of the same crystals.

Thus, on the basis of the foregoing results, we may consider that in the luminescence of phosphors based on the alkali iodides CsI, RbI, KI, and NaI the activator luminescence is accompanied by two unactivated types of emission. The first of these emits in a shorter-wave part of the spectrum than the other forms of luminescence and its intensity falls with increasing temperature; the second occupies an intermediate position between the short-wave and activator

---

*The apparatus was designed and created by E. E. Bukke.

emission, and its behavior varies slightly, depending on the presence of other anions or micro-defects. In addition to these we have seen that in indium phosphors the intensity of the activator luminescence increases with rising temperature at the expense of a reduction in the unactivated luminescence. We also note that in the ultraviolet emission bands single crystals of the alkali iodides store no light, whereas in the intermediate band the storage is considerable.

All this reliably confirms the existence of two different types of luminescence not associated with the activator in alkali iodides; however, at the moment it is hard to guess the nature of these. We consider that a study of the spectral excitation characteristics and a comparison of these with the known absorption spectra of the alkali iodides (the bands of which are fully treated in the literature) may help in understanding the nature of the unactivated types of luminescence in question.

STUDY OF THE SPECTRAL CHARACTERISTICS OF THE PHOTOEXCITATION OF PHOSPHORS BASED ON ALKALI IODIDES

1. Review of Published Data Relating to the Absorption of Alkali Halides

A study of the excitation spectra is of very great importance in understanding the nature of various emission bands. When light of a specific spectral composition is absorbed, individual emission bands may be excited; this offers the possibility of separating these out and controlling their behavior in relation to various external factors. By excitation spectrum we mean the relation between the active absorption (active in respect of a given form of luminescence) and the wavelength of the exciting light [77, 78]. However, before studying the active part of the absorption, we shall review published data relating to the absorption of alkali halide salts.

Hilsch and Pohl [66] measured the absorption spectra of alkali halides, using thin layers of crystals obtained by sublimation in high vacuum. The absorption spectra of chlorides, bromides, and iodides were measured up to 160 mμ. All the iodides and bromides showed similar absorption spectra. The absorption of the crystals was affected by the nature of the anions and type of lattice as well as the method of obtaining the samples; this may explain the differences between various authors as regards the positions of the maxima, for example, in [66] the first absorption maximum of KCl occurred at 162.5 mμ and in [79] at 158.1 mμ.

In order to determine the position of the band maxima, Hilsch and Pohl proposed the following empirical formula:

$$h\nu_{\max} = \alpha \frac{e^2}{4\pi\varepsilon_0 r} + E - A, \tag{1}$$

where E is the electrical affinity of the halides, A is the work of ionization in the alkali atom, α is the Madelung coefficient, r is the distance between the ions, and ε_0 is the dielectric constant. Measurement of the positions of the bands at room temperature agrees closely with the values calculated from the formula proposed.

In order to obtain absorption spectra in the shorter-wave part of the spectrum, Schneider and O'Bryan [67] used LiF as substrates, this salt being transparent up to 105 mμ. This made it possible to obtain the absorption spectra of alkali halides right up to the characteristic absorption of LiF. The absorption spectra of the alkali halides showed a number of narrow bands, and the first long-wave maximum determined their absorption edge. The behavior of the maximum may be described by an exponential law for temperatures up to 1000°K and for ν, varying over many orders:

Table 3. Position of Absorption Bands at 20°K

Crystal	I_a		I_b		Step		II_a	
	mμ	eV	mμ	eV	mμ	eV	mμ	eV
NaI	221	5.59	184	6.72	213	5.80	—	—
KI	212	5.83	185	6.68	199	6.19	—	—
RbI	216	5.72	190	6.49	201	6.13	—	—
			186	6.65				
CsI	212	5.81	181	6.80	—	—	208	5.926
							206	5.993

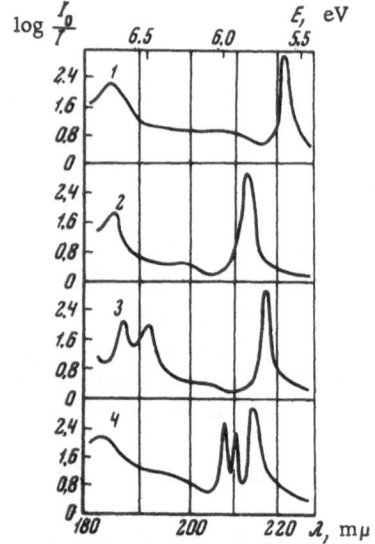

Fig. 11. Absorption spectra of sublimed layers of alkali iodides at 20°K (according to [74]). 1) NaI; 2) KI; 3) RbI; 4) CsI.

$$K = K_0 e^{-\frac{\sigma h(\nu_{max} - \nu)}{kT}}, \qquad (2)$$

where K is the absorption coefficient for frequency ν, K_0 is the absorption coefficient at the band maximum, k is Boltzmann's constant, T is the absolute temperature, σ is a constant, and ν_{max} is the frequency at the maximum of the absorption band.

For zero temperature this law leads to the degeneration of the band into an absorption line. Frenkel [58] and Wennier [80] explain the relative narrowness of the absorption band by considering that the absorption of light leads to the formation of an exciton. The absorption spectra of the alkali iodides at low temperatures were studied by Martienssen [74]. On cooling the samples to 20°K, the absorption bands become sharper and stronger. The absorption spectra of each of the iodides contain: a long-wave band in the region of 210 to 225 mμ, called, according to Hippel [81], the I_a band, and a I_b band in the region of 180-195 mμ (in the case of RbI this band is split); between these bands there is a clear absorption step in all the salts studied (Fig. 11). On passing from the light to the heavy iodides, the long-wave band changes its position in an irregular manner.

Cesium iodide has a lattice differing from those of the other iodides and also constitutes an exception in the series of absorption spectra. Schulz showed by electron microscopy [82] that the absorption band with $\lambda_{max} \sim 206$ mμ was a structural band of CsI; this was resolved at 20°K into two very narrow bands II_a with $\lambda_{max} \sim 206$ and 208 mμ. The long-wave band, having a half-width of 1 mμ (0.03 eV) was the sharpest of all bands found in the alkali halides. The distance between the components of the doublet was 0.065 eV. The positions of the absorption-band maxima of the alkali iodides of present interest are shown in Table 3.

As temperature varies, the absorption spectrum of the alkali iodides changes monotonically. With increasing temperature both bands move in the long-wave direction; at the same time, the half-width increases and the height diminishes. At temperatures above 120°K there is a linear displacement of band maxima

$$h\nu_{max, T} = h\nu_0 - \beta T, \qquad (3)$$

where $\nu_{max,T}$ is the position of the absorption band maximum at T°K, ν_0 is the position of the absorption band maximum at 120°K, and β is a quantity constant for the crystal in question.

The effect of crystal-lattice defects on the behavior of the absorption bands is analogous with the effect of temperature. Fisher [75] obtained very defective layers of KI by simultaneous sublimation with KF on a substrate, at 9°K. The absorption spectrum at this temperature was completely identical with that of uncontaminated KI at 480°K. The addition of KF (10% in the present case) stabilized the defects in the KI lattice, while the absorption of the KF itself failed to appear in this region. Thus, it is quite immaterial whether the crystal contains defects "frozen in" at low temperatures or defects caused by thermal vibrations of the lattice at high temperatures. This point of view was first expressed by Pohl [83]. It might be thought that, roughly speaking, a microphotograph of a damaged layer taken at low temperature should not differ from an instantaneous photograph of the damaged layer at high temperature, taken with an extremely short exposure; however, the displacement of the lattice elements from their normal position in the case of frozen defects is considerably greater than the amplitude of thermal vibrations at 480°K.

In order to discover the nature of the absorption bands of alkali halide salts, the photoconductivity and photoemission spectra were studied [84-86, 86a]. It was found experimentally that the absorption of light in the region of the first characteristic absorption band produced no photoconductivity in the crystals. This indicated that optical absorption in the region of the first band created no free electrons and holes in the crystals, but simply led to the formation of excited states: excitons. Free electrons and holes arise after the absorption of light of a shorter wavelength, such that its photon energy corresponds to a zone—zone transition; this corresponds to the absorption step in the absorption spectra of alkali halide salts [86, 86a, 87]. This principle was confirmed by the Tartu school when studying recombination phosphorescence spectra [88, 89].

When studying the absorption spectra of mixed crystals a number of authors [90-94] came to the conclusion that the first absorption maximum was associated with an exciton localized near the anion, the so-called anion exciton.

Thus it is now considered as proved that the longest-wave maximum in the absorption spectrum at alkali halide salts is due to the anion exciton, while the absorption in the region of the step corresponds to a zone—zone transition.

2. Method of Studying Excitation Spectra

Excitation spectra may be studied in two ways. The first of these lies in comparing the emission intensity of the samples under consideration with the emission of luminophores having a constant quantum yield on excitation by the same light source. A constant quantum yield characterizes fluorescent substances, the emissions of which obey Vavilov's law, which says that the quantum yield of the fluorescence is independent of the wavelength of the exciting light up to a certain limiting wavelength, after which it starts falling sharply [95]. Substances with a constant quantum yield include a whole series of fluorescent solutions and such solids as α-naphthol, light-yellow phosphorogen, anthracene, sodium salicylate, etc.

In studying excitation spectra by the intensity-comparison method, when the intensity of the radiation may be recorded from the side of the incident exciting radiation, fairly thick phosphorogen screens were selected; the quantum yield of these, according to the data presented in [96], was constant for excitation by short-wavelength light (according to unpublished data of Alentsev and Morgenshtern this constancy is preserved up to a wavelength of 450 mμ).*

---

*According to recent data [97], the light-yellow phosphorogen is an insufficiently standardized

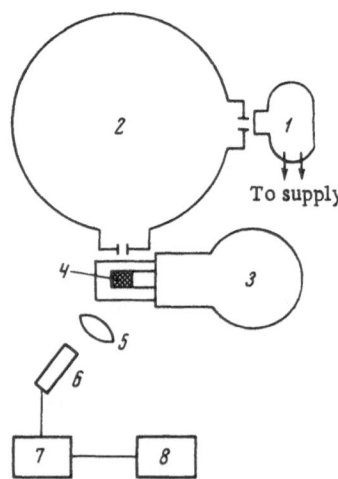

Fig. 12. Arrangement of apparatus for measuring excitation spectra at low temperatures. 1) Hydrogen lamp; 2) monochromator with fluorite prism; 3) Dewar; 4) sample; 5) condenser lens; 6) photomultiplier; 7) amplifier; 8) recording system.

The second method of measuring excitation spectra lies in determining the distribution of the spectral energy density of the excitation source. This distribution is also measured by reference to the emission of a substance with a constant quantum yield. When we used the second method of measuring the excitation spectra, we chose sodium salicylate, having a quantum yield constant over a wavelength range of 90–250 mμ [98]; semitransparent screens of this material are easily obtained by deposition from an alcohol solution on quartz or glass substrates. The measurements could be made either by transmission or with the help of an integrating sphere.

When measuring in transmission we used semitransparent sodium salicylate screens, choosing the thickness so that, on the one hand, they were thin enough for the visible light of the radiation from the sodium salicylate ($\lambda_{max} \sim 410$ mμ according to [99]) to pass through them, and on the other hand thick enough to absorb the exciting light completely. We were interested in the ultraviolet part of the spectrum, the absorption in which is much greater than the absorption in the sodium salicylate range of emission, so that screens a few microns thick ensured complete absorption of the ultraviolet light, remaining transparent for the radiation of the sodium salicylate itself.

The use of an integrating sphere offers the possibility of making more accurate measurements, since, in this case, all the light of the luminescence propagating within a solid angle of 4π is recorded, whereas on measuring with the help of screens only the flow of luminescence in one arbitrarily chosen direction is measured. The inner surface of the integrating sphere was covered with a layer of magnesium oxide, the reflection coefficient of which is very high and depends very little on the wavelength of the exciting light in the range of optical frequencies of present interest; the reflection coefficient of MgO is in fact 90–95% in the wavelength range 200–700 mμ [100, 100a]. Data relating to the energy distribution of the radiation from a hydrogen lamp obtained by the two methods just described agree completely for any one lamp.

We used both methods to study the excitation spectra of the alkali iodides. At room temperature the measurements of the excitation spectra in the wavelength range greater than 220 mμ was carried out with an SF-4 spectrophotometer having a special attachment designed and created by E. E. Bukke.

The spectrometer was modified in that the photomultiplier recording the luminescence of the sample was set at right angles to the ray falling on the latter. In order to avoid recording the scattered exciting light, a liquid filter (an aqueous solution of sodium nitrite) completely absorbing light of wavelength shorter than 400–420 mμ was placed at the entrance to the photomultiplier [63]. The use of this filter made it possible to cut off the ultraviolet part of the scat-

---

product; its relative yield varies from batch to batch. However, in these experiments we used the first-batch phosphorogen prepared in the Institute of Organic Chemistry (by V. K. Matveev), which exhibited a fairly good constancy in its quantum yield.

tered exciting light. Correspondingly, the sample itself was set on an angle differing from 45° so as to avoid recording the light reflected from the sample. Measurements in the range $\lambda >$ 220 mμ were carried out with this apparatus.

Measurements at shorter wavelengths (to $\lambda = 175$ mμ) were carried out at room temperature in a vacuum monochromator with a concave diffraction grating of the type developed by Tonsey et al. [101]. A monochromator with a diffraction grating, of course, has a linear dispersion, which is particularly valuable for investigations in the ultraviolet part of the spectrum.

The excitation spectra at low temperatures were measured by means of an apparatus employing a standard laboratory monochromator with a fluorite prism. The arrangement of the optical part of the apparatus is shown in Fig. 12. The dispersion curve of the monochromator is drawn out considerably in the ultraviolet part of the spectrum, so that the energy distribution of the radiation from the hydrogen lamp at the exit from the monochromator falls sharply in this region.

As excitation sources we used hydrogen lamps of the DVS-200 type (developed by an electric-lamp factory). These were low-voltage arc lamps with a heated cathode and a window made of fused optical quartz, transparent to $\lambda \sim 155$ mμ. Such lamps have a high emission intensity in the ultraviolet part of the spectrum [102].

The radiation from the samples studied was recorded with the help of photoelectron multipliers of the FEU-19M and FEU-29 types. Depending on the requirements of the experiment, various light filters were placed between the photomultiplier and sample.

3. Experimental Results. Photoexcitation Spectra of the Ultraviolet Fluorescence of Alkali Iodides

As indicated earlier, excitation by hard radiation offers the possibility of revealing all forms of luminescence in each of the samples, thus facilitating the classification of existing samples by reference to their luminescence. If special conditions were not imposed while the crystals were being formed, samples were very frequently obtained with several bands of luminescence. It was interesting to study the excitation spectra of each band separately. In the case of a composite luminescence spectrum, the excitation spectra of various emission bands may overlap, so that sometimes it may even be impossible to determine how many excitation bands there are in a given spectral range.

Since the spectral separation of the emission bands is not always possible, the first problem confronting us was that of determining the excitation spectra of alkali iodides giving only one band each in their emission spectrum, i.e., in this case we did not have to consider the interaction of the emission bands.

Let us first consider the excitation spectra of the short-wave bands in the alkali iodides in question. Figures 13-15 (curves b and c) show the excitation spectra of the shortest emission bands of CsI, RbI, and KI. Two excitation spectra are given for each substance, using different samples. The excitation spectra of the emission bands in question constitute complex bands with four maxima in the case of CsI and three in the case of RbI and KI. For each given substance the position of the maxima on the excitation bands changes very little on passing from sample to sample; the intensity ratios of the bands do change, and this is apparently related to the prehistory of the crystals.

It is interesting to see how the positions of the excitation bands change on passing from light cations to heavy, and also to note how the positions of the maxima change with changing

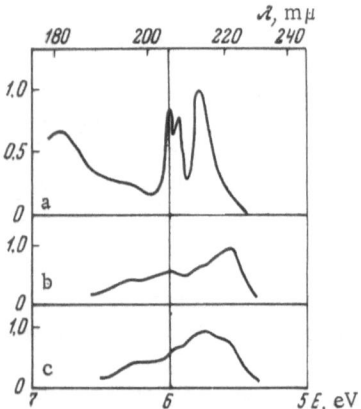

Fig. 13. Excitation spectra of the short-
wave emission band of CsI. a) Absorption
at 20°K from the results of [74]; b) excitation of CsI at 77°K; c) excitation of CsI—
Tl (crystal No. 1) at 77°K (UG-1 filter).

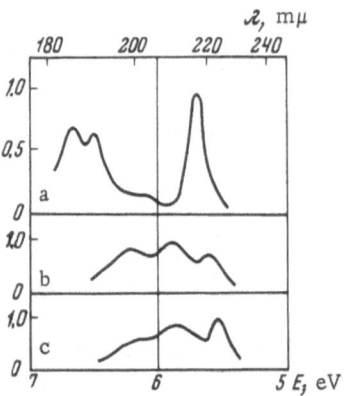

Fig. 14. Excitation spectra of the short-
wave emission band of RbI. a) Absorption
at 20°K from the results of [74]; b,c) excitation of various samples grown from solution at liquid-nitrogen temperature.

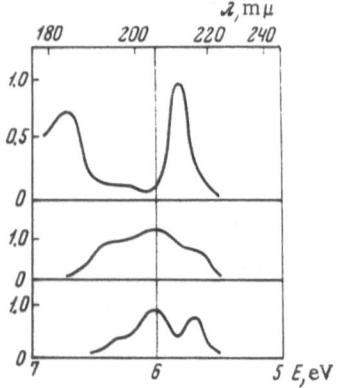

Fig. 15. Excitation spectra of the short-
wave emission band of KI. a) Absorption
at 20°K from the results of [74]; b,c) excitation of various samples of crystals
grown from solution at liquid-nitrogen
temperature.

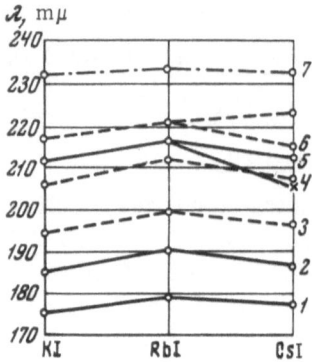

Fig. 16. Positions of the maxima on the absorption and excitation bands of various alkali
iodides. 1,2,5) Absorption at 20°K from
results of [74]; 3,4,6) excitation of the short-
wave band at liquid-nitrogen temperature;
7) excitation of the intermediate band at
liquid-nitrogen temperature.

positions of the maxima in the absorption bands. Figure 16 presents a graph of this relationship,
from which we see that the positions of the excitation bands follow those of the absorption bands
quite strictly. On passing from KI to RbI there is a uniform shift of all the absorption bands in
the long-wave direction (curves 1, 2, 5); the excitation bands also move in the long-wave direction by the same amount as the corresponding absorption bands (curves 3, 4, 6).

On passing from RbI to CsI the picture becomes more complex; the absorption spectrum
of CsI shows a new maximum at 206 mμ due to the change in crystal structure on passing from
RbI to CsI (cross in Fig. 16) while the maxima of the absorption bands move in the short-wave
direction. An analogous picture is seen on comparing the excitation spectra of CsI and RbI:

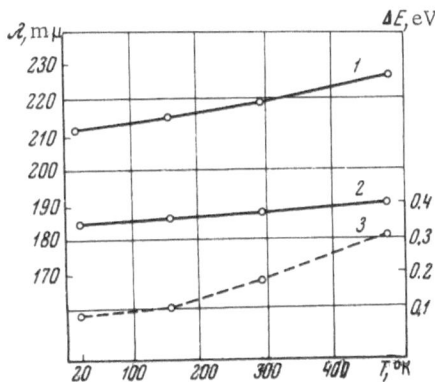

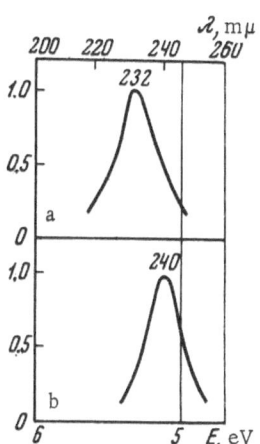

Fig. 17. Temperature dependence of the position and half-width of the absorption bands (from [74]). 1) I_a band; 2) I_b band; 3) half-width of I_a band (right-hand scale).

Fig. 18. Excitation spectra of the intermediate emission band in CsI crystals. a) At liquid-nitrogen temperature; b) at room temperature.

A new band appears with a maximum at ~214 mμ and the shortest band moves in the short-wave direction by the same amount as the I_b band in the absorption spectrum (curve 3). This comparison was made for excitation at liquid-nitrogen temperature and for absorption at 20°K. In order to show that the comparison was valid, let us consider the temperature dependence of the position of the absorption bands I_a and I_b and the half-width of the band I_a for KI (Fig. 17, taken from [74]). The axis of abscissas represents the temperature in °K and the ordinate axis the position of the maxima on the I_a and I_b bands (curves 1 and 2) and the half-width of the I_a band (curve 3, right-hand scale). The change in all these characteristics between 20 and 155°K is insignificant, of the order of 0.5-1.5 mμ for the displacement of the bands and 0.02 eV for the change in half-width. On this basis we considered it justifiable to compare the excitation spectra at liquid-nitrogen temperature (77°K) with the absorption spectra at 20°K.

Figures 13-15 show the absorption spectra of CsI, RbI, and KI (curve a) in addition to the excitation spectra. On considering these figures we see that all the excitation spectra of the ultraviolet bands lie in the region of the first absorption band; the excitation falls to zero in the region of the second absorption band in all cases considered.

4. Photoexcitation Spectra of the Intermediate

Emission Band of Alkali Iodides

Let us consider the excitation spectra of the intermediate emission band. Figures 18-20 show the corresponding spectra for CsI, RbI, and KI crystals. In all cases considered the excitation band of the luminescence in question is situated at the characteristic absorption-edge of the crystals, its position changing very little on passing from one alkali iodide to another; the half-widths of the bands also vary very little (Table 4). On raising the temperature the excitation bands move in the red direction, in the same way as the absorption spectra. Even at room temperature the relative change in the positions of the excitation bands changes little on changing the cation of the alkali iodide.

These laws may be seen very clearly from Fig. 16, in which the positions of the absorption maxima are shown graphically in conjunction with the excitation of the short-wave emission bands (curves 3, 4, 6) and the intermediate emission bands (curve 7).

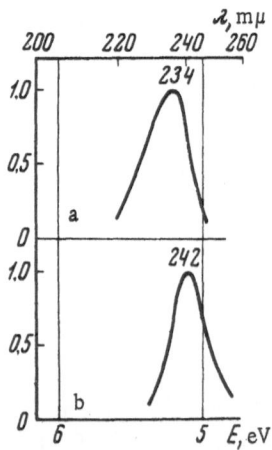

Fig. 19. Excitation spectra of the intermediate emission band in RbI crystals.
a) At liquid-nitrogen temperature; b) at
room temperature.

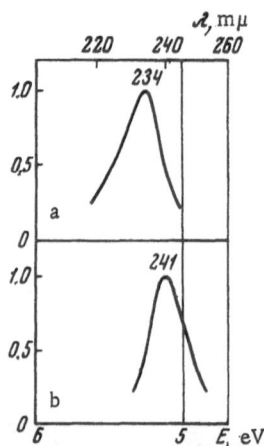

Fig. 20. Excitation spectra of the intermediate emission band in KI crystals.
a) At liquid-nitrogen temperature; b) at
room temperature.

Table 4. Positions of the Maximum and Half-Widths of the Excitation Spectrum
of the Intermediate Emission Band of Alkali Iodides

Crystal	Liquid-nitrogen temperature			Room temperature		
	first absorp-tion max., mμ	excitation of intermed. emission band, mμ	half-width	first absorp-tion max., mμ	excitation of intermed. emission band, mμ	half-width
CsI	212	232	0.34	218	240	0.30
RbI	216	234	0.35	223	242	0.30
KI	212	232	0.32	219	241	0.30

5. Photoexcitation Spectra of Alkali Iodides

Subjected to Heat Treatment

As indicated earlier, emission of the intermediate band may be induced in crystals not
formerly possessing this kind of luminescence if these are subjected to special heat treatment.
In CsI crystals the ultraviolet and intermediate emission bands are spectrally well separated,
so that the appearance of the intermediate band in the course of heat treatment of the crystal
may easily be detected by reference to the emission spectrum of the sample. After 54 h heating
and subjection to rapid quenching, the crystals only gave one intermediate band with $\lambda_{max} \sim$
400 mμ in the emission spectrum. In order to verify that this luminescence did in fact correspond to the intermediate emission band, we measured its excitation spectrum. The excitation
spectrum of the emission thus arising had a maximum at about 240 mμ at room temperature,
i.e., it coincided with the excitation spectrum of the intermediate band of CsI. The half-width
of the excitation band of the intermediate luminescence of the heated CsI crystal was 0.50 eV
and that of a CsI crystal not subjected to heat treatment, 0.30 eV.

When the other alkali iodides (RbI and KI) are subjected to heat treatment, the appearance
of the intermediate band in the emission spectra of samples not formerly containing it is very
difficult to detect, since it lies very close to the ultraviolet band, and the broadening of the latter

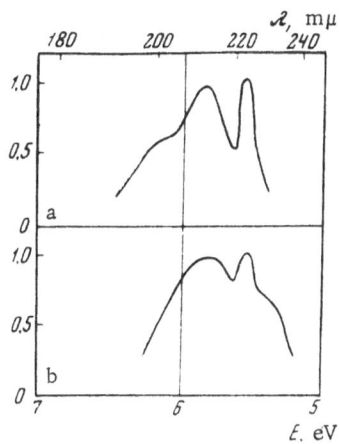

Fig. 21. Excitation spectra of RbI crystals grown from solution at liquid-nitrogen temperature. a) Before heat treatment; b) after 25-h heating at about 500°C and subsequent quenching.

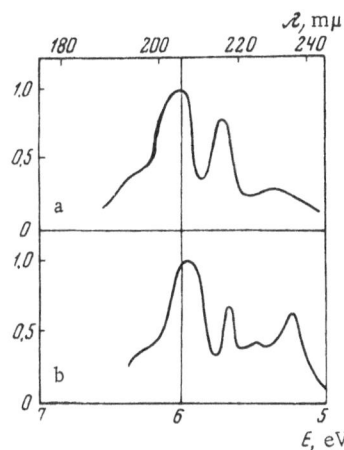

Fig. 22. Excitation spectra of KI crystals grown from solution at liquid-nitrogen temperature. a) Before heat treatment; b) after 7-h heating at about 500°C and subsequent quenching.

after heat treatment may be caused either by the appearance of the second emission band or by damage to the crystal structure. A study of the excitation spectra of heat-treated samples, however, showed that an intermediate band appeared in the luminescence spectra of the other alkali iodides in the same way as in CsI.

Thus, after heating an RbI crystal not having an intermediate band for 25 h in a sealed quartz ampoule and then rapidly cooling (Fig. 21a), a maximum appeared on the falling part of the excitation spectrum of the ultraviolet emission band; this maximum corresponded to the excitation of the intermediate emission band (Fig. 21b). As we should expect, the excitation spectrum of the ultraviolet band became diffuse as a result of the heat treatment and its maxima tended to merge.

If, however, the intermediate band was already present in the emission spectrum of the crystal, heat treatment increased its intensity; this may be seen more clearly for the case of KI (Fig. 22), since KI was more sensitive to heat treatment. Thus, heating the samples for 7 h at 500°C raised the intensity of the intermediate band by a factor of 2-3, and its intensity in the excitation spectrum reached a value half that of the ultraviolet-band excitation maximum. Heating RbI for the same time under the same conditions produced no marked change in the luminescence of the crystals, only a 25-h heating period having any appreciable effect on the excitation spectrum.

An explanation for this phenomenon may be found on considering the energy of the cohesive forces of the substances studied. Thus the energy of the cohesive forces of ionic crystals, referred to their original state in the form of the monatomic gaseous components, is as follows [103]: for NaI, 120.8 kcal/mole; for KI, 124.3 kcal/mole; for RbI, 125.5 kcal/mole; and for CsI, 128.3 kcal/mole. The force of cohesion between the ions forming the KI lattice is the smallest of all the alkali iodides studied in the present investigation, and this correspondingly facilitates the formation of structural defects. This evidently explains the fact that only in the case of KI were crystals having an intermediate band in the emission spectrum obtained from solution.

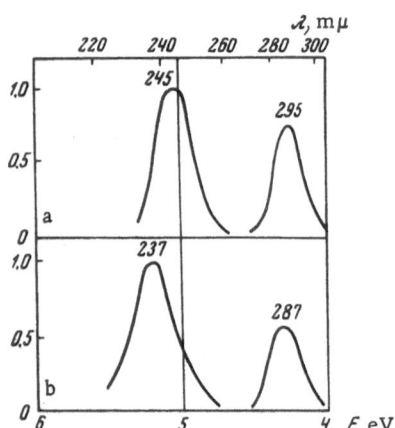

Fig. 23. Excitation spectra of the activator band of RbI—Tl phosphor. a) At room temperature; b) at liquid-nitrogen temperature.

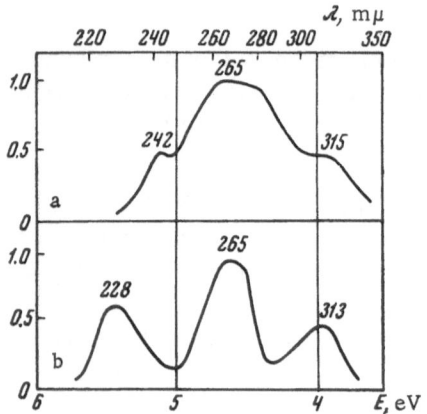

Fig. 24. Excitation spectra of the activator luminescence of RbI—In phosphor. a) At room temperature; b) at liquid-nitrogen temperature.

6. Photoexcitation of the Activator Fluorescence of Phosphors Based on Alkali Iodides

For large quantities of activator in the crystals, only the activator luminescence occurs in the emission spectrum; the excitation of this coincides with the position of the activator absorption bands. Figures 23 and 24 show the excitation spectra of Tl and In in RbI. The excitation spectrum of Tl consists of two separated bands with $\lambda_{max} \sim 245$ and 295 mμ, respectively; this agrees with the results of other authors [73, 104]. The excitation in the long-wave band is smaller than in the short-wave band; this difference increases slightly with falling temperature. As temperature falls, the shape of the bands changes little, but there is a slight increase in the intensity of the short-wave band at the expense of the long-wave; both bands move together in the short-wave direction.

On considering the excitation spectra of RbI—In, however, the picture becomes more complex. At room temperature the spectrum consists of three bands with $\lambda_{max} \sim 242$, 265, and 315 mμ, which agrees closely with the absorption of the activator In in RbI as indicated in [76]. Of the three maxima observed in the excitation spectrum at room temperature, only the shortest undergoes any marked change in position as temperatures fall further. The long-wave band moves very little: from 315 mμ at room temperature to 313 mμ at the temperature of liquid nitrogen. It is hard to judge the change in the position of the middle excitation band, since this is very diffuse at room temperature. Although the bands become narrower and are well resolved with falling temperature, the whole excitation spectrum is extended owing to the different ways in which temperature affects the position of the individual bands.

We thought it interesting to study the excitation spectra of the activator luminescence in crystals also possessing other forms of luminescence. For this purpose we studied CsI—Tl and RbI—In crystals. Figure 25 shows two CsI—Tl excitation spectra: the excitation of the ultraviolet emission band and the excitation of the activator luminescence for the same crystal. The whole band of activator excitation lies more to the long-wave side than the excitation of the ultraviolet band, while in the region corresponding to the onset of self-absorption in the crystal, in which the ultraviolet band is excited, the activator excitation diminishes; no marked luminescence is observed on excitation in this region.

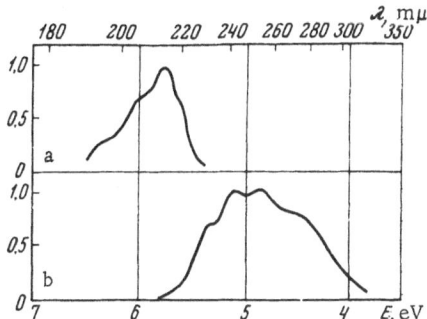

Fig. 25. Excitation spectra of various bands of CsI—Tl phosphor (crystal No. 1) at liquid-nitrogen temperature. a) Excitation of the short-wave band (UG-4 filter); b) activator excitation (OS-13 filter).

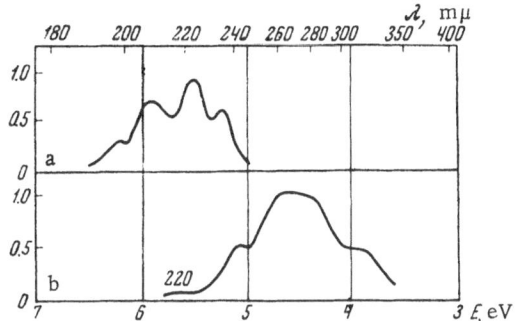

Fig. 26. Excitation spectra of various emission bands of RbI—In phosphor at room temperature. a) Excitation of the short-wave band (FS-7 filter); b) excitation of the activator luminescence (ZhS-18 filter).

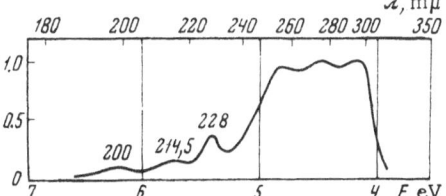

Fig. 27. Excitation spectrum of the activator luminescence in a fresh cleavage of CsI—Tl phosphor (crystal No. 1) at room temperature, measured in a monochromator with a diffraction grating.

An analogous picture is obtained for RbI—In (Fig. 26). The activator excitation also lies to the long-wave side of the excitation band of the ultraviolet radiation. The excitation falls on moving in the short-wave direction, but fails to reach zero; in the cases studied one very weak excitation band still remains. The intensity of the excitation at the maximum of this band, which lies at about 220 mμ, is 2-5% of the excitation in the activator band.

We made an attempt at discovering an analogous band in the excitation spectrum of Tl in CsI—Tl. We took a fresh cleavage (roughly through the middle of the sample) from a CsI—Tl crystal with an activator concentration of 0.0011 mole Tl/mole CsI and immediately measured the excitation in a vacuum monochromator with a diffraction grating. In this way we were able to observe activator excitation bands with $\lambda \sim$ 228, 215, and 200 mμ (Fig. 27). The excitation in the 215-mμ band maximum was 10-15%, and that in the 200-mμ band 5-10% of that at the maximum of the activator band. We were unable to observe such excitation in crystals after long storage; this was apparently because of changes taking place in the surface layer of the crystals.

CONCENTRATION DEPENDENCE OF THE TIME CHARACTERISTICS OF THE PHOSPHORS CsI-Tl AND KI-Tl

1. Attenuation Time of CsI—Tl and KI—Tl Phosphors as a Function of Activator Concentration

As already indicated in the introduction, the scintillation characteristics of alkali iodides have been studied by a number of authors as functions of activator concentration, temperature, exciting-particle energy, and exciting-particle ionization density. The attenuation time of the scintillations in NaI—Tl was studied in [29] and found to vary in a nonmonotonic manner; the growth time was also determined for this phosphor.

We studied the duration of the scintillations and their growth times for two series of phosphors: CsI—Tl and KI—Tl [105, 106]. The activator concentration in the first series varied from $0.14 \cdot 10^{-4}$ to $15.75 \cdot 10^{-4}$ mole Tl/mole CsI and in the second from $0.70 \cdot 10^{-4}$ to $5.45 \cdot 10^{-4}$ mole Tl/mole KI (determined polarographically by the method described in [9]).

The attenuation of the scintillations was studied by the single oscillograph-record method developed by Plyavin' [24]. The crystals were excited by a Co^{60} γ-source. The pulse repetition frequency could be varied over a wide range by varying the distance between the source and sample. The system is shown schematically in Fig. 28.

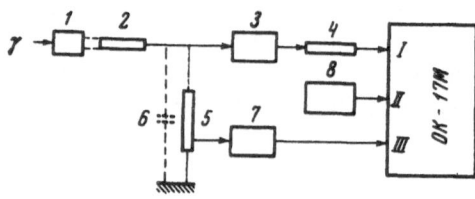

Fig. 28. Arrangement of apparatus for studying the attenuation of scintillations by a single oscillograph record. 1) Crystal; 2) photomultiplier; 3) preamplifier; 4) delay line; 5) load resistance R; 6) parasitic capacity C; 7) amplifier; 8) standard-signal generator GSS-6. I) First oscillograph input; II) second oscillograph input; III) sweep trigger input.

The γ quantum of exciting radiation falls on the crystal, which is placed in a chamber impenetrable to light; the light from the sample scintillations falls on a photomultiplier. We used photomultipliers of the FEU-29 type designed for spectrometric use. The signals taken from the resistance of the photomultiplier through a preamplifier and delay line passed to the first-beam input of an OK-17M oscillograph, the amplifier of which had a uniform frequency characteristic up to 10 Mc/sec. As delay line we used an RK-3 coaxial cable 700 m long, which gave a delay of about 3.5 μsec. The preamplifier, operating at maximum amplification (50 times), also had a uniform frequency characteristic up to 10 Mc/sec. The output of the preamplifier was matched with the wave impedance of the RK-3 cable.

The pulses for triggering the sweep were taken from a small section of the load resistance of the photomultiplier and after passing through a linear pulse amplifier reached the sweep-trigger input of the OK-17M oscillograph.

A sinusoidal voltage from a GSS-6 standard signal generator was applied to the second beam input of the oscillograph in order to give time calibration.

To an accuracy limited by the spread of the time of flight of the electrons in the photomultiplier ($\sim 10^{-8}$ sec), the current in the photomultiplier reproduces the shape of the light pulse in the crystal; however, the voltage taken from the load resistance of the photomultiplier is affected by parasitic capacities. If the light scintillation in the crystal follows an exponential law with attenuation time τ, the voltage on the load resistance U_R is according to [107] determined by the relation

$$U_R = \frac{i_0 R \tau}{1 - RC} (e^{-\frac{t}{\tau}} - e^{-\frac{t}{RC}}), \qquad (4)$$

where i_0 is the maximum current in the pulse, τ is the attenuation time of the light pulse, R is the load resistance of the photomultiplier, and C is the parasitic capacity.

The load resistance was chosen in such a way that the condition

$$RC \ll \tau \qquad (5)$$

was satisfied. For this relationship between RC and τ the attenuation of the luminescence pulse will be accurately represented by the attenuation of the voltage pulse taken from the photomultiplier. On the other hand, considering that the intensity of the light pulse was very low, the value of R was taken as high as possible, subject to satisfying condition (5). Under the conditions of the experiment RC was kept constant at 10^{-8} sec.

In order to determine the apparatus error we studied the scintillation of tolan, the attenuation time of which ($\tau \sim 10^{-8}$ sec) may be neglected [108], and found that the spread of the pulse on the oscillogram may be regarded as being simply due to the apparatus employed.

On excitation by γ rays, it was inevitable that there should be considerable statistical fluctuations in the time characteristics of the light pulse. These fluctuations, or light noise, were due to the fact that the intensity of the luminescence in an individual pulse was so small as to yield a comparatively small number of photoelectrons from the photocathode of the photomultiplier during the resolving time of the apparatus. The inevitable statistical fluctuations in this number of electrons, or, what is the same thing, the quantum fluctuations in the number of photons, tends to disrupt the smooth run of the pulse. Plyavin' [24] found that for an $\sim 10\%$ yield of γ luminescence (favorable case), a luminescence pulse length of $\sim 3 \cdot 10^{-7}$ sec, a photocathode efficiency of 10%, and a light-gathering factor of 10%, the statistical fluctuations in the number of photoelectrons leaving the photocathode of the photomultiplier during a period of $2 \cdot 10^{-8}$ sec, equal to the resolving time of the apparatus, were about 20%. Hence, in determining the time characteristics of the scintillations, averaging was carried out first over each pulse individually and then over all pulses in the series of experiments relating to a given crystal (15-20 photographs). Figures 29 and 30 present photographs of typical oscillograms of KI—Tl and CsI—Tl scintillations. The averaged pulses were corrected graphically by reference to formula (4). Figure 31 presents such corrected pulses from the scintillations of CsI—Tl (crystal No. 7) and KI—Tl (crystal No. 3) phosphors.

The experimental apparatus described enabled us to study the attenuation of photoscintillations over a 20- to 50-times fall in intensity. Within these limits the graph of log (luminescence intensity) versus scintillation attenuation time is a straight line, which indicates that the fall in

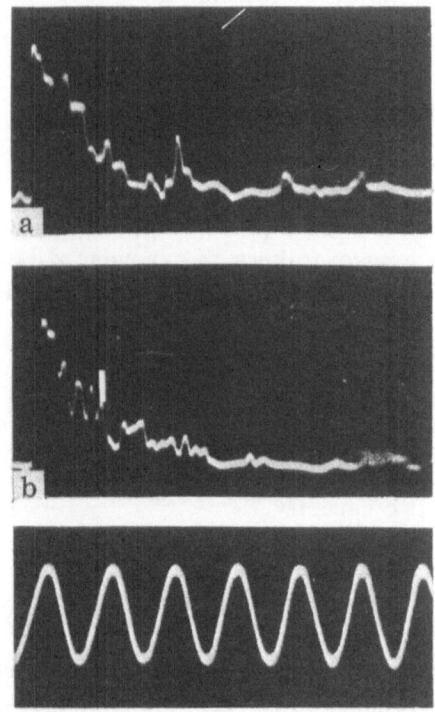

Fig. 29. Oscillograms of the scintillations of KI—Tl crystals with activator
(nominal) concentrations: a) 0.01 wt.%;
b) 3.0 wt.%. Lower curve represents
the timing scale (signal from a GSS-6,
frequency 2 Mc/sec).

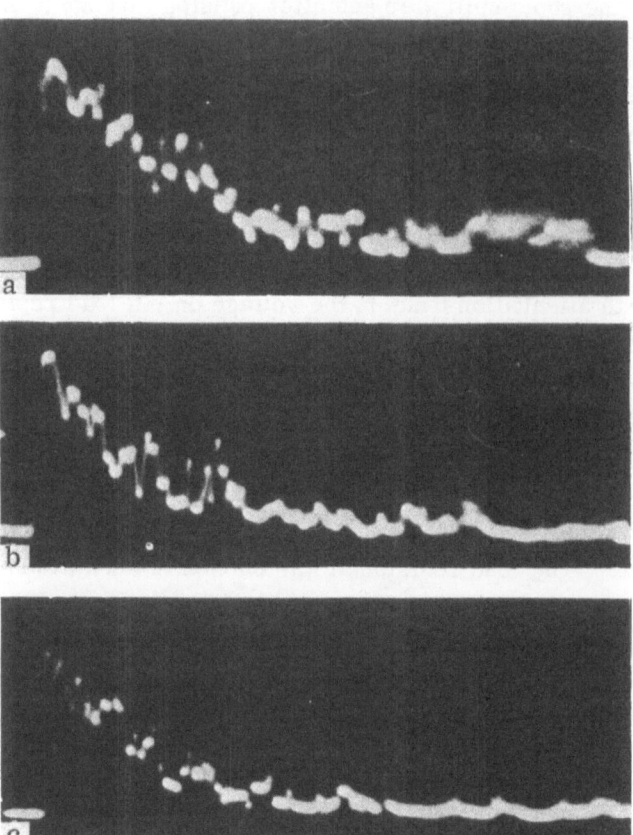

Fig. 30. Oscillograms of the scintillations of
CsI—Tl crystals with activator concentrations:
a) 0.001 mol.%; b) 0.048 mol.%; c) 0.124 mol.%.
Sweep as in Fig. 29.

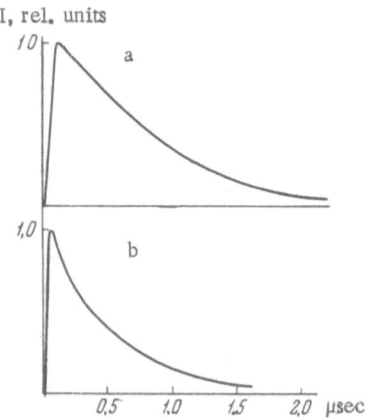

Fig. 31. Averaged and corrected [by
formula (4)] scintillations of the phosphors: a) CsI—Tl (crystal No. 7); b)
KI—Tl (crystal No. 3).

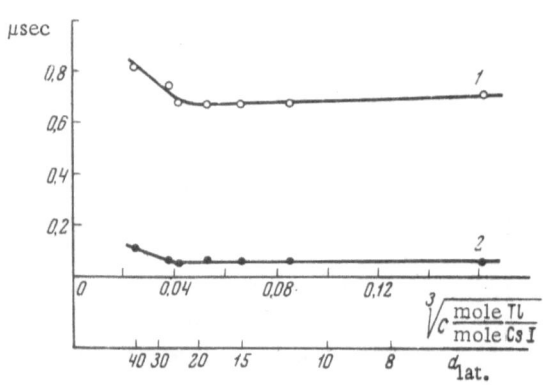

Fig. 32. Concentration dependence of the time
characteristics of the scintillations of CsI—Tl
phosphor. 1) Attenuation time of the scintillations; 2) growth time of the scintillations.

Table 5. Attenuation and Growth Time Constants of CsI—Tl Phosphors

Crystal No.	C, $\dfrac{\text{mole Tl}}{\text{mole CsI}} \cdot 10^4$	$\sqrt[3]{C\,\dfrac{\text{mole Tl}}{\text{mole CsI}}}$	Mean distance between Tl ions, Å	τ att. μsec	t_{gr}, μsec
1	0.140	0.0242	191	0.810	0.11±0.01
2	0.584	0.0387	117	0.736	0.06
3	0.726	0.0416	108	0.672	0.05
4	1.56	0.0537	85	0.666	0.07
5	2.88	0.0660	69	0.670	0.065
6	6.15	0.0850	54	0.678	0.06
7	15.75	0.1610	27	0.710	0.06

Table 6. Attenuation and Growth Time Constants of KI—Tl Phosphors

Crystal No.	Charge Tl conc., %	Tl conc. by polarography [9], g Tl/g KI	Tl ions per cm^3, $1/\text{cm}^3$	$C\,\dfrac{\text{mole Tl}}{\text{mole KI}}$	$\sqrt[3]{C\,\dfrac{\text{mole Tl}}{\text{mole KI}}}$	τ att, μsec	t_{gr}, μsec
1	0.1	$0.86 \cdot 10^{-4}$	$8.0 \cdot 10^{17}$	$7.0 \cdot 10^{-5}$	0.041	0.49±0.01	0.025±0.010
2	0.2	$1.5 \cdot 10^{-4}$	$1.4 \cdot 10^{18}$	$1.22 \cdot 10^{-4}$	0.049	0.48	0.015
3	1.0	$3.0 \cdot 10^{-4}$	$2.8 \cdot 10^{18}$	$2.44 \cdot 10^{-4}$	0.062	0.45	0.030
4	2.0	$6.7 \cdot 10^{-4}$	$6.3 \cdot 10^{18}$	$5.45 \cdot 10^{-4}$	0.082	0.44	0.020

luminescence of these phosphors obeys an exponential kinetic law and may therefore be characterized by an attenuation time τ.

The results obtained by this method for the scintillations of CsI—Tl and KI—Tl phosphors are given in Tables 5 and 6 and Figs. 32 and 33. As the activator concentration in the series of CsI—Tl phosphors increases, the attenuation time first falls to a certain minimum value, and then starts rising again, though only a little. Similar behavior was observed in the case of NaI—Tl by Eby and Jentschke [29] (Fig. 34). In the case of KI—Tl there was no bend on the concentration-dependence curve.

This kind of concentration dependence may be explained from the following considerations. The attenuation time of the scintillations is characterized by the thermal release of the activator from fine metastable levels. A number of authors [22, 29, 32, 109] consider that there are at least two excited levels for the Tl ion, of which one is nonradiating; the probabilities of a transition between these levels may change on changing the activator concentration owing to the interaction of the activator ions with each other and with the lattice of the base. This evidently explains the different behavior of the concentration curves of the attenuation time of CsI—Tl and KI—Tl phosphors: In the case of CsI—Tl the attenuation time first falls to a minimum and then rises, while in KI—Tl the attenuation time of the scintillations falls continuously with increasing activator concentration.

We see from Fig. 32 that as the activator concentration varies not only the attenuation time but also the growth time of the scintillations does likewise; however, the variation of the latter lies within the limits of sensitivity of the apparatus. In view of the fact that this experimental result appeared very interesting, we made a special further study of the scintillation growth times of the concentration series of CsI—Tl and KI—Tl in question.

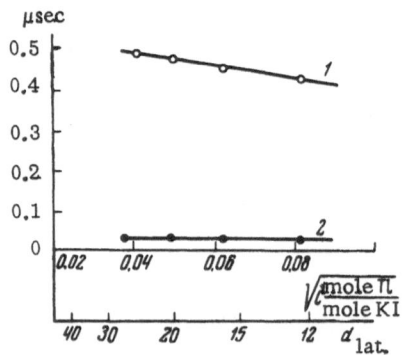

Fig. 33. Concentration dependence of the time characteristics of the scintillations of KI—Tl phosphor. 1) Scintillation attenuation time; 2) scintillation growth time.

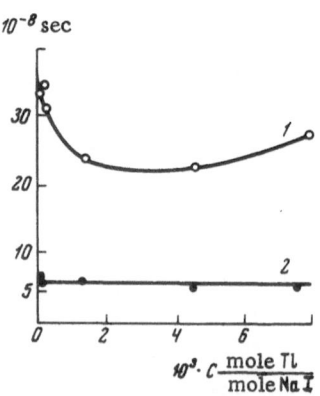

Fig. 34. Concentration dependence of the time characteristics of the scintillations of NaI—Tl phosphor [29]. 1) Scintillation attenuation time; 2) scintillation growth time.

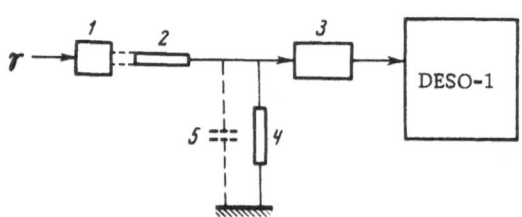

Fig. 35. Arrangement of apparatus for studying the growth time of the scintillations. 1) Sample crystal; 2) photomultiplier; 3) wide-band amplifier IM-1; 4) load resistance of photomultiplier R; 5) parasitic capacity C.

2. Concentration Dependence of the Growth Time of Scintillations in a Number of CsI—Tl and KI—Tl Phosphors

In order to study the relation between the scintillation ignition time and the activator concentration in the phosphor, we used a DESO-1 oscillograph having a uniform frequency characteristic in the range 60 cps to 60 Mc/sec; this enabled us to measure the growth times of the phosphors very accurately. The growth times were of the order of 0.06-0.8 μsec for CsI—Tl and 0.02 μsec for KI—Tl.

The arrangement of the apparatus employed is shown schematically in Fig. 35. A γ quantum from the radioactive Co^{60} source falls on the sample crystal 1, situated in a lightproof chamber; the light from the scintillations of the sample is recorded with an FEU-33 photomultiplier; the voltage from the load resistance of the photomultiplier 4 passes to the input of the DESO-1 oscillograph either directly or through a wide-band IM-1 amplifier 3, having a uniform frequency characteristic up to 60 Mc/sec. As in the case of the attenuation time, we studied the growth time of the scintillations at room temperature. As in the previous case, the apparatus error was determined by studying the scintillations of tolan.

When studying CsI—Tl phosphors with the lowest activator concentration (sample No. 1), we found two components, a short-term and a long-term one, in the emission. These components could be separated with light filters. The short-term component was easily separated with a UFS-1 light filter. However, nothing could be said as to the corresponding attenuation or growth times, since these were of the same order as the photomultiplier noise. We consider that this component was associated with the emission of the ultraviolet band of pure CsI. Enz and Rossel [31] studied the temperature dependence of the cold component of CsI corresponding to the emission of the ultraviolet band. The duration of this component fell very sharply with increasing temperature, and if the corresponding curve is extrapolated to room temperature then the attenuation time should not exceed 10^{-8} sec; this in no way contradicts our own data.

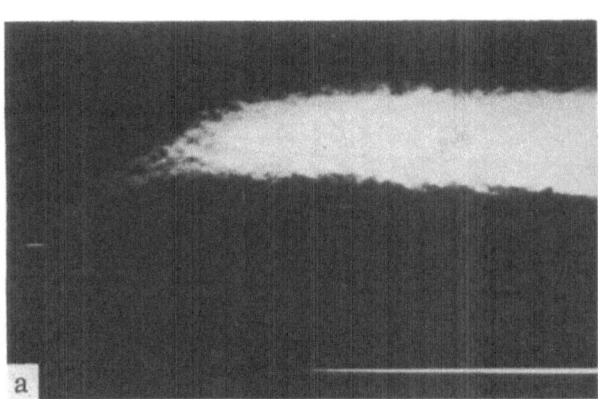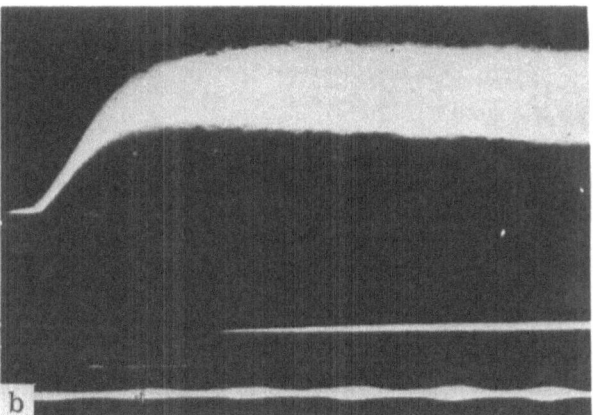

Fig. 36. Oscillograms of the scintillations of activator emission from CsI—Tl crystals. a) Crystal No. 1 ($C_{Tl} = 1.4 \cdot 10^{-5}$ mole Tl/mole CsI); b) crystal No. 2 ($C_{Tl} = 5.84 \cdot 10^{-5}$ mole Tl/mole CsI). Timing marks 0.05 μsec.

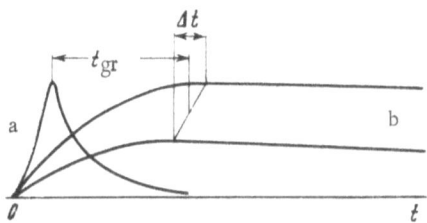

Fig. 37. Diagram to explain the method of determining the growth time of the CsI—Tl scintillations. a) Scintillation of tolan; b) scintillations of CsI—Tl; Δt) error in determining the maximum point in the growth of the CsI—Tl scintillations owing to the fluctuating background; t_{gr}) growth time of CsI—Tl scintillations, measured as the difference between the times required to reach the maxima of the CsI—Tl and tolan scintillations, respectively.

The long component of CsI—Tl scintillation corresponds to the emission of Tl and is easily separated with an OS-12 light filter. Oscillograms of the scintillations of CsI—Tl are shown in Fig. 36.

In order to eliminate the apparatus error, the scintillations of the sample crystals were compared with those of tolan. It was considered that the growth time of the tolan scintillations was simply due to the apparatus effect. The growth time of the sample scintillations was then given by the period elapsing between the moment at which the tolan scintillations (identical in intensity with those of the sample) reached their maximum and the moment at which the sample phosphor did likewise (Fig. 37). The error in determining the growth time of the phosphor scintillation was mainly due to the fact that the scintillation maximum was very shallow, which made it more difficult to estimate the moment at which it had been reached. The error due to the diffuseness of the tolan maximum may be neglected, as it was very small compared with the error committed in determining the maximum of the sample scintillation.

Some measured scintillation growth times obtained for a CsI—Tl concentration series are presented in Fig. 38. The scintillation growth time diminishes with increasing concentration of the activator, and after reaching a minimum value of 0.06 μsec thereafter remains unaltered with further increase in the Tl concentration. A break in the curve relating the time characteristic to the activator concentration occurs at $0.8 \cdot 10^{-4}$ mole Tl/mole CsI (or $\sqrt[3]{C \text{ mole Tl/mole}}$ $\overline{CsI} = 0.0416$), which corresponds to a mean distance between the activator ions of ~108 Å (see Table 5). For this concentration, as indicated in Chapter II, both forms of unactivated luminescence in CsI vanish.

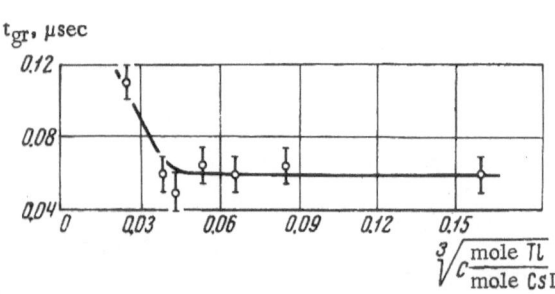

Fig. 38. Concentration dependence of the growth time of a CsI—Tl phosphor.

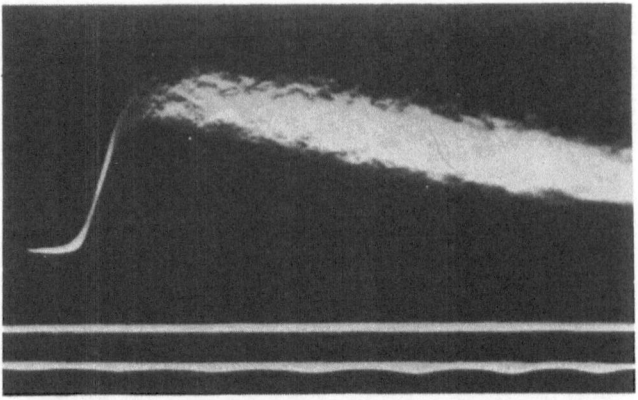

Fig. 39. Oscillogram of the scintillations of a KI—Tl crystal (0.01 wt.%). Timing marks 0.05 μsec.

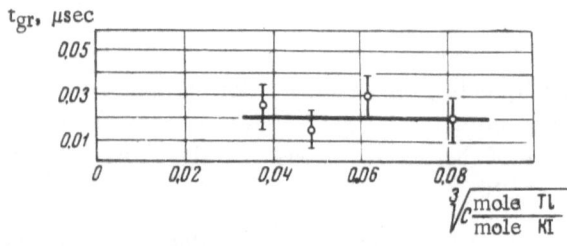

Fig. 40. Concentration dependence of the growth time of a KI—Tl phosphor.

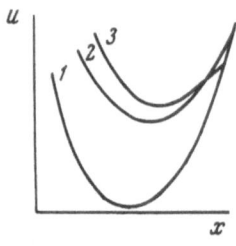

Fig. 41. Potential curves of a thallium center [24]. 1) Ground state; 2,3) excited states.

In the case of KI—Tl, the growth time of the scintillations is very short (Fig. 39), the relative errors in its determination are large, and no variation in their concentration dependence can be detected (Fig. 40). [The figure shows the dependence of the growth time on $\sqrt[3]{C}$ (mole Tl/mole KI); the corresponding Tl concentrations are given in Table 6.]

When studying the temperature dependence of the scintillation characteristics of alkali-halide phosphors activated with Tl, Plyavin' also observed a specific growth time of the CsI—Tl scintillations, increasing with falling temperature.

A study of the temperature dependence of τ and the use of the composite model of the thallium center given in [109-111] suggested that, in the excited state, the Tl ion had two closely contiguous levels, the distance between these being a few hundredths of an electron volt; moreover the level 2 was metastable (Fig. 41). The activator ions were able to become thermally redistributed between the levels 2 and 3 during the lifetime of the excited state. The attenuation of the scintillations corresponds to the thermal transition 2 → 3, while the lifetime in state 3 should characterize the growth time of the scintillations. Usually, one considers that the time for the energy to be transferred from the base in which the γ quantum is absorbed to the activator is much shorter than the lifetime of the activator in the excited state, and therefore has no effect on the measured scintillation growth time. It is logical to consider, however, that the energy-transfer time will differ for different distances between the activator ions, being larger for longer distances. As soon as the transfer time becomes comparable with the activator lifetime in the excited state (this should occur for a small activator content), it begins introducing its own contribution into the scintillation growth time.

Thus we consider that the increase in the luminescence growth time of Tl with falling Tl concentration in CsI may be explained by the fact that the γ-quantum energy absorbed in the lattice is transferred to the activator, not instantaneously, but after a finite time dt depending on the Tl concentration. (For $C_{Tl} = 0.14 \cdot 10^{-4}$ mole Tl/mole CsI, dt = 0.05 μsec.) The longer the period dt, the greater will be the probability of a transfer of energy to a structural center of the CsI (giving blue luminescence) or else its conversion into heat; at low temperatures there will be a greater probability of the development of ultraviolet luminescence.

The absence of any break on the concentration curve of the scintillation growth time for KI—Tl may be explained by the fact that in our sample crystals the Tl concentration was such that dt had already become too small and could not be observed experimentally.

Moreover, according to [112, 113], the Tl concentration in KI for which luminescence due to exciton annihilation might be observed should not exceed 10^{15} Tl ions/cm^3, which corresponds to $0.88 \cdot 10^{-7}$ mole Tl/mole KI. In our own case, the Tl concentration in the samples exceeded this value by more than two orders (see Table 6).

CHAPTER V

DISCUSSION OF RESULTS

1. Discussion of the Nature of the Fluorescence in the Ultraviolet Emission Band of Pure Alkali Iodides

As a result of the foregoing investigation we have established that the luminescence of pure alkali iodides contains two emission bands with different characteristics, suggesting the existence of two independent luminescence mechanisms.

Let us first discuss the characteristics obtained for the ultraviolet emission band. The maximum of this band moves monotonically in the long-wave direction on passing to heavier base cations, i.e., from NaI to KI and RbI, occupying positions of 303, 370, and 400 mμ, respectively, at liquid-nitrogen temperature. On passing to CsI, the band maximum moves sharply in the short-wave direction, apparently as a result of a change in crystal structure; all the alkali iodides have a fcc lattice except for CsI, which has a bcc structure.

In all cases the ultraviolet emission band occurs mainly at low temperatures and the intensity of its luminescence falls sharply with increasing temperature, diminishing by a factor of several tens before reaching room temperature in the case of NaI and CsI, or no longer appearing at all in the case of RbI.

The presence of foreign impurities or structural defects has a bad effect on the ultraviolet emission band of alkali iodides.

Our own results for CsI and KI, and those of [38] for NaI, show that a Tl concentration of the order of $2 \cdot 10^{-4}$ mole Tl/mole CsI in CsI—Tl and 10^{-5} mole Tl/mole NaI in NaI—Tl, and an In concentration of the order of $1.5 \cdot 10^{-5}$ mole In/mole KI in KI—In completely eliminates the unactivated emission.

Unactivated emission appears not only on excitation by γ rays but also on photoexcitation, when light is absorbed in the region of the first absorption maximum of the alkali iodide; on reducing the wavelength of the exciting light the emission intensity fell, and we were quite unable to observe any marked luminescence in the region of the second absorption maximum.

On studying the thermoluminescence of alkali iodides in the ultraviolet emission band, we found no trace of appreciable light storage, i.e., the luminescence was not of a recombination character. This conclusion is supported by the fact that the attenuation in this band obeys an exponential law [38]. Data relating to the luminescence of pure alkali iodides are given in Table 7.

Luminescence of pure alkali iodides analogous to that found in our own experiments was later observed by a number of authors; thus, when studying NaI activated with traces of Tl, Vishnevskii et al. [114] found that the long-wave radiation was accompanied by a short-wave

Table 7. Data Relating to the Luminescence of Pure Alkali Iodides

Band	Characteristics of the emission	CsI	RbI	KI	NaI***
Exciton (short wave)	Emission λ_m, mμ	340** 320**	400	370	300
	Absorption* λ_m, mμ	183, 208, 212	190, 201, 216	185, 199, 213	184, 213, 221
	Excitation λ_m, mμ	196, 206, 214, 222	199, 211, 221	194, 206, 217	Shorter than 225
	Temperature, °C	$\sim(-196)$ ~ 20	$\sim(-196)$	$\sim(-196)$	$\sim(-196)$
	Quenching conc. of activator	$C_{Tl} \sim 10^{-4} \dfrac{\text{mole Tl}}{\text{mole CsI}}$	$C_{In} > 10^{-5} \dfrac{\text{mole In}}{\text{mole RbI}}$	$C_{In} \sim 10^{-5} \dfrac{\text{mole In}}{\text{mole KI}}$	$C_{Tl} \sim 10^{-5} \dfrac{\text{mole Tl}}{\text{mole NaI}}$
	Attenuation	Exponential	–	Exponential	Exponential****
	Stored light	Extremely little	Practically no storage	Not observed	Not observed
Structural (intermed.)	Emission λ_m, mμ	405**	404	410	410
	Excitation λ_m, mμ	232 240	234 242	234 241	–
	Temperature, °C	$\sim(-196)$ ~ 20	$\sim(-196)$ ~ 20	$\sim(-196)$ ~ 20	–
	Quenching conc. of activator	$C_{Ti} \sim 10^{-4} \dfrac{\text{mole Tl}}{\text{mole CsI}}$	$C_{In} > 10^{-5} \dfrac{\text{mole In}}{\text{mole RbI}}$	$C_{In} \sim 10^{-5} \dfrac{\text{mole In}}{\text{mole KI}}$	$C_{Tl} > 10^{-5} \dfrac{\text{mole Tl}}{\text{mole NaI}}$
	Attenuation	–	–	–	Complex law
	Stored light	Stores	Stores	Stores	Stores

*At 20°K [74].
**From [42].
***From [21, 22, 29].
****From [31].

band with $\lambda_{max} \sim 302$ mμ. In [115, 116] (x-ray fluorescence of KI at liquid-nitrogen temperature) luminescence with an emission maximum at about 376 mμ was observed; in [91, 93, 117, 118] (mixed KCl—KI salts) an emission band of KI at about 363 mμ was also found. Japanese authors [119], studying KBr—KI mixed crystals, also found a KI emission band, with a maximum occupying various positions between 360 and 400 mμ for different samples. The ultraviolet emission band also occurs in the luminescence of CsI crystals [120] purified by zone melting, with an emission maximum at about 350 mμ. An ultraviolet emission band of CsI with a maximum $\lambda_{max} \sim 330$ mμ was also found in [33, 121, 122].

Although there can no longer be any doubt regarding the luminescence of pure alkali iodides, there has not yet been any single, unified explanation of the nature of this luminescence. Thus some authors consider that the luminescence of pure salts is due to their activation by uncontrollable impurities [33]. Other authors associate the luminescence of pure salts simply with their own crystal lattice, while some attribute the ultraviolet band of luminescence to the direct annihilation of an exciton captured as a result of the polarization of the crystal [38], and others associate the ultraviolet band with the recombination of an electron and a self-captured [121] or localized [122] hole. However, the authors of [122] do not deny the possibility of the radiative annihilation of an exciton. It is pointed out in [43] that a localized exciton may radiate on annihilation.

We consider that the ultraviolet emission band is associated with the annihilation of an anion exciton. This point of view is supported by a whole series of investigations [93, 117-119] on mixed KCl—KI and KBr—KI crystals, in which the ultraviolet emission band of KI appears in the luminescent spectrum. In these crystals the migration of energy from one iodine ion to another is impossible, since the I$^-$ is not in resonance with its surroundings, although its nearest neighbors in the mixed crystals are the same as in KI. If an exciton localized at an I$^-$ emits in the mixed crystal, then of course an exciton in KI itself will also have a radiation configuration, appearing in both absorption and luminescence. Since the excitation of the ultraviolet band which we observed in all alkali iodides occurs in the region of the first absorption maximum, which is ascribed to the anion exciton, we consider it most reasonable to associate this band with the annihilation of the anion exciton.

If in fact the ultraviolet emission of the alkali iodides is associated with the annihilation of excitons, then the law governing the attenuation of its luminescence should have an exponential character, and there should be no light storage in its emission bands. These characteristics are in fact observed experimentally, and this in our view confirms the foregoing theory; the ultraviolet emission band is in fact associated with the annihilation of the anion exciton, and we shall therefore call it the "exciton emission band" of the alkali iodides.

From the point of view of the mechanism under consideration, the luminescence of the ultraviolet band may be easily explained by the fact that an increase in the foreign impurities and structural defects in the crystal reduces the intensity of luminescence in the ultraviolet band. On the one hand, an increase in foreign impurities and any disruption of the crystal lattice interferes with the creation of excitons in the crystal, and on the other there is a competition between the radiative annihilation of excitons and the capture of these by impurity centers and structural defects.

2. Discussion of the Nature of the Fluorescence
in the Intermediate Emission Band of Alkali Iodides

Let us now consider the nature of the intermediate emission band of the alkali iodides. As indicated earlier, the intermediate emission band is observed in all the iodides at 400-415 mμ at the temperature of liquid nitrogen and at 425-437 mμ at room temperature, slightly changing its position on passing from one alkali iodide to another. The photoexcitation spectrum of this band itself constitutes a narrow band situated on the falling part of the characteristic absorption of the alkali iodides, with a maximum at about 230-235 mμ at the temperature of liquid nitrogen and 240-245 mμ at room temperature. With increasing temperature, the intensity of luminescence of the intermediate band falls (like that of the ultraviolet emission), in some cases almost vanishing, as in RbI, but in others remaining appreciable even after the vanishing of the ultraviolet emission, as in CsI. We noted a single case in which the intensity of the luminescence from the intermediate band rose with increasing temperature; this was the case of the activation of CsI by thallium nitrate.

We also found that the addition of Tl and In reduced the luminescence of the intermediate and ultraviolet emission bands to the same extent, but the addition of activator Tl to CsI in the form of Tl_2NO_3 stimulated the luminescence of the intermediate band.

There was a rise in the luminescent intensity of the intermediate band (or even an initiation of luminescence in samples not previously possessing this) after special heat treatment of the crystals. On studying thermoluminescence, light storage was observed in this band, and the attenuation of the luminescence was governed [22, 38] by a complicated law. All the characteristics of the intermediate band of luminescence in the alkali iodides are given in the composite Table 7.

Emission bands of pure salts analogous to our own intermediate band have been observed by other authors. Thus, an emission band with a maximum at about 400 mμ was observed by Fieschi and Spinolo [115] when studying the x-ray fluorescence of KI after heating the sample in vacuum, i.e., after heat treatment similar to that employed by ourselves in order to increase the luminescent intensity of the intermediate band. A certain stimulation of this band was found on activating KI with copper [116]. The band was also observed in absorption; thus a band with a maximum at $\lambda_{max} \sim 235$ mμ was observed in [123] at the characteristic absorption edge of KI at liquid-nitrogen temperature.

An intermediate band was also observed for CsI in [120]; the maximum lay at about 437 mμ. Tomura and Kaifu [119] found a displacement of the emission maxima in KI, different samples having their maxima between 360 and 400 mμ; this was probably because two bands appeared in the emission of these samples (an ultraviolet band and an intermediate band) with different relative intensities, and this appeared in the form of a displacement of the overall maximum.

Regarding the nature of the luminescence in the intermediate band, there is quite a well-supported view that this is associated with cation vacancies [120, 123, 124]. It is found that the presence of divalent ions in the crystal stimulates this radiation, and that the number of cation vacancies is in equilibrium with the divalent ions [120]. Another point of view is expressed in [33], where it is considered that the luminescence of pure alkali iodides is due to the presence of uncontrollable impurities.

Timusk and Martienssen [124], studying analogous luminescence in KCl, came to the following conclusion regarding the nature of the luminescence in the band with $\lambda_{max} \sim 430$ mμ. The cation vacancy plays the part of a recombination center, capturing first an electron and then a hole, which recombine with the accompaniment of emission (the so-called α luminescence).

The fact that the intermediate emission band of alkali iodides fails to appear in the most perfect crystals grown from solution, and also the fact that this emission may be created or intensified by heat treatment, compels us to consider that the band in question is associated with structural defects, and we shall call it the "structural emission band." However, whether Timusk and Martienssen's mechanism is the true one or whether an iodine ion emits near an anion vacancy is a problem very difficult to solve at the present time.

3. Discussion of the Mechanism of Energy Transfer from the Main Substance to the Activator

The mechanism underlying the transfer of energy from the main (base) substance to the activator on absorption of the exciting radiation by the former during γ luminescence is not yet clear. A study of the photoexcitation of the activator emission (by excitation in the absorption band of the base substance) might facilitate an understanding of this problem.

Even in 1935, Terenin and Klement [125] showed that in a number of alkali halide sublimate phosphors with large activator concentrations the excitation spectrum extended to 186 mμ (the limit of investigation in the paper in question). Later, Alderson and Williams [126] observed that the energy absorbed by the base was transferred to the activator even on reducing the wavelength of the exciting light to 110 mμ. In discussing the mechanism underlying the transfer of energy from the base substance to the activator, Curie [127] considered two possible means of transfer: 1) the transfer of energy by the transport of holes, and, 2) transfer by the diffusion of excitons; he also showed to a first approximation that in excitation processes accompanied by photoconduction the main contribution to transfer came from the charge carriers. The same point of view was expressed in [128]. Studying the luminescence of the activator excitation of KI—Tl in the region of 175 mμ, where photoconductivity begins to appear, the authors determined the quantum yield in this region as being equal to 0.8, while the quantum yield in the activator absorption band was close to unity [129, 130]. The authors considered that the emission in this region was due to the transfer of energy by means of holes, and the subsequent recombination of these with electrons at thallium ions.

In the region of the first (exciton) absorption maximum, however, we and other authors [124, 126, 130, 131] observed a fall in the photoexcitation spectrum of activator luminescence, usually not quite reaching zero; this latter fact suggested a possible exciton mechanism for the transfer of energy from the base to the activator. This point of view was also expressed earlier by other authors [46, 55, 87, 89, 121, 132, 133].

The considerable excitation of activator emission which we observed in CsI—Tl on ir-radiation in the region of the exciton absorption maximum for fresh crystal cleavages also tends to favor the exciton mechanism of energy transfer. Activator luminescence following exciton excitation was also observed in the emission of KI—Sn crystals [134]. The fact that the yield of activator emission in the exciton absorption band was much smaller than that obtained by excitation of the activator luminescence may be explained by considering that the excitons arising after the absorption of light may themselves be annihilated, to the accompaniment of radiation. In the present case, various competing processes take place; on the one hand we may have the capture of excitons by thallium ions, with subsequent de-excitation of the energy in the activator emission band, and on the other hand the radiative or radiationless annihilation of excitons; the probability of one process or the other taking place is a function of temperature, activator con-centration, and the concentration of lattice defects which may capture excitons. Tomura and Kaifu [135] observed luminescence of the activator in KI—Tl on excitation in the first absorption band of the base, the efficiency of the exciton excitation falling with temperature. On the ab-sorption of light in the region between the exciton band and the band in which absorption gives rise to photoconductivity, both transfer mechanisms will occur, i.e., both the exciton and the hole type [131].

Thus we see from all these results that the excitons formed in alkali halide crystals on the absorption of light in the first absorption band may give their energy to activator luminescence centers.

What kind of mechanism is involved in the γ excitation of alkali iodides? We showed earlier that after γ irradiation excitons were formed in the crystals of alkali iodides and that these were annihilated, with radiation in the ultraviolet luminescence band. We may suppose that some of these will transfer the primary excitation energy to activator luminescence cen-ters, and with increasing concentration of the latter the probability of their capture of excitons will increase, to the detriment of the radiative annihilation of excitons.

Antonov-Romanovskii, when studying the character of the electron motion inside the crys-tal lattice, defined the volume of the phosphor as consisting of three regions, A, B, and C [136]. In region A (free zone) the electrons only experienced diffusion-type displacements, in region B they were captured, and in region C, lying in immediate proximity to the ionized centers, they recombined almost instantaneously.

If we transfer this idea to the motion of excitons in a crystal, we find that in the free zone the excitons only experience diffusive motion, and in the region lying close to the activator cen-ters, characterized by an effective interaction cross-section radius r_0, the excitons are annihil-ated almost instantaneously, with the excitation of an activator center. The diffusion of the ex-citons in the free zone will determine the scintillation growth time, which depends on the dimen-sions of the free zone, i.e., on the activator concentration.

Since there is a clear relationship between the radius of the effective cross section and the diffusion coefficient for two interacting particles [137], a knowledge of the diffusion coeffi-cient of the excitons offers the possibility of determining the radius of the effective cross sec-tion of the interaction between the excitons and the activator. A formula for determining the diffusion coefficient of excitons in alkali halide salts was derived by Trlifay [112], who based his calculations on the following assumptions: The transfer of the anion exciton energy from one ion to another is effected by way of the resonance migration of the excitation energy, and the absorption and emission bands are symmetrical.

Trlifay's formula has the form

$$D = 0.463 \frac{e^4 \, | \, (S_2 \, | \, \mathbf{r}_s \, | \, S_1) \, |^4}{\hbar \varepsilon_0^2 \delta_A} \sum \frac{1}{| \, \mathbf{R}_{li} \, |^4} e^{\frac{2 \log 2 \, (\Delta E)^2}{\delta_A^2}}, \tag{6}$$

where D is the diffusion coefficient of the excitons, $|(S_2 \, | \, \mathbf{r}_S \, | S_1) |^2$ is the square of the absolute value of the matrix element associated with the emission of light, $|\mathbf{R}_{li}|$ is the absolute distance between the i and l halide ions, δ_A is the half-width of the absorption band, ΔE is the Stokes displacement, and ε_0 is the dielectric constant. The square of the absolute value of the matrix element is obtained from the probability of the spontaneous emission of light [138]:

$$\frac{1}{\tau} = \frac{4e^2 E_s^3}{3\hbar c^3} \, | \, (S_2 \, | \, \mathbf{r}_s \, | \, S_1) \, |^2, \tag{7}$$

where τ is the exciton lifetime and E_S is the energy of the maximum of the emission band.

The quantity $\sum \frac{1}{| \, \mathbf{R}_{li} \, |^4}$ is a function of the smallest distance a between the alkali halide ions, and for an fcc lattice this sum equals $6.335/a^4$.

The Stokes displacement of the emission band was determined from the formula derived by Pekar [139] on the assumption that the half-widths of the absorption and emission bands equalled

$$\delta_\omega = 2 \sqrt{2 \ln 2} \, \sqrt{\frac{\Delta \omega}{\hbar} kT}, \tag{8}$$

where δ_ω is the half-width of the absorption band, $\Delta \omega$ is the Stokes shift of the emission band relative to the absorption band, and T is the absolute temperature.

At room temperature the half-widths of the absorption bands of KI and CsI are equal and, according to [66], constitute 0.18 eV, which gives a value of 0.23 eV for the Stokes shift in both crystals. It should be noted that the value of the Stokes shift derived by Pekar differs from the value obtained in our own experiments. This difference is evidently due to the fact that Trlifay, like Pekar, calculated the Stokes shift for a free exciton; we observed the radiation of a self-localized exciton, and the Stokes shift of the radiation from this should exceed the values obtained for a free exciton. However, for calculating the exciton diffusion coefficient in the crystal we should consider a free exciton and use the value of the Stokes shift obtained for this.

In order to obtain a numerical value for the exciton diffusion coefficient in KI, Trlifay used the following values of the quantities coming into formulas (6) and (7):

$$\tau = 10^{-9} \text{ sec}, \quad \Delta E = 0.23 \text{ eV}, \quad E_s = E_A - \Delta E = 5.40 \text{ eV},$$

where E_A = 5.63 eV is the energy at the maximum of the absorption band, δ_A = 0.18 eV, a = 5.53 Å; he obtained $1.4 \cdot 10^{-1}$ cm^2/sec for the diffusion coefficient of the excitons in KI.

We feel, however, that Trlifay's exciton lifetime (as mentioned earlier) is too low. A value of 10^{-8} sec is given for the lifetime of excitons in [140, 141]. We also consider that the lifetime of the free exciton is no shorter than 10^{-8} sec; otherwise it could not appear in the growth time of the scintillations. This change reduces the exciton diffusion coefficient in KI by two orders, and its value should not in fact exceed $1.4 \cdot 10^{-3}$ cm^2/sec.

For calculating the exciton diffusion coefficient in CsI we used the following values of the quantities involved in formulas (6)-(8):

$$\tau = 10^{-8} \text{ sec,}$$
$$\Delta E = 0.23 \text{ eV.}$$

E_A = 5.63 eV (the positions of the absorption maxima of CsI and KI coincide, λ_{max} = 218 mμ),

$$\delta_A = 0.18 \text{ eV,}$$
$$a = 3.95\text{Å} \quad \text{(reported in [142, 143]),}$$
$$\varepsilon_0 = 5.65.$$

The value of $\sum \frac{1}{|R_{li}|^4}$ will differ for different types of lattice; hence, we must introduce a change into the value used when calculating the diffusion coefficient of excitons in CsI, since the CsI lattice is bcc (I), whereas that of KI is fcc (II). As to the values of these sums, the following relation is given in [144]:

$$\frac{1}{z_I^{4/3}} \sum \frac{1}{|R_{li}|_I^4} = \frac{1}{z_{II}^{4/3}} \sum \frac{1}{|R_{li}|_{II}^4} , \tag{9}$$

where z is the number of molecules in the crystal cell, equal to 4 for an fcc lattice and 1 for a bcc lattice [142]. From relation (9) we obtain

$$\sum \frac{1}{|R_{li}|_I^4} = \frac{1}{6.35} \sum \frac{1}{|R_{li}|_{II}^4} = \frac{0.977}{a^4}.$$

Putting the resultant values into formulas (6)-(8), we obtain a value of D = 1.0 · 10^{-4} cm^2/sec for the exciton diffusion coefficient in CsI.

Smoluchowski [137], considering the rate of precipitation of colloidal particles from an originally homogeneous solution on to an absorbing spherical surface of radius R, obtained the following formula for determining this rate:

$$\frac{M}{t} = 4\pi DRC \left(1 + \frac{2R}{\sqrt{t\pi D}} \right), \tag{10}$$

where M is the mass of the precipitated material after a time t and D is the diffusion coefficient of the colloidal particles.

If we neglect the second term and refer this idea to the interaction of excitons initially distributed uniformly in a crystal containing stationary activator ions, we obtain the following formula for the rate of recombination of the excitons:

$$p = 4\pi D r_0 C_{Tl}, \tag{11}$$

where p is the specific rate of recombination of the excitons with Tl ions, D is the exciton diffusion coefficient, r_0 is the radius of the effective cross section of the interaction between the excitons and activator ions, and C_{Tl} is the concentration of Tl ions in the crystal.

Earlier it was indicated that in the case of CsI—Tl (crystal No. 1) the incremental scintillation growth time dt which had to be added to the growth time associated with the lifetime of the activator in the excited state was determined by the average time required for the exciton to pass through the distance from its place of origin to the activator ion, this increment being 0.05 μsec in the present case; the specific recombination rate was thus p = 1/dt = 0.2 · 10^8

sec^{-1}. The activator concentration in crystal No. 1 was $1.5 \cdot 10^{17}$ cm^{-3}; the exciton diffusion coefficient in CsI obtained by the Trlifay formula was $1.0 \cdot 10^{-4}$ cm^2/sec.

Putting these values into formula (11), we obtain a value of 10.6 Å for the radius of the effective interaction between the excitons and the Tl ions in the CsI—Tl; this equals ~2d (the lattice-constant of CsI is 4.56 Å according to [142]). Considering the approximate nature of the calculations, the value obtained for r_0 must be regarded as very reasonable, and in our view this fact supports the mechanism assumed for the transfer of energy from the base to the activator, namely, that in the case of alkali iodides this is effected by the exciton mechanism.

The same conclusion was reached by Lomonosov and Nemilov [35] after studying the effect of an electric field on scintillation processes in CsI—Tl. These authors showed that individual scintillations did not depend on the electric field. In the case of the electron—hole mechanisms, however, an external electric field would change the intensity of the observed luminescence. A confirmation of this is the fact that in the case of indium phosphors the energy was, as it were, "thrown" from the bands of unactivated luminescence to the activator centers with increasing temperature.

CONCLUSION

As a result of the foregoing investigation, we have shown that there are two forms of unactivated luminescence in the luminescence spectra of pure alkali iodides and of those activated with slight traces of foreign impurities: exciton luminescence (this has the shortest wavelength and is situated in the ultraviolet part of the spectrum), and structural luminescence (occupying an intermediate position between the exciton and activator emission and lying in the blue part of the spectrum).

The photoexcitation spectrum of the exciton emission band is situated in the region of the first exciton maximum of the absorption spectrum of the base material. There is hardly any light storage in this emission band and the corresponding attenuation characteristic is exponential.

The photoexcitation spectrum of the structural emission band is itself a narrow band situated on the falling part of the characteristic absorption of the crystal. Light storage does occur in the structural band, and the attenuation law is of a complex nature.

Increasing the concentration of foreign impurities in the crystals leads to a fall in the intensity of both forms of unactivated luminescence, while increasing the number of structural defects eliminates the exciton luminescence and intensifies the structural band.

Our study of the photoexcitation spectra of the activator luminescence shows that the transfer of energy from the base material to the activator may take place by way of an exciton mechanism. This point of view is supported by our study of the concentration dependence of the time characteristics of CsI—Tl phosphors. Using the diffusion coefficient in the crystal determined from the Trlifay formula, we have obtained a value for the radius of effective interaction between the excitons and the activator ions in the case of CsI—Tl.

In conclusion, the author wishes to thank Z. L. Morgenshtern for directing the research, M. D. Galanin and V. V. Antonov-Romanovskii for interest in the work and discussion of the results, and N. V. Kostina for help with the experiments.

LITERATURE CITED

1. R. W. Pohl, Naturwissenschaften, 16:477 (1928); Proc. Phys. Soc., Vol. 49, Extra Part, page 3 (1938); Ann. Phys., 29:239 (1937).
1a. R. Pohl and E. Rupp, Ann. Phys., 81:1161 (1926).
2. H. Kallmann, Naturwiss. und Technik, July, 1947.
3. P. R. Bell, Phys. Rev., 73:1405 (1948).
4. R. Hofstadter, Phys. Rev., 74:100 (1948).
5. W. Van Sciver and R. Hofstadter, Phys. Rev., 84:1062 (1951); 87:522 (1952).
5a. J. C. D. Milton and R. Hofstadter, Phys. Rev., 75:1289 (1949).
6. J. Bonanomi and J. Rossel, Helv. Phys. Acta, 24:310 (1951).
7. I. Broser, H. Kallman, and U. Martins, Z. Naturforsch., 4a:204 (1949).
8. W. Hanle and H. Schneider, Z. Naturforsch., 6a:290 (1951).
9. L. M. Belyaev, M. D. Galanin, Z. L. Morgenshtern, and Z. A. Chizhikova, Dokl. Akad. Nauk SSSR, 99:691 (1954); 105:57 (1955); Izv. Akad. Nauk SSSR, Ser. Fiz., 21:548 (1957).
10. Z. L. Morgenshtern, Dokl. Akad. Nauk SSSR, 105:250 (1955).
11. J. Bonanomi and J. Rossel, Helv. Phys. Acta, 25:75 (1952).
12. S. K. Allison and H. Casson, Phys. Rev., 90:880 (1953).
13. M. L. Halbert, Phys. Rev., 107:647 (1957).
14. A. Colansky, C. H. Johnson, and C. D. Moak, Rev. Sci. Instr., 27:58 (1956).
15. S. Bashkin, R. K. Carlson, K. A. Douglas, and J. A. Jacobs, Phys. Rev., 109:434 (1958).
16. R. S. Storey, W. Jack, and A. Ward, Proc. Phys. Soc., 72:1 (1958).
17. J. C. Robertson and A. Ward, Proc. Phys. Soc., 73:523 (1959).
18. J. C. Robertson and J. G. Lynch, Proc. Phys. Soc., 77:751 (1961).
19. K. Heilig, Exptl. Tech. Phys., 10:187 (1962).
20. Z. L. Morgenshtern, Zh. Eksper. i Teor. Fiz., 29:903 (1955).
21. W. Van Sciver and R. Hofstadter, Phys. Rev., 97:1181 (1955).
22. W. Van Sciver, IRE Trans. Nucl. Sci., NS-3(4):39 (1956).
23. J. C. Robertson, J. G. Lynch, and W. Jack, Proc. Phys. Soc., Vol. 78, Pt. 1, p. 1188 (1961).
24. I. K. Plyavin', Dissertation, Fiz. Inst. Akad. Nauk, Moscow (1958).
25. S. N. Komnik, V. I. Startsev, and Yu. A. Tsirlin, Opt. i Spektroskopiya, 4:411 (1958).
26. Yu. A. Tsirlin, S. N. Komnik, and L. M. Soifer, Opt. i Spektroskopiya, 6:422 (1959).
27. V. I. Startsev, Z. B. Baturicheva, and Yu. A. Tsirlin, Opt. i Spektroskopiya, 8:541 (1960).
28. F. S. Eby, W. K. Jentschke, and G. De Pasquali, Phys. Rev., 91:495 (1953).
29. F. S. Eby and W. K. Jentschke, Phys. Rev., 96:911 (1954).
30. G. A. Mikhal'chenko and M. Ya. Shukan, in collection: Transactions of the Seventh Conference on Luminescence (Crystal Phosphors), Tartu (1959), p. 191.
31. H. Enz and J. Rossel, Helv. Phys. Acta, 31:25 (1958).
32. I. K. Plyavin', Opt. i Spektroskopiya, 2:384 (1957); 4:266 (1958); 7:71 (1959).
33. Yu. A. Tsirlin, V. I. Startsev, and L. M. Soifer, Opt. i Spektroskopiya, 8:537 (1960).

34. R. Fieschi, R. Oggioni, and G. Spinolo, Phys. Stat. Solidi, 3:1207 (1963).

35. I. I. Lomonosov and Yu. A. Nemilov, Fiz. Tverd. Tela, 2:1629 (1960).

36. B. Hahn, Phys. Rev., 91:772 (1953).

37. Muelhause, der Mateosian, and McKeown, Phys. Rev., A95:598 (1954).

38. W. Van Sciver, Nucleonics, 15:50 (1956); Phys. Rev., 120:1193 (1960).

39. H. Knöpfel, E. Loepfe, and P. Stoll, Helv. Phys. Acta, 29:241 (1956); 30:521 (1957); Z. Naturforsch., 12a:348 (1957).

40. B. Hahn and J. Rossel, Helv. Phys. Acta, 26:803 (1957).

41. K. J. Teegarden, Phys. Rev., 105:1222 (1957).

42. Z. L. Morgenshtern, Opt. i Spektroskopiya, 7:231 (1959); 8:672 (1960).

43. H. Rüchardt, Z. Phys., 140:547 (1955).

44. U. Fano, Phys. Rev., 57:564 (1940); 58:544 (1940).

45. F. Seitz, Imperfections in Nearly Perfect Crystals, New York (1954).

46. Ch. B. Lushchik, N. E. Lushchik, G. G. Liid'ya, and L. A. Teiss, Tr. Inst. Fiz., Akad. Nauk EstSSR, No. 6, p. 63 (1957).

47. Ch. B. Lushchik, F. N. Zaitov, and G. G. Liid'ya, Photoelectric and Optical Processes in Semiconductors, Kiev (1959), p. 180.

48. F. Seitz, Rev. Mod. Phys., 26(7):29 (1954).

49. L. R. Apker and E. A. Taft, Phys. Rev., 79:964 (1950); 81:968 (1951).

50. H. R. Philipp and E. A. Taft, J. Phys. Chem. Solids, 1(3):159 (1956); Phys. Rev., 106:671 (1957).

51. N. Inchauspé, Phys. Rev., 106:898 (1957).

52. V. E. Lashkarev and G. A. Fedorus, Izv. Akad. Nauk SSSR, Ser. Fiz., 16:81 (1962).

53. V. P. Zhuze and S. M. Ryvkin, Izv. Akad. Nauk SSSR, Ser. Fiz., 16:93 (1952).

54. Ch. B. Lushchik and G. G. Liid'ya, Trudy IFA Akad. Nauk EstSSR, No. 7, p. 193 (1958); Transactions of the Seventh Conference on Luminescence, Tartu (1959), p. 101.

55. G. G. Liid'ya, Tr. Inst. Fiz., Akad. Nauk EstSSR, No. 11, p. 187 (1960); Dissertation, Tartu (1962).

56. M. Ueta, M. Hirai, and H. Watanabe, J. Phys. Soc. Japan, 14:253 (1952).

57. N. F. Mott, Proc. Phys. Soc., A167:384 (1938).

58. J. Frenkel, Phys. Rev., 37(17):1276 (1931).

59. J. Franck and E. Teller, J. Chem. Phys., 6:861 (1938).

60. G. Diemer and W. Hoogenstraaten, J. Chem. Phys. Solids, 2:119 (1957).

61. Yu. V. Vorob'ev and Yu. M. Karkhanin, Fiz. Tverd. Tela, 3:206 (1961).

62. D. Chanvy and J. Rossel, Helv. Phys. Acta, 32:481 (1959).

63. Luminescence Analysis, Handbook, Fizmatgiz, Moscow (1961), pp. 84, 95.

63a. A. A. Babushkin, P. A. Bazhulin, F. A. Korolev, L. V. Levshin, V. K. Prokof'ev, and A. R. Striganov, Methods of Spectral Analysis, Izd. Mosk. Univ. (1962), p. 89.

64. S. L. Mandelstam, Introduction to Spectral Analysis, Gostekhizdat, Moscow (1946).

65. Z. L. Morgenstern (Morgenshtern) and N. N. Vasiljeva (Vasil'eva), Czech. J. Phys., B13:226 (1963).

66. R. Hilsch and R. W. Pohl, Z. Phys., 59:812 (1930).

67. E. G. Schneider and H. N. O'Bryan, Phys. Rev., 51:293 (1937).

68. Scintillators and Scintillation Materials, 1960, p. 70. Informats. Byull, No. 10(22):3 (1960).

69. N. N. Vasil'eva, Transactions of the Second Conference on the Physics of Alkali Halide Crystals, Riga (1962), p. 325.

70. N. N. Vasil'eva and Z. L. Morgenshtern, Opt. i Spektroskopiya, 9:676 (1960); 12:86 (1962).

71. N. N. Vasil'eva and Z. L. Morgenshtern, Izv. Akad. Nauk SSSR, Ser. Fiz., 25:47 (1961).

72. L. M. Shamovskii and Yu. I. Zhvanko, Izv. Akad. Nauk SSSR, Ser. Fiz., 21:557 (1957).

73. M. Forro, Z. Phys., 58:613 (1929).

74. W. Martienssen, J. Phys. Chem. Solids, 2:257 (1957).
75. F. Fischer, Z. Phys., 139:328 (1954).
76. A. A. Dunina, Z. L. Morgenshtern, and L. M. Shamovskii, Opt. i Spektroskopiya, 4:105 (1958).
77. V. L. Levshin, Photoluminescence of Liquids and Solids, Gostekhizdat, Moscow (1951).
78. P. Pringsheim, Fluorescence and Phosphorescence [Russian translation], IL, Moscow (1951). [Interscience, New York.]
79. A. H. Pfund, Phys. Rev., 32:39 (1928).
80. G. Wennier, Phys. Rev., 52:191 (1937).
81. A. V. Hippel, Z. Phys., 101:680 (1936).
82. L. G. Schulz, Acta Cryst., 4:487 (1951).
83. R. W. Pohl, Naturwissenschaften, 16:477 (1928).
84. J. Apper and E. Taft, Phys. Rev., 79:964 (1950); 81:698 (1951); 82:814 (1951).
85. E. Taft and L. Apker, J. Chem. Phys., 20:1648 (1952).
86. E. A. Taft and H. R. Philipp, J. Phys. Chem. Solids, 3:1 (1959).
86a. H. R. Philipp and E. A. Taft, J. Phys. Chem. Solids, 1:159 (1956); Phys. Rev., 106:671 (1957).
87. J. Taylor and P. Hartmann, Phys. Rev., 113:1421 (1959).
88. I. V. Yaék and G. G. Liid'ya, Opt. i Spektroskopiya, 8:142 (1960).
89. Ch. B. Lushchik, G. G. Liid'ya, I. V. Yaék, and É. S. Tiisler, Opt. i Spektroskopiya, 9:70 (1960).
89a. Ch. B. Lushchik, Physics of Alkali Halide Crystals, Transactions of the All-Union Conference, Riga (1961), p. 245.
90. K. Teegarden, Phys. Rev., 108:660 (1957).
90a. J. E. Eby, K. J. Teegarden, and D. B. Dutton, Phys. Rev., 116:1099 (1959).
91. K. Nakamura, K. Fukuda, R. Kato, A. Matsui, and Y. Uchida, J. Phys. Soc. Japan, 16:1262 (1961).
92. K. Fukuda, R. Kato, K. Nakamura, and Y. Uchida, J. Phys. Soc. Japan, 15:1344 (1960).
93. H. Mahr, Phys. Rev., 122:1464 (1961); 125:1510 (1962); 130:2257 (1963).
94. D. L. Dexter, Phys. Rev., 83:435 (1951).
95. S. I. Vavilov, Collected Papers, Vol. 1, Izd. Akad. Nauk SSSR, Moscow (1954), p. 222; Izv. Akad. Nauk SSSR, Ser. Fiz., 9:283 (1945).
96. V. B. Neustruev, Diploma Work, Fiz. Inst. Akad. Nauk, Moscow (1961).
97. Z. L. Morgenshtern, V. B. Neustruev, and M. I. Épshtein, Zh. Prikl. Spektroskopiya (in press).
98. K. Watanabe and C. J. Inn, J. Opt. Soc. Am., 43:31 (1953).
99. D. H. Thurnau, J. Opt. Soc. Am., 46:346 (1956).
100. M. Luchiesh and A. H. Taylor, J. Opt. Soc. Am., 36:227 (1946).
100a. F. Benford, G. P. Lloyd, and S. Schwarz, J. Opt. Soc. Am., 38:964 (1948).
101. R. Tonsey, F. S. Johnson, J. Richardson, and N. Toran, J. Opt. Soc. Am., 41:696 (1951).
102. Electric Lamp Apparatus. Spectral Hydrogen Lamps Types DVS-200 and DVS-201. Time Technical Conditions No. SUO. 337. 108. TU.
103. F. R. Bichowsky and F. D. Rossini, The Thermochemistry of the Chemical Substances, New York (1936).
104. K. V. Shalimova, Dokl. Akad. Nauk SSSR, 97:437 (1954).
105. N. N. Vasil'eva, Opt. i Spektroskopiya, 16:851 (1964).
106. N. N. Vasil'eva, Zh. Prikl. Spektroskopiya, 3:470 (1965).
107. J. Broser, H. Kallmann, and O. Rember, Z. Phys., 59:79 (1950).
108. Z. A. Chizhikova, Dissertation, Fiz. Inst. Akad. Nauk, Moscow (1959).
109. Ch. B. Lushchik, N. E. Lushchik, and I. V. Yaék, Izv. Akad. Nauk SSSR, Ser. Fiz., 26:488 (1962).

110. K. V. Shalimova, Dissertation, Moscow (1954).

111. N. E. Lushchik and Ch. B. Lushchik, Tr. Inst. Fiz. Akad. Nauk EstSSR, No. 6, p. 5 (1957).

112. M. Trlifay, Czech. J. Phys., 5:463 (1955).

113. J. Toyazawa, Progr. Theoret. Phys., 12:421 (1954).

114. V. N. Vishnevskii, O. B. Liskovich, M. S. Pidzirailo, and Z. P. Chornii, Ukr. Fiz. Zh., 7(10):1101 (1962); 7(12):1292 (1962).

115. R. Fieschi and G. Spinolo, Nuovo Cimento, 23 (4):738 (1962).

116. R. Oggioni and G. Spinolo, Phys. Rev., 131:1114 (1963).

117. H. Mahr, International Symposium on Color Centers in Alkali Halides, Stuttgart (1962).

118. K. Fukudo, R. Kato, K. Nakamura, and Y. Uchida, J. Phys. Soc. Japan, 15:1344 (1960).

119. M. Tomura and Y. Kaifu, J. Phys. Soc. Japan, 15:314 (1960).

120. H. Bessou, D. Chanvy, and J. Rossel, Helv. Phys. Acta, 35(3):211 (1962).

121. R. Gwin and R. B. Murray, Phys. Rev., 131:501, 508 (1963).

122. K. Teegarden and R. Weeks, J. Phys. Chem. Solids, 10(2-3):211 (1959).

123. C. J. Delbecq, P. Pringsheim, and P. Yuster, J. Chem. Phys., 19:574 (1951).

124. T. Timusk and W. Martienssen, International Symposium on Color Centers in Alkali Halides, Stuttgart (1962); Phys. Rev., 128:1656 (1962).

125. A. N. Terenin and F. D. Klement, Uch. Zap. Leningr. Gos. Univ., Ser. Fiz., No. 1 (1935).

126. J. E. Alderson and S. E. Williams, Australian J. Phys., 14:386 (1961).

127. D. Curie, J. Chim.-Phys. et Phys.-Chim. Biol., 55:607 (1958).

128. M. Ueta and T. Ishii, J. Phys. Soc. Japan, 14:857 (1959).

129. N. Bünger, Z. Phys., 66:311 (1930).

130. T. Timusk, Phys. Chem. Solids, 18(2/3):265 (1961).

131. B. Smaller and E. Avery, Phys. Rev., 92:232 (1953).

132. G. G. Liid'ya, Tr. Inst. Fiz. Akad. Nauk EstSSR, No. 17, p. 93 (1961).

133. Ch. B. Lushchik, G. G. Liid'ya, T. A. Soovik, and I. V. Yaěk, Tr. Inst. Fiz. Akad. Nauk EstSSR, No. 15, p. 103 (1961).

134. G. K. Vale, Tr. Inst. Fiz. Akad. Nauk EstSSR, No. 21, p. 281 (1962).

135. M. Tomura and Y. Kaifu, J. Phys. Soc. Japan, 15:1295 (1960).

136. V. V. Antonov-Romanovskii, Doctor's Dissertation, Fiz. Inst. Akad. Nauk, Moscow (1942).

137. A. Einstein and M. Smoluchowski, The Brownian Motion, Collection of Articles [Russian translation], ONTI, Moscow (1936), p. 332.

138. W. Heitler, Quantum Theory of Radiation [Russian translation], IL, Moscow (1956). [Third edition, Oxford University Press, New York, 1956.]

139. S. I. Pekar, Usp. Fiz. Nauk, 50:197 (1953).

140. D. L. Dexter and W. R. Heller, Phys. Rev., 84:377 (1951).

141. M. Tomura, J. Phys. Soc. Japan, 15:1508 (1960).

142. C. Kittel, Introduction to Solid-State Physics [Russian translation], Gostekhizdat, Moscow (1957). [Second edition, Wiley, New York, 1956.]

143. F. Seitz, Modern Theory of Solids [Russian translation], Gostekhizdat, Moscow (1949). [McGraw-Hill, New York, 1940.]

144. T. Förster, Ann. Phys., 2:55 (1948).